DEVELOPING TECHNOLOGIES IN FOOD SCIENCE

Status, Applications, and Challenges

DEVELOPING TECHNOLOGIES IN FOOD SCIENCE

Status, Applications, and Challenges

Edited by
Murlidhar Meghwal, PhD
Megh R. Goyal, PhD, PE

Apple Academic Press Inc.
3333 Mistwell Crescent
Oakville, ON L6L 0A2 Canada

Apple Academic Press Inc.
9 Spinnaker Way
Waretown, NJ 08758 USA

© 2017 by Apple Academic Press, Inc.

First issued in paperback 2021

Exclusive worldwide distribution by CRC Press, a member of Taylor & Francis Group
No claim to original U.S. Government works

ISBN 13: 978-1-77-463041-9 (pbk)
ISBN 13: 978-1-77-188447-1 (hbk)

Library and Archives Canada Cataloguing in Publication

Developing technologies in food science / edited by Murlidhar Meghwal, PhD, Megh R. Goyal, PhD, PE.
(Innovations in agricultural and biological engineering)
Includes bibliographical references and index.
Issued in print and electronic formats.
ISBN 978-1-77188-447-1 (hardcover).--ISBN 978-1-315-36597-8 (PDF)
1. Food industry and trade--Technological innovations. I. Meghwal, Murlidhar, editor II. Goyal, Megh Raj, editor III. Series: Innovations in agricultural and biological engineering

TP372.5.D48 2017	664	C2017-901099-9	C2017-901100-6

Library of Congress Cataloging-in-Publication Data

Names: Meghwal, Murlidhar, editor. | Goyal, Megh Raj, editor.
Title: Developing technologies in food science : status, applications, and challenges / editors, Murlidhar Meghwal, PhD, Megh R. Goyal, PhD, PE.
Description: Toronto ; New Jersey : Apple Academic Press, 2017. | Series: Innovations in agricultural & biological engineering | Includes bibliographical references and index.
Identifiers: LCCN 2017005305 (print) | LCCN 2017007904 (ebook) | ISBN 9781771884471 (hardcover : alk. paper) | ISBN 9781315365978 (ebook)
Subjects: LCSH: Food--Biotechnology.
Classification: LCC TP248.65.F66 D48 2017 (print) | LCC TP248.65.F66 (ebook) | DDC 664--dc23
LC record available at https://lccn.loc.gov/2017005305

Apple Academic Press also publishes its books in a variety of electronic formats. Some content that appears in print may not be available in electronic format. For information about Apple Academic Press products, visit our website at **www.appleacademicpress.com** and the CRC Press website at **www.crcpress.com**

CONTENTS

    Processing.. 99

    Gabriela John Swamy, Sangamithra Asokapandian,
    and Chandrasekar Veerapandian

6. Minimal Processing of Fruits and Vegetables..................................... 119

    Jagbir Rehal, Deepika Goswami, Hradesh Rajput,
    and Harshad M. Mandge

PART III: BIOPROCESS ENGINEERING AND
         BIOTECHNOLOGY: APPLICATIONS AND PRODUCTS............. 141

7. Application of Biocomposite Polymers in Food Packaging:
    A Review... 143

    Arijit Nath, Arpia Das, and Chiranjib Bhattacharjee

8. Value Added Product Recovery From Fruit and Vegetable Waste
    Through Biotechnological Intervention 175

    Anshu Singh

9. An Overview of Bioprocessing and Biorefinery Approach
    for Sustainable Fisheries... 193

    Vegneshwaran V. Ramakrishnan, Winny Routray, and Deepika Dave

PART IV: FOODS FOR BETTER HUMAN HEALTH.............................. 279

10. Fortification of Paddy and Rice Cereals 281

    Shruti Pandey and A. Jayadeep

11. Functional Foods from the Indian Subcontinent 305

    Arpita Das, Arijit Nath, and Runu Chakraborty

12. Phytochemistry and Anti-Oxidative Effects of *Aspalathus linearis*
    Herbal Tea – A Review... 319

    Ademola Ayeleso, MutiuIdowu Kazeem, Taiwo Ayeleso,
    and Emmanuel Mukwevho

    Appendix A: Food Processing Equipments and Their Applications 333

    Index 339

LIST OF CONTRIBUTORS

Sangamithra Asokapandian, PhD
Assistant Professor, Department of Food Technology, Kongu Engineering College, Perundurai, Tamil Nadu – 638052, India. E-mail: asokmithra@gmail.com

Ademola Olabode Ayeleso, DTech
Postdoctoral Research Fellow, Department of Biological Sciences, North-West University, Private Bag X2046, Mmabatho, 2735, South Africa. E-mail: ademola.ayeleso@gmail.com

Taiwo Betty Ayeleso, MTech
PhD Student, Department of Biological Sciences, North-West University, Private Bag X2046, Mmabatho, 2735, South Africa. E-mail: taiwo.ayeleso@gmail.com

Rintu Banerjee, PhD
Professor, Microbial Biotechnology and Downstream processing lab, Agricultural and Food Engineering Department, Indian Institute of Technology, Kharagpur, West Bengal, India. E-mail: rb@agfe.iitkgp.ernet.in

Sunil Manohar Behera
PhD Research Scholar, Department of Agricultural and Food Engineering, Indian Institute of Technology, Kharagpur 721302, West Bengal, India. Mobile: +91-9434371464. Tel.: +91-3222-281673, E-mail: sunilmanoharb@gmail.com

Chiranjib Bhattacharjee, PhD
Professor, Chemical Engineering Deptartment, Jadavpur University, Kolkata, 700032, West Bengal, India.

Sudhanshi Billoria, MSc
PhD Research Scholar, Department of Agricultural and Food Engineering, Indian Institute of Technology, Kharagpur 721302, West Bengal, India. Mobile: +91-8768126479. E-mail: sudharihant@gmail.com

Runu Chakraborty, PhD
Professor, Department of Food Technology and Biochemical Engineering, Jadavpur University, Kolkata – 700032, India, E-mail: crunu@hotmail.com

Arpita Das, PhD
Postdoc Fellow, Faculty of Chemistry and Chemical Engineering, Babes Bolyai University, 400028 Cluj-Napoca, Romania. E-mail: arpita_84das@yahoo.co.in

Deepika Dave, PhD
Research Scientist and Adjunct Professor, Marine Bioprocessing Facility, Centre of Aquaculture and Seafood Development, Fisheries and Marine Institute, Memorial University of Newfoundland, St. John's, Newfoundland, P.O. Box 4920, St. John's, NL, A1C5R3, Canada. Tel.: +001 (709) 7570732; E-mail: Deepika.Dave@mi.mun.ca

Deepika Goswami, MSc
Scientist, Division of Food Grains and Oilseeds Processing, Central Institute of Post-Harvest Engineering & Technology, Ludhiana – 141004, India. Mobile: +91-9592317693; E-mail: deepika007@gmail.com

Megh R. Goyal, PhD, PE
Retired Faculty in Agricultural and Biomedical Engineering from General Engineering Department, University of Puerto Rico – Mayaguez Campus; and Senior Technical Editor-in-Chief in Agriculture Sciences and Biomedical Engineering, Apple Academic Press Inc., USA. E-mail: goyalmegh@gmail.com

A. Jayadeep, PhD
Principal Scientist, Department of Grain Science and Technology, CSIR – Central Food Technological Research Institute (CFTRI), Mysore, 570020, Karnataka, India. Mobile: +91-9449807228; E-mail: jayadeep@cftri.res.in

Mandira Kopari, MSc
Department of Food Technology, Maharshi Dayanand University, Rohtak, Haryana – 124001, India, Mobile: +91-9051245266. E-mail: kaprimandira@gmail.com

Gagan Jyot Kaur
Assistant Research Engineer, Department of Processing and Food Engineering, Punjab Agricultural University, Ludhiana, 141004, Punjab, India. Mobile: +91-9878114400; E-mail: engg-gagan@pau.edu

Mutiu Idowu Kazeem, PhD
Lecturer, Department of Biochemistry, Lagos State University, P.M.B 0001, Lagos, Nigeria. E-mail: mikazeem@gmail.com

Ajesh Kumar
PhD Research Scholar, Department of Agricultural and Food Engineering, Indian Institute of Technology, Kharagpur 721302, WB, India. Mobile: +91-8900479007. Tel.: +91-3222-281673, E-mail: ajeshmtr@gmail.com

Dipendra Kumar Mahato, MSc
Senior Research Fellow, Division of Food Science and Post-harvest Technology, Indian Agricultural Research Institute (IARI), New Delhi, India. Mobile: +91-9911891494, E-mail: kumar.dipendra2@gmail.com

Harshad M. Mangde, MSc
Assistant Professor, Post-Harvest Technology, College of Horticulture, Banda University of Agriculture and Technology, Banda – 210001, Uttar Pradesh, India. Mobile: +91-9780724557; E-mail: mandgeharshad@gmail.com

Vinay Mannam, PhD
Assistant Professor, Department of Chemistry and Food Science, Framingham State University, 100 State St, Framingham, MA 01701, USA Tel.: +001 5086201220, E-mail: vmannam@framingham.edu

Murlidhar Meghwal, PhD
Assistant Professor, Department of Food Science and Technology, National Institute of Food Technology Entrepreneurship & Management, Kundli - 131028, Sonepat, Haryana, India; Mobile: +91 9739204027; Email: murli.murthi@gmail.com

Emmanuel Mukwevho, PhD
Associate Professor, Department of Biological Sciences, North West University, Private Bag X2046, Mmabatho, 2735, South Africa. E-mail: emmanuel.mukwevho@nwu.ac.za

Arijit Nath, PhD
Post Doctoral Research Fellow, Food Engineering Department, Faculty of Food Science, Corvinus University of Budapest, Hungary.

Shruti Pandey, PhD
Scientist, Department of Grain Science and Technology, CSIR – Central Food Technological Research Institute, Mysore, 570020, Karnataka, India. Mobile: +91-9449859778, E-mail: shruti@cftri.res.in

P. K. Prabhakar, MSc
PhD Research Scholar, Department of Agricultural and Food Engineering, Indian Institute of Technology, Kharagpur 721302, WB, India. Mobile: +91-9973112421. Tel.: +91-3222-281673, Fax: +91-3222-282224. E-mail: prabhatseeds@yahoo.co.in

Hradesh Rajput, MSc
Research Fellow, Department of Food Science & Technology, Punjab Agricultural University, Ludhiana – 141004, India. Mobile: +91-9454183802; E-mail: hrdesh802@gmail.com

Vegneshwaran V. Ramakrishnan, MSc
Research Fellow, Marine Biotechnologist, Marine Bioprocessing Facility, Centre of Aquaculture and Seafood Development, Fisheries and Marine Institute, Memorial University of Newfoundland, St. John's, Newfoundland, P.O. Box 4920, St. John's, NL, A1C 5R3, Canada

Jagbir Rehal
Assistant Fruit and Vegetable Technologist, Department of Food Science and Technology, Punjab Agricultural University, Ludhiana,141004, Punjab, India. Mobile: +91-9417751567; E-mail: jagbir@pau.edu

Winny Routray, PhD
Research Associate, Marine Bioprocessing Facility, Centre of Aquaculture and Seafood Development, Fisheries and Marine Institute, Memorial University of Newfoundland, St. John's, Newfoundland, P.O. Box 4920, St. John's, NL, A1C 5R3, Canada

Anshu Singh, PhD
Research Associate, Microbial Biotechnology and Downstream Processing Lab, Agricultural and Food Engineering Department, Indian Institute of Technology, Kharagpur, WB, India. Mobile: +91-9378096583; E-mail: anshu@agfe.iitkgp.ernet.in

Prem Prakash Srivastav, PhD
Associate Professor, Department of Agricultural and Food Engineering, Indian Institute of Technology, Kharagpur 721 302, West Bengal, India. Mobile: +91-9434043426.Tel.:+91-3222–283134, E-mail: pps@agfe.iitkgp.ernet.in

Gabriela John Swamy, PhD
Research Assistant, Department of Agriculture and Biosystems Engineering, South Dakota State University, Brookings, SD, 57006, USA. E-mail: gabrielafoodtech@gmail.com

Chandrasekar Veerapandian, PhD
Scientist (ICAR)-cum-Researcher, Department of Food and Agricultural Process Engineering, Tamil Nadu Agricultural University, Coimbatore, Tamil Nadu, India. E-mail: chandrufpe@gmail.com

Deepak Kumar Verma, MTech
PhD Research Scholar, Department of Agricultural and Food Engineering, Indian Institute of Technology, Kharagpur 721302, West Bengal, India. Mobile: +91-7407170260; +919335993005. Tel.: +91-3222–281673, Fax: +91-3222–282224. E-mail: deepak.verma@agfe.iitkgp.ernet.in

LIST OF ABBREVIATIONS

5-HIAA	5-hydroxyindoleacetic acid (5-HIAA)
AA	ascorbic acid
ABS	acrylonitrile butadiene styrene
ADI	acceptable daily intake
AITC	allyl isothiocyanate
Al-NMR	aluminum-NMR
APEDA	Agricultural and Processed Food Development Authority
ATP	adenosine triphosphate
BOD	biochemical oxygen demand
CA	controlled atmospheres
$CaCO_3$	calcium carbonate
CAGR	compound annual growth rate
CCl_4	carbon tetrachloride
CDC	Center for Disease Control
CFU	colony forming units
CNS	central nervous system
CO_2	carbon dioxide
COD	chemical oxygen demand
CP	carrier protein
CP-MAS	cross-polarization-magic angle of spinning
CP-MAS NMR	cross polarization magic angle spinning-NMR
DHA	cis-4,7,10,13,16,19-docosahexaenoic acid
DNA	deoxyribonucleic acid
DSS	dextran sodium sulfate
EC	electrode electrochemical
ECNMR	electrode electrochemical NMR
EDTA	ethylenediaminetetraacetic acid
EFSA	European Food Safety Authority
EMAP	equilibrium modified atmosphere packaging
EPA	Environmental Protection Agency
EPA-EE	EPA-ethyl esters

EVA	ethylene vinyl acetate
FAME	fatty acid methyl ester
FDA	Food and Drug Administration
FDI	foreign direct investment
FFA	free fatty acids
FSSAI	food Safety and Standards Authority of India
FTIR	Fourier transform infrared
FWP	fresh water prawns
FY	financial year
GAG	glycosoaminoglycans
GAPs	good agricultural practices
GC-MS	gas-chromatography mass spectroscopy
GMPs	good manufacturing practices
GOI	Government of India
GPC	gel permeation chromatography
GPx	glutathione peroxidase
GR	glutathione reductase
GSH	reduced glutathione
GSSG	oxidized glutathione
H_2O_2	hydrogen peroxide
HACCP	hazard analysis and critical control points
HDPE	high density polyethylene
HFFS/VFFS	horizontal/vertical form-fill-seal
HIPS	high impact polystyrene
HMFP	high moisture fruit products
HNO_2	nitrous oxide
HO	hydroxyl radical
HOCl	hydrochlorous acid
HPLC	high performance liquid chromatography
HRO_2-	hydroperoxyl radical
Hz	hertz
IFI	Indian Food Industry
IRS	infrared spectroscopy
LC-MS	liquid chromatography–mass spectrometry
LDPE	low density polyethylene
LF-NMR	low field nuclear magnetic resonance
MAP	modified atmosphere packaging

MAS-NMR	magic angle spinning-NMR
MDA	malondialdehyde
MHO	menhaden oil
MIT	Massachusetts Institute of Technology
MMT	million metric ton
MPEDA	Marine Products Export Development Authority
MPPO	modified polyphenylene oxide
MRI	magnetic resonance imaging
MT	metric ton
MUFA	monounsaturated fatty acids
NADPH	nicotinamide adenine dinucleotide phosphate
NaOH	sodium hydroxide
NHB	National Horticultural Board
NMR	nuclear magnetic resonance
NO^{-}	nitric oxide
NO_2^{-}	nitrogen dioxide
O_2^{-}	superoxide anion
OML	overall migration limit
ONNO	peroxynitrite
ORAC	oxygen radical absorbance capacity
PB	olybutadiene
PBAT	poly-butylenes adipate terephthalate
PBSA	poly-butylene succinate adipate
PCL	polycaprolactone
PE	polyethylene
PHA	polyhydroxylalkanoate
PHB	poly-3-hydroxybutyrate
Pi	pigments
PLA	phospholipase A
PLA	polylactic acid
PLC	phospholipase C
PME	pectin methylesterase
PP	polyethylene
PPO	poly phenol oxidase
PS	polystyrene
PTLF	preformed tray and lidding film
PUASE	pulsed ultrasound-assisted solvent extraction

PUFA	polyunsaturated fatty acids
PVC	polyvinyl chloride
PVDC	polyvinyl dichloride
PVOH	polyvinyl alcohol
RF	radiofrequency
RH	rice husk
RNS	nitrogen oxygen species
RO_2-	peroxyl radical
RONOO	alkyl peroxynitrates
ROS	reactive oxygen species (ROS)
RTE	ready to eat
$SC\text{-}CO_2$	super critical CO_2
SCF	Scientific Committee on Food
SCP	single cell protein
SDS	sodium dodecyl sulfate
SFA	saturated fatty acid
SFE	supercritical fluid extraction
SML	specific migration limit
SOD	superoxide dismutase
SSI	small scale industry
SSOPs	Sanitation Standard Operating Practices
TAG	triglycerides
TBARS	thiobarbituric acid reactive substances
TDI	tolerable daily intake
TLC	thin layer chromatography
TMS	tetramethylsilane
TS	total solids
TSS	total soluble solids
TWTFFS	three-web thermoform-fill-seal
UAUF	ultrasound-assisted ultrafiltration
USA	United States of America
USDA	United States Department of Agriculture
UV	ultraviolet
VSS	volatile suspended solids
W/W	weight/weight
WHC	water holding capacity
WSBP	waste sugar beet pulp
WVP	water vapor permeability

LIST OF SYMBOLS

a and *b*	empirical constants
a_w	water activity
D_e	diffusion co-efficient [m²/s]
D_{ef}	effective diffusivity [m²/s]
D_{eff}	effective diffusion coefficient [m²/s]
D_o	pre-exponential factor [m²/s]
E_a	activation energy [KJ mol⁻¹]
F_C	mean force in the cutting sequence
F_F	mean force in friction test
g	gravity [m/s²]
k_1	constant related to the initial rate of sorption [hour per % weight of dry solids to water]
k_1 and k_2	permeability parameters
k_2	constant related to equilibrium moisture content [% weight of dry solids to water]
K_{cat}	specific activity
K_{cat}/K_m	specificity constants
kg	kilogram
K_m	substrate affinity
M_0	initial moisture content [% dry basis]
M_e	equilibrium moisture content [% dry basis]
M_s	surface moisture content [% dry basis]
m_s and m_0	saturation and initial moisture contents [kg/kg, dry basis]
M_t	moisture content [% dry basis] at time, t [s]
R	equivalent radius (radius of the sphere having the same volume as the grain) [m]
r	pore radius [m]
R_g	universal gas constant [8.314 KJ mol⁻¹K⁻¹]
S	surface area [m²]
T	absolute temperature [K]
t	time [s]

V	volume [m³]
V_{max}	maximum velocity
y	distance traveled by the liquid at time t [m]
ΔM_0	initial moisture gain [% dry basis]
δ	contact angle [°]
γ	surface tension [N/m]
μ	viscosity [N s/m²]

FOREWORD BY
TRIDIB KUMAR GOSWAMI

I feel very delighted and honored to write this foreword to the book ***Developing Technologies in Food Science: Status, Applications, and Challenges*** published under the book series *Innovations in Agricultural and Biological Engineering* edited by Murlidhar Meghwal and Megh R. Goyal.

Food science is the applied science dedicated to the study of food, nutrition, and their health effect. It is a discipline in which the engineering, biological, chemical, principles of food processing and physical sciences basic knowledge are used to study the nature of foods, the causes of deterioration, the principles underlying food processing, and the improvement of foods for the consuming public.

In this book, first three sections cover many important topics on food science and technology.

The last fourth section covers fortification of paddy and rice cereals, functional foods from Indian subcontinent, and a update on phytochemistry and anti-oxidative effects in biological systems of aspalathus linearis herbal tea.

I congratulate the editors for the timely decision of bringing out this book for use by scientists, engineers, professionals and students. I am sure it will be a very useful reference book for professionals working in food science and technology, food processing and nutrition.

Dr. Tridib Kumar Goswami,
Professor
Agricultural and Food Engineering Department,
I.I.T., Kharagpur–721302, India

Tridib Kumar Goswami, PhD

PREFACE 1 BY MURLIDHAR MEGHWAL

Right from birth to till death one has to eat, drink, chew or suck food to survive. The food eating habits of an individual depend on the available food sources, cultural background, specific nutritional requirements, and health. Nowadays, governments, private organizations, NGOs, food scientists and technologists are trying their best to provide sufficient, safe, cheaper and healthier food to all citizens of their countries. The World Health organization (WHO) and the World Bank, etc. are actively working in this direction. WHO is focusing on research and innovative projects to feed the disadvantaged sector and make the world a healthy and safer place to live.

Of course, the field of food science is not new but there is not much literature found which shows the emerging trends, applications and challenges in food science and technology. Because these aspects have taken on new importance and dimensions in today's fast food and high level and precise research world, there is lot of scope and career opportunities in this focus area. This subject demands a very broad educational background in the sciences due to the wide variety of problems and the number of chemical, physical, and biological sciences required to solve them.

Physicochemical characterization of food comes under food science whereas application of basic knowledge to come up with a valuable food product or food processing operation from the selection of raw materials through processing, preservation, and distribution comes under food technology. The current book not only considers food processing, preservation and distribution's, but it also taken into account the consumer's wants and needs.

I hope and believe that the subject matter in this book will be of great interest to a wide audience of those who are professionally engaged in some aspects of food production, marketing, and distribution. The targeted audience for this book includes food scientists, food technologists, practicing food process engineers, researchers, lecturers, teachers, professors, food professionals, food industry, students of these fields and all those who have inclination for food processing sector. The book not only covers the practical aspects of the subject, but also includes a lot of basic information. Therefore students in undergraduate, graduate courses, postgraduate and post-doctoral research programs will also receive benefits. In order for the book to be useful to engineers, coverage of each topic is comprehensive enough to serve as an overview of the most recent and relevant research and technology. Numerous references are included at the end of each chapter. A wide range of reference material has been included. Readers are encouraged to study widely and intensively from as many of these references as much time and resources permits.

The editors wish to acknowledge the professionals who have contributed to this book volume.

For any technical, typographical and other human error in this book, I am extremely sorry and welcome your valuable suggestions.

—Murlidhar Meghwal, PhD

PREFACE 2 BY MEGH R. GOYAL

So are you to my thoughts as food to life,
Or as sweet seasoned showers are to the ground;
And for the peace of you I hold such strife,
As 'twixt a miser and his wealth is found.
Now proud as an enjoyer, and anon
Doubting the filching age will steal his treasure;
Now counting best to be with you alone,
Then bettered that the world may see my pleasure;
Sometime all full with feasting on your sight,
And by and by clean starved for a look;
Possessing or pursuing no delight
Save what is had, or must from you be took.
Thus do I pine and surfeit day by day,
Or gluttoning on all, or all away.
*—**William Shakespeare***

The Institute of Food Technologists (http://www.ift.org) defines Food Science (FS) as *"the discipline in which the engineering, biological, and physical sciences are used to study the nature of foods, the causes of deterioration, the principles underlying food processing, and the improvement of foods for the consuming public."* The textbook ***Food Science*** defines food science as *"the application of basic sciences and engineering to study the physical, chemical, and biochemical nature of foods and the principles of food processing."* Food science brings together multiple scientific disciplines. It incorporates concepts from fields such as microbiology, chemical engineering, and biochemistry.

https://en.wikipedia.org/wiki/Food_science indicates that *"Food science is the applied science devoted to the study of food."* The focus areas of food science are: food chemistry, food physical chemistry, food engineering, food microbiology, food packaging, food preservation, food substitution, food technology, molecular gastronomy, new product development, quality control, and sensory analysis. Activities of food scientists include: the development of new food products, design of processes to produce these foods, choice of packaging materials, shelf-life studies, sensory evaluation of products using panels or potential consumers, as well as microbiological and chemical testing. Food scientists may study more fundamental phenomena that are directly linked to the production of food products and its properties.

http://www.ift.org/ defines Food Technology (FT) as *"the application of food science to the selection, preservation, processing, packaging, distribution, and use of safe food."* Related fields include analytical chemistry, biotechnology, engineering, nutrition, quality control, and food safety management. The "http://www.amity.edu/aift/AboutUs.asp" mentions that *"Food Technology is the application of food science to the selection, preservation, processing, packaging, distribution and consumption of delicious, safe, nutritious and wholesome food. Food Science is a discipline concerned with all the technical aspects of food, beginning with harvest or slaughtering or milking or catching and ending with its processing and consumption. Food Science is in fact highly inter-disciplinary applied science, incorporating concepts and involving topics from Chemistry, Microbiology, Life Sciences, Chemical Engineering, Biochemistry, Nutrition, Biotechnology, Mechanical Engineering, Agricultural, Dairy and Veterinary Sciences. India's agricultural production base is strong, kudos to our farmers and agricultural scientists but at the same time wastage of agricultural produce is massive mostly due to lack of postharvest technological and food processing facilities. The wastage of fruits and vegetables alone is estimated at about 35%, the value of which is approximately Rs. 370 billion annually. This is mainly because the processing level is very low, i.e., around 2.2% for fruits and vegetables compared to countries like USA (65%), Philippines (78%) and China (23%); 26% for marine, 6% for poultry and 20% for buffalo meat, as against 60–70% in developed countries. The share of India's export of processed food in global trade is only 1.5% at present. There exists, therefore, tremendous scope of expansion of food processing sector and as a result, increasing employment opportunities for the qualified food technologists in the times to come."*

The journal *Trends in Food Science and Technology* covers product development issues in the science and the technology related to food production. The journal is the official publication of the European Federation of Food Science and Technology and of the International Union of Food Science and Technology.

At 49th annual meeting of the Indian Society of Agricultural Engineers at Punjab Agricultural University (PAU) during February 22–25 of 2015, a group of ABEs and FEs convinced me that there is a dire need to publish book volumes on focus areas of agricultural and biological engineering (ABE). This is how the idea was born for new book series titled Innovations in Agricultural & Biological Engineering. This book volume under the AAP book series sheds light on different technological aspects of food science and technology; and it contributes to the ocean of knowledge on food science.

The *Global Confederation for Higher Education Association for Agriculture and Life Sciences* (GCHERA) named Dr. R. Paul Singh the "2015 Laureate of the World Agriculture Prize." Dr. Singh, my class fellow at Punjab Agricultural University and Professor Emeritus in Food Engineering (FE) at University of California, Davis, is a distinguished Food Engineer, and a pioneer in FE to improve lives world over. The award not only recognizes his research and teaching experience, but also grants importance to the technological advances in FE, food science and technology [Resource, November/December issue 2015 22(6), 29 by ASABE].

The contributions by the contributing authors to this book volume have been most valuable in the compilation. Their names are mentioned in each chapter and in the list of contributors. I appreciate you all for having patience with my editorial skills. This book would not have been written without the valuable cooperation of these investigators, many of whom are renowned scientists who have worked in the field of food engineering throughout their professional careers. I am glad to introduce Dr. Murlidhar Meghwal, who is an Assistant Professor in the Food Technology, Center for Emerging Technologies at Jain University – Jain Global Campus in District Karnataka, India. With several awards and recognitions, including from the President of India, Dr. Meghwal brings his expertise and innovative ideas in this book series. Without his support and leadership qualities as editor of the book volume and his extraordinary work on food technology applications, readers will not have this quality publication.

I will like to thank editorial staff, Sandy Jones Sickels, Vice President, and Ashish Kumar, Publisher and President at Apple Academic Press, Inc., for making every effort to publish the book when the diminishing water resources are a major issue worldwide. Special thanks are due to the AAP Production Staff also.

I request that the reader offers his constructive suggestions that may help to improve the next edition.

I express my deep admiration to my family and colleagues for understanding and collaboration during the preparation of this book volume. Can anyone live without food or milk?

As an educator, there is a piece of advice to one and all in the world: *"Permit that our almighty God, our Creator, provider of all and excellent Teacher, feed our life with Healthy Food Products and His Grace; and Get married to your profession."*

—*Megh R. Goyal, PhD, PE*
Senior Editor-in-Chief

WARNING/DISCLAIMER

User Must Read It Carefully

The goal of this book volume, *Developing Technologies in Food Science: Status, Applications, and Challenges*, is to guide the world community on how to manage efficiently for technology available for different processes in food science and technology. The reader must be aware that dedication, commitment, honesty, and sincerity are important factors for success. This is not a one-time reading of this compendium.

The editors, the contributing authors, the publisher and the printer have made every effort to make this book as complete and as accurate as possible. However, there still may be grammatical errors or mistakes in the content or typography. Therefore, the content in this book should be considered as a general guide and not a complete solution to address any specific situation in food engineering. For example, one type of food process technology does not fit all cases in food engineering/science/technology.

The editors, the contributing authors, the publisher and the printer shall have neither liability nor responsibility to any person, any organization or entity with respect to any loss or damage caused, or alleged to have caused, directly or indirectly, by information or advice contained in this book. Therefore, the purchaser/reader must assume full responsibility for the use of the book or the information therein.

The mention of commercial brands and trade names are only for technical purposes. No particular product is endorsed over another product or equipment not mentioned. The author, cooperating authors, educational institutions, and the publishers Apple Academic Press Inc., do not have any preference for a particular product.

All weblinks that are mentioned in this book were active on December 20, 2016. The editors, the contributing authors, the publisher and the printing company shall have neither liability nor responsibility, if any of the weblinks are inactive at the time of reading of this book.

ABOUT LEAD EDITOR

 Murlidhar Meghwal, PhD, is a distinguished researcher, engineer, teacher and professor in Food Technology, Centre for Emerging Technology, Jain Global Campus, Jain University, Bangalore, India. He is the lead editor for this book volume. He received his BTech degree (Agricultural Engineering) in 2008 from the College of Agricultural Engineering Bapatla, Acharya N. G. Ranga Agricultural University, Hyderabad, India; his MTech degree (Dairy and Food Engineering) in 2010, his PhD degree (Food Process Engineering) in 2014 from the Indian Institute of Technology Kharagpur, West Bengal, India.

He worked for one-year as a research associate at INDUS Kolkata for the development of a quicker and industrial-level parboiling system for paddy and rice milling. In his PhD research, he worked on ambient and cryogenic grinding of fenugreek and black pepper by using different grinders to select a suitable grinder.

Currently, Dr. Meghwal is working on developing inexpensive, disposable and biodegradable food containers using agricultural wastes; quality improvement, quality attribute optimization and storage study of kokum (*Garcinia indica* Choisy); and freeze drying of milk. At present, he is actively involved in research and is the course coordinator for MTech (Food Technology) courses. He also teaches at the Food Science and Technology Division, Jain University Bangalore, India. He has written two books and many research publications on food process engineering. He has attended many national and international seminars and conferences. He is reviewer and member of editorial boards of reputed journals.

He is recipient of the Bharat Scout Award from the President of India as well as the Bharat Scouts Award from the Governor. He received a meritorious "Foundation for Academic Excellence and Access (FAEA-New Delhi) Scholarship" for his full undergraduate studies from 2004 to 2008.

He also received a Senior Research Fellowship awarded by the Ministry of Human Resources Development (MHRD), Government of India, during 2011–2014; and a Scholarship of Ministry of Human Resources Development (MHRD), Government of India research during 2008–2010.

Readers may contact him at: murli.murthi@gmail.com

ABOUT SENIOR EDITOR-IN-CHIEF

 Megh R. Goyal, PhD, PE, is a Retired Professor in Agricultural and Biomedical Engineering from the General Engineering Department in the College of Engineering at University of Puerto Rico–Mayaguez Campus; and Senior Acquisitions Editor and Senior Technical Editor-in-Chief in Agriculture and Biomedical Engineering for Apple Academic Press Inc.

He has worked as a Soil Conservation Inspector and as a Research Assistant at Haryana Agricultural University and Ohio State University. He was first agricultural engineer to receive the professional license in Agricultural Engineering in 1986 from College of Engineers and Surveyors of Puerto Rico. On September 16, 2005, he was proclaimed as "Father of Irrigation Engineering in Puerto Rico for the twentieth century" by the ASABE, Puerto Rico Section, for his pioneer work on micro irrigation, evapotranspiration, agroclimatology, and soil and water engineering. During his professional career of 45 years, he has received many prestigious awards. A prolific author and editor, he has written more than 200 journal articles and textbooks and has edited over 48 books. He received his BSc degree in engineering from Punjab Agricultural University, Ludhiana, India; his MSc and PhD degrees from Ohio State University, Columbus; and his Master of Divinity degree from Puerto Rico Evangelical Seminary, Hato Rey, Puerto Rico, USA.

BOOK ENDORSEMENTS

This book encompasses versatile aspects of emerging scientific technologies associated with the food industry; such as the application of nuclear magnetic resonance and ultrasound processing to food processing. It addresses niche topics like value-added product recovery from food wastes and cereal fortification with ease and emphasis. The food industry will find this book to be an intellectually enriched handbook incorporating various aspects of food processing in a concise and unembellished manner. This book will prove to be valuable resource on pertinent food technology for students aspiring to be associated with food processing industry.

— Harita R. Desai, PhD
Research Fellow, Department of Pharmaceutical Sciences
and Technology (DPST),
Institute of Chemical Technology (formerly UDCT), Nathalal Parekh
Marg, Mumbai, India

The current technology and research of food science are rapidly expanding in the current food production scenario. Innovation is a must for incorporating better food process technology, food safety and food and health nutrition. The book chapters are highly innovative and written by experienced research experts for improvement and innovation in research.

— R. K. Raigar, PhD
Research Scholar
Agricultural and Food Engineering Department,
Indian Institute of Technology Kharagpur, West Bengal, India

My heartiest congratulations to Professors Murlidhar Meghwal and Megh R. Goyal for bringing out this book which covers most of the topics relevant to those in food processing. This will be a great help for academicians and non-academicians interested in food systems.

— *Abhinav Mishra, PhD*
Research Scholar
Department of Nutrition and Food Science and Center for Food Safety
and Security Systems,
University of Maryland, College Park, Maryland, USA

The book is a nicely written and surely will be very useful for researchers, industry people and students.

— *Kacoli Banerjee, PhD*
Assistant Professor
Department of Zoology, Maharaja Sayajirao University of Baroda,
Gujarat, India

The book volume on *Developing Technologies in Food Science: Status, Applications, and Challenges* under the book series *Innovations in Agricultural and Biological Engineering* covers almost all the emerging issues in the field of food science and technology. The book is well written and shares important knowledge of the novel engineering and technology.

— *Bhupendra M. Ghodki, PhD*
Research Scholar in Food Process Engineering
Agricultural and Food Engineering Department, Indian Institute of
Technology Kharagpur, Kharagpur, India

The book is an authoritative, comprehensive, conceptually sound, and highly informative compilation of recent advances in various important emerging areas of food technology. It contains outstanding chapters written by scientists working in this field for a good number of years. The in-depth literature review and state-of-the-art treatment of the scientific and technological information means that the book can serve as an essential reference source to students and researchers in universities and research institutions as well act as a guideline for the food processors.

*— **J. Chitra, PhD Scholar***
Researcher
Agricultural and Food Engineering Department, IIT Kharagpur,
Kharagpur, West Bengal, India

I am fortunate to read "*Developing Technologies in Food Science: Status, Applications, and Challenges*" and happy to see the comprehensive content covered in it. The book provides an overview and background of recent advances in food process engineering, and will be a good introductory guide for students and professionals in the field of food process engineering.

*— **Deepak Kumar Garg, PhD***
Postdoctoral Research Associate
ADM Institute for the Prevention of Postharvest Loss,
University of Illinois at Urbana–Champaign, USA

OTHER BOOKS ON AGRICULTURAL AND BIOLOGICAL ENGINEERING BY APPLE ACADEMIC PRESS, INC.

Management of Drip/Trickle or Micro Irrigation
Megh R. Goyal, PhD, PE, Senior Editor-in-Chief

Evapotranspiration: Principles and Applications for Water Management
Megh R. Goyal, PhD, PE, and Eric W. Harmsen, Editors

Book Series: Research Advances in Sustainable Micro Irrigation
Senior Editor-in-Chief: Megh R. Goyal, PhD, PE

Volume 1: Sustainable Micro Irrigation: Principles and Practices
Volume 2: Sustainable Practices in Surface and Subsurface Micro Irrigation
Volume 3: Sustainable Micro Irrigation Management for Trees and Vines
Volume 4: Management, Performance, and Applications of Micro Irrigation Systems
Volume 5: Applications of Furrow and Micro Irrigation in Arid and Semi-Arid Regions
Volume 6: Best Management Practices for Drip Irrigated Crops
Volume 7: Closed Circuit Micro Irrigation Design: Theory and Applications
Volume 8: Wastewater Management for Irrigation: Principles and Practices
Volume 9: Water and Fertigation Management in Micro Irrigation
Volume 10: Innovation in Micro Irrigation Technology

Book Series: Innovations and Challenges in Micro Irrigation
Senior Editor-in-Chief: Megh R. Goyal, PhD, PE

Volume 1: Principles and Management of Clogging in Micro Irrigation

Book Series: Innovations in Agricultural and Biological Engineering
Senior Editor-in-Chief: Megh R. Goyal, PhD, PE

- Modeling Methods and Practices in Soil and Water Engineering
- Food Engineering: Modeling, Emerging issues and Applications.
- Emerging Technologies in Agricultural Engineering
- Dairy Engineering: Advanced Technologies and Their Applications
- Food Process Engineering: Emerging Trends in Research and Their Applications
- Soil and Water Engineering: Principles and Applications of Modeling
- Developing Technologies in Food Science: Status, Applications, and Challenges
- Agricultural and Biological Engineering Practices
- Soil Salinity Management in Agriculture: Emerging Technologies and Applications
- Engineering Practices for Agricultural Production and Water Conservation: An Interdisciplinary Approach
- Flood Assessment: Modeling and Parameterization
- Food Technology: Applied Research and Production Techniques
- Processing Technologies for Milk and Milk Products: Methods, Applications, and Energy Usage
- Engineering Interventions in Agricultural Processing
- Technological Interventions in Processing of Fruits and Vegetables

- Technological Interventions in Management of Irrigated Agriculture
- Engineering Interventions in Foods and Plants
- Technological Interventions in Dairy Science: Innovative Approaches in Processing, Preservation, and Analysis of Milk Products
- Novel Dairy Processing Technologies: Techniques, Management, and Energy Conservation
- Sustainable Biological Systems for Agriculture: Emerging Issues in Nanotechnology, Biofertilizers, Wastewater, and Farm Machines
- State-of-the-Art Technologies in Food Science: Human Health, Emerging Issues and Specialty Topics

EDITORIAL

Apple Academic Press Inc., (AAP) will be publishing various book volumes on the focus areas under book series titled *Innovations in Agricultural and Biological Engineering*. Over a span of 8 to 10 years, Apple Academic Press, Inc., will publish subsequent volumes in the specialty areas defined by *American Society of Agricultural and Biological Engineers* (http://asabe.org).

The mission of this series is to provide knowledge and techniques for agricultural and biological engineers (ABEs). The series aims to offer high-quality reference and academic content in Agricultural and Biological Engineering (ABE) that is accessible to academicians, researchers, scientists, university faculty, and university-level students and professionals around the world. The following material has been edited/modified and reproduced from: *"Megh R. Goyal, 2006. Agricultural and biomedical engineering: Scope and opportunities. Paper Edu_47 Presentation at the Fourth LACCEI International Latin American and Caribbean Conference for Engineering and Technology (LACCEI' 2006): Breaking Frontiers and Barriers in Engineering: Education and Research by LACCEI University of Puerto Rico – Mayaguez Campus, Mayaguez, Puerto Rico, June 21–23."*

WHAT IS AGRICULTURAL AND BIOLOGICAL ENGINEERING (ABE)?

"Agricultural Engineering (AE) involves application of engineering to production, processing, preservation and handling of food, fiber, and shelter. It also includes transfer of technology for the development and welfare of rural communities," according to http://isae.in. *"ABE is the discipline of engineering that applies engineering principles and the fundamental concepts of biology to agricultural and biological systems and tools, for the safe, efficient and environmentally sensitive production, processing, and management of agricultural, biological, food, and natural resources*

systems," according to http://asabe.org. "*AE is the branch of engineering involved with the design of farm machinery, with soil management, land development, and mechanization and automation of livestock farming, and with the efficient planting, harvesting, storage, and processing of farm commodities,*" definition by: http://dictionary.reference.com/browse/agricultural+engineering.

"*AE incorporates many science disciplines and technology practices to the efficient production and processing of food, feed, fiber and fuels. It involves disciplines like mechanical engineering (agricultural machinery and automated machine systems), soil science (crop nutrient and fertilization, etc.), environmental sciences (drainage and irrigation), plant biology (seeding and plant growth management), animal science (farm animals and housing), etc.,*" (Source: http://www.bae.ncsu.edu/academic/agricultural-engineering.php)

According to https://en.wikipedia.org/wiki/Biological_engineering: "*Biological engineering (BE) is a science-based discipline that applies concepts and methods of biology to solve real-world problems related to the life sciences or the application thereof. In this context, while traditional engineering applies physical and mathematical sciences to analyze, design and manufacture inanimate tools, structures and processes, biological engineering uses biology to study and advance applications of living systems.*"

SPECIALTY AREAS OF ABE

Agricultural and Biological Engineers (ABEs) ensure that the world has the necessities of life including safe and plentiful food, clean air and water, renewable fuel and energy, safe working conditions, and a healthy environment by employing knowledge and expertise of sciences, both pure and applied, and engineering principles. Biological engineering applies engineering practices to problems and opportunities presented by living things and the natural environment in agriculture. BA engineers understand the interrelationships between technology and living systems, have available a wide variety of employment options. The http://asabe.org indicates that "*ABE embraces a variety of following specialty areas.*" As new technology and information emerge, specialty areas are created, and many overlap with one or more other areas.

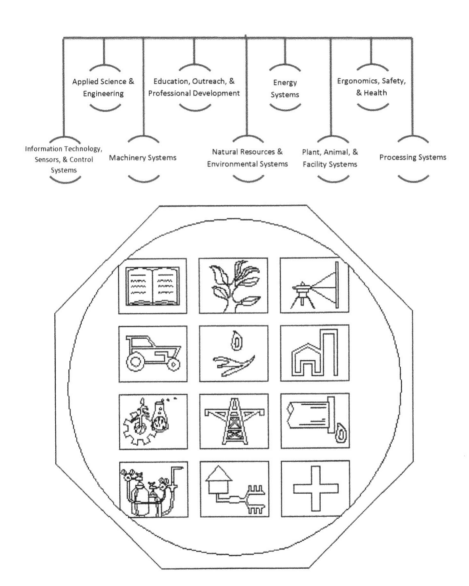

1. **Aqua Cultural Engineering**: ABEs help design farm systems for raising fish and shellfish, as well as ornamental and bait fish. They specialize in water quality, biotechnology, machinery, natural resources, feeding and ventilation systems, and sanitation. They seek ways to reduce pollution from aqua cultural discharges, to

reduce excess water use, and to improve farm systems. They also work with aquatic animal harvesting, sorting, and processing.

2. **Biological Engineering** applies engineering practices to problems and opportunities presented by living things and the natural environment.

3. **Energy:** ABEs identify and develop viable energy sources – biomass, methane, and vegetable oil, to name a few – and to make these and other systems cleaner and more efficient. These specialists also develop energy conservation strategies to reduce costs and protect the environment, and they design traditional and alternative energy systems to meet the needs of agricultural operations.

4. **Farm Machinery and Power Engineering**: ABEs in this specialty focus on designing advanced equipment, making it more efficient and less demanding of our natural resources. They develop equipment for food processing, highly precise crop spraying, agricultural commodity and waste transport, and turf and landscape maintenance, as well as equipment for such specialized tasks as removing seaweed from beaches. This is in addition to the tractors, tillage equipment, irrigation equipment, and harvest equipment that have done so much to reduce the drudgery of farming.

5. **Food and Process Engineering**: Food and process engineers combine design expertise with manufacturing methods to develop economical and responsible processing solutions for industry. Also food and process engineers look for ways to reduce waste by devising alternatives for treatment, disposal and utilization.

6. **Forest Engineering**: ABEs apply engineering to solve natural resource and environment problems in forest production systems and related manufacturing industries. Engineering skills and expertise are needed to address problems related to equipment design and manufacturing, forest access systems design and construction; machine-soil interaction and erosion control; forest operations analysis and improvement; decision modeling; and wood product design and manufacturing.

7. **Information and Electrical Technologies Engineering** is one of the most versatile areas of the ABE specialty areas, because it is applied to virtually all the others, from machinery design to soil

testing to food quality and safety control. Geographic information systems, global positioning systems, machine instrumentation and controls, electromagnetics, bioinformatics, biorobotics, machine vision, sensors, spectroscopy: These are some of the exciting information and electrical technologies being used today and being developed for the future.

8. **Natural Resources**: ABEs with environmental expertise work to better understand the complex mechanics of these resources, so that they can be used efficiently and without degradation. ABEs determine crop water requirements and design irrigation systems. They are experts in agricultural hydrology principles, such as controlling drainage, and they implement ways to control soil erosion and study the environmental effects of sediment on stream quality. Natural resources engineers design, build, operate and maintain water control structures for reservoirs, floodways and channels. They also work on water treatment systems, wetlands protection, and other water issues.

9. **Nursery and Greenhouse Engineering**: In many ways, nursery and greenhouse operations are microcosms of large-scale production agriculture, with many similar needs – irrigation, mechanization, disease and pest control, and nutrient application. However, other engineering needs also present themselves in nursery and greenhouse operations: equipment for transplantation; control systems for temperature, humidity, and ventilation; and plant biology issues, such as hydroponics, tissue culture, and seedling propagation methods. And sometimes the challenges are extraterrestrial: ABEs at NASA are designing greenhouse systems to support a manned expedition to Mars!

10. **Safety and Health**: ABEs analyze health and injury data, the use and possible misuse of machines, and equipment compliance with standards and regulation. They constantly look for ways in which the safety of equipment, materials and agricultural practices can be improved and for ways in which safety and health issues can be communicated to the public.

11. **Structures and Environment**: ABEs with expertise in structures and environment design animal housing, storage structures, and

greenhouses, with ventilation systems, temperature and humidity controls, and structural strength appropriate for their climate and purpose. They also devise better practices and systems for storing, recovering, reusing, and transporting waste products.

CAREERS IN AGRICULTURAL AND BIOLOGICAL ENGINEERING

One will find that university ABE programs have many names, such as biological systems engineering, bioresource engineering, environmental engineering, forest engineering, or food and process engineering. Whatever the title, the typical curriculum begins with courses in writing, social sciences, and economics, along with mathematics (calculus and statistics), chemistry, physics, and biology. Student gains a fundamental knowledge of the life sciences and how biological systems interact with their environment. One also takes engineering courses, such as thermodynamics, mechanics, instrumentation and controls, electronics and electrical circuits, and engineering design. Then student adds courses related to particular interests, perhaps including mechanization, soil and water resource management, food and process engineering, industrial microbiology, biological engineering or pest management. As seniors, engineering students' work in a team to design, build, and test new processes or products.

For more information on this series, readers may contact:

Ashish Kumar, Publisher and President
Sandy Sickels, Vice President
Apple Academic Press, Inc.,
Fax: 866-222-9549
E-mail: ashish@appleacademicpress.com
http://www.appleacademicpress.com/
publishwithus.php

Megh R. Goyal, PhD, PE
Book Series Senior Editor-in-Chief
Innovations in Agricultural and Biological Engineering
E-mail: goyalmegh@gmail.com

PART I

FOOD SCIENCE AND TECHNOLOGIES: STATUS, EMERGING APPLICATIONS AND CHALLENGES

STATUS OF AGRICULTURAL PRODUCTION AND FOOD PROCESSING INDUSTRIES IN INDIA WITH NATIONAL AND INTERNATIONAL PERSPECTIVE

GAGAN JYOT KAUR and JAGBIR REHAL

CONTENTS

1.1   INTRODUCTION

Indian food processing industry is widely recognized as a 'sunrise industry' having huge potential for uplifting agricultural economy, creation of large scale processed food manufacturing and food chain facilities, of employment and export earnings. The industry's estimated worth is around US\$ 67 billion, employing about 13 million people directly and about 35 million people indirectly. The food-processing sector in India is geared to meet the International Standards. FSSAI has the mandate to develop standards, harmonize them with International Standards consistent with food hygiene and food safety requirement as per India's food industry conditions.

Two nodal agencies, APEDA and MPEDA, promote exports from India. MPEDA is responsible for overseeing all fish and fishery product exports; and APEDA holds responsibility for the exports of other processed food products.

This chapter discusses current statues of Indian Food Processing Industry.

1.2   PRESENT SCENARIO

In 2011, India was amongst the world's top five largest producers of milk (121.8 million tons per annum), fruits and vegetables (170 million tons per annum) and food grain (230 million tons per annum). It also has one of highest livestock population (485 million). Its high productivity/yield is due to excellent climatic conditions with 52% cultivable land compared to 11% (world average), 46 different types of soil types (out of 60), in 20 agro-climatic regions. India is the world's second largest producer of food and holds the potential to acquire the top most status with sustained efforts. India is amongst the top producers of wheat, rice paddy, mangoes, guavas, sugarcane, cotton lint, bananas, potatoes, lentils, etc. [7].

Considering the manufacturing industries, the share of food processing industries is just 9% out of which only 25% is organized (Figure 1.1b). In this organized sector, the major stakeholders are dairy products, grain mill products, beverages, fruits, vegetables, meat, fish, oils and other as shown in Figure 1.1a. [6].

Table 1.1 indicates that the number of rice mills top the processing industry in India in comparison to other processing units [4]. Even though India tops in the production of fruits and vegetables, milk and meat but the number of units involved in the processing are less. The large number of

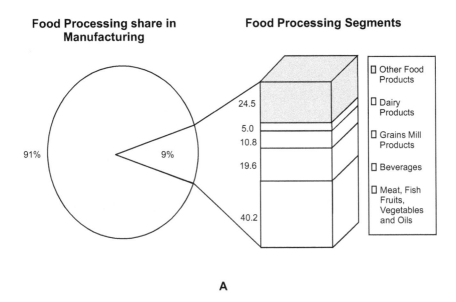

Food Processing share in Manufacturing

Food Processing Segments

91%   9%

24.5
5.0
10.8
19.6
40.2

☐ Other Food Products

☐ Dairy Products

☐ Grains Mill Products

☐ Beverages

☐ Meat, Fish Fruits, Vegetables and Oils

A

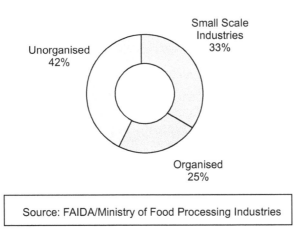

Structure of the Indian Food Processing Industry

Small Scale Industries
33%

Unorganised
42%

Organised
25%

Source: FAIDA/Ministry of Food Processing Industries

B

FIGURE 1.1   Percentage share of (a) different food processing industries (b) organized/ unorganized and small-scale industries.
(Source: Unpublished report by the Ministry of Food Processing Industries, Annual Report 2005-06, [4])

TABLE 1.1   Food Processing Units in Organized Sector

Name	Units
Fish processing units	568 (482 cold Storage)
Flour mills	516
Fruits and vegetable processing units	5293
Meat processing units	171
Milk products units	266
Modernized rice mills	35,088
Rice mills	139,208
Solvent extract units	725
Sugar mills	429
Sweetened and aerated water units	656

Source: Ministry of Food Processing Industries, Annual Report 2005–2006 [4].

rice mills can be due to various reasons like: profitability, demand, export of rice/basmati, ease in handling, less spoilage, and above all clarity in Government procedures for its establishment and export.

1.3   MAJOR PLAYERS

1.3.1   GRAIN PROCESSING

India is the world's second largest producer of rice, wheat and other cereals. The huge demand for cereals in the global market is creating an excellent environment for the export of Indian cereal products. In 2008, India had imposed ban on export of rice and wheat to meet its domestic needs, which was later lifted so that the global market demand can be met from the surplus production. The allowed marginal quantity of export cereals could not make any significant impact either on domestic prices or the storage conditions.

India is not only the largest producer of cereals but also the largest exporter of cereal products in the world. India's cereal export stood at Rs. $300,911.08 \times 10^6$ (Rs. 60.00 = 1.00 US$) during the year 2011–2012, which included rice 78% (including Basmati and Non Basmati) and 22% of other cereals (wheat and milled products). The 75% of the export share is from grains [5], which are mainly exported to the neighboring countries like UAE, Saudi Arabia, Iran, Indonesia and Nigeria. Looking at the availability of resources there is huge potential in processing of cereals and its export. A number of commercial units involved in primary and secondary processing of grains (handling, cleaning, grading, drying, storage, treatment and bagging of cereals for seed and food applications) are operational but limited number of units is operational in tertiary processing. Analyzing the market, it is observed that the demand for processed products is on the increase due to social, economic and behavior change worldwide which is responsible for the increased demand of bread, biscuits, flakes, breakfast cereals, etc. over traditional products like *chappati, pura,* etc.

1.3.2 DAIRY PROCESSING

India is the second largest producer as well as a major consumer of milk in the world with a total production of 121.8 million tons in 2010–2011 [5]. A major portion of milk is consumed either in the form of milk (46%) and traditional products (54%) like *ghee, dahi, butter milk, khoya, traditional sweets, channa, and milk powder.* It is estimated that 50–55% of the milk production is converted by traditional sector (*halwais*). Market size of traditional Indian milk products is estimated at more than one billion rupees with an annual growth at 50 billion rupees [11]. This underlines the significance of Indian milk sweets in national economy. The exports of milk and milk products from India form less than 1% of the total international trade. The export market is concentrated to the neighboring countries due to low cost of the end product. Concentrated milk powder (skimmed milk powder) form bulk of exports from India [2]. A niche global market has strongly emerged for ethnic Indian dairy products. Indigenous dairy products have a market worth INR 5 billion in North America, Canada, Europe and Middle East [10].

1.3.3 FRUITS AND VEGETABLES

India ranks second in production of vegetables (111.77 MMT) and fruits (57.73 MMT) [8], accounting for nearly 11.90% and 10.90% of country's share in the world production. Even though the production is high, but the processing units are in numbers only. It is recorded that actual postharvest losses are estimated at 25–40%. The major contributory factors of postharvest loss of fruits and vegetables are lack: of awareness about harvesting of fruits and vegetables, improper packaging and transportation and inadequate market infrastructure [9]. Amongst the top contributors to the total production are apple (2.40%), banana (23%), mango (51%), pineapple (9%), grapes (1.20), oranges (4%), guava (3%), onion (11.38%), tomato (5.65%), cauliflower (37.70%), cabbage (7.70%), green peas (36%) and potatoes (6.08%). In export market, the percentage contribution of grapes, fresh mangoes and vegetables has increased over the last three years 2009–2012 (Figure 1.2).

1.3.4 MEAT AND POULTRY PROCESSING

Meat and poultry production in India has been about 4.6 MT per year, where sheep (54%), cattle (26%), poultry (13%) and piggery (7%) is recorded. The available meat is mostly used fresh. Efforts are on to develop infrastructure for export of both fresh and processed meat and poultry. It is recorded that production of buffalo meat and poultry is increasing with every year.

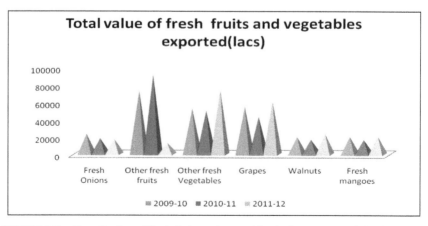

FIGURE 1.2   Contribution of fresh fruits and vegetables in the export pool (2009–2012).

1.4 FOOD PROCESSING INDUSTRY

Food processing is defined as set of techno-economic activities, applied to all the produces, originating from agricultural farm, livestock, aquaculture sources and forests for their conservation, handling and value-addition to make them usable as food, feed, fiber, fuel or industrial raw materials. It is regarded as the sunrise sector of the Indian economy in view of its large potential for growth and likely socio economic impact specifically on employment and income generation. It is critical to India's development, for it establishes a vital linkage and synergy between the two pillars of the Economy-Industry and Agriculture. The broad classification of the processing unit can be done as under (Figure 1.3).

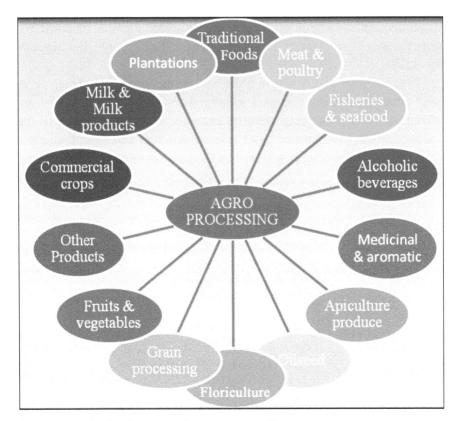

FIGURE 1.3 Major food processing industry in India.

MOFPI is a body, which looks after the matters of food processing industry, establishment, incentives and dispersal of funds. It has categorized the industry into following segments (Table 1.2) [3].

Considering the abundance in availability of resources, the percentage share of processing industry is just 9% but it is encouraging to see the annual growth rate of 9% in packaged food products (Table 1.1). This could be due to abundance of raw material, shift from conventional farming of food grains to horticulture, low production cost, change in consumption patterns and growing demand with increase in population. Along with it, the fiscal incentives given by Government of India are encouraging the investors to invest in establishment of processing industry (both SSI and mega food projects). Accounting for about 32% of country's food market, the food processing industry is one of the largest industries in India and is ranked top fifth (Table 1.2) in terms of production, consumption, export and expected growth. Its attractive growth rate of 9% (Table 1.3) has encouraged the Foreign Direct Investment (FDI).

The total food production in India is likely to double in next ten years with country's domestic market estimated to reach US$258 billion by 2015. The food processing industry forms an important segment of the Indian economy in terms of GDP, employment and investment, and is a major driver in the country's growth in the near future [1].

TABLE 1.2 Segmentation in Food Processing Industry

Sectors	Products
Consumer Foods	Snack foods, namkeens, biscuits, ready to eat food, alcoholic and non-alcoholic beverages
Dairy	Whole milk powder, Skimmed milk powder, Condensed milk, Ice cream, Butter, Ghee and Cheese
Fisheries	Frozen and canned products mainly in fresh form
Fruits and Vegetables	Beverages, juices, concentrates, Pulp, slices, frozen and dehydrated products, potato wafers/chips, etc.
Grains and cereals	Flour, bakeries, starch, glucose, cornflakes, malted foods, vermicelli, beer and malt extracts, grain based alcohol
Meat and Poultry	Frozen and packed – mainly in fresh form, egg powder

Source: Ministry of Food Processing, India, Annual report 2004 [3].

TABLE 1.3   Status of the Food Processing Industry

Rank	5th
Percentage share of packaged food in Indian exports	63%
Annual growth of packaged food	9%
Food processing industry worth (2012)	$121 billion
Foreign Direct Investment (2000–2013)	$1,811.06 million

1.5   INDIA VERSUS REST OF THE WORLD

While the country's agricultural produce is significant, the food processing in India is still at nascent stage. In spite of abundant resources and favorable climatic conditions, the levels of processing are not comparable to the rest of the world. Table 1.4 shows the percentage share of processed produce in India in comparison to the top processing industries [5].

Only 35% of the total milk produced is processed in India. Comparing to the lead countries, it is considerably low. On an average, the top processing countries are processing 60–65% of the commodity whereas as the India only processes 2.2 to 35% depending upon the commodity. India is processing way below its potential signifying a huge scope and bright future of the industry. India's share of processed food in global trade is around 1.5%, and it is expected to reach USD 255 billion by FY2016 at 13% growth rate per annum. The growth is due to its strategic geographic positioning (surrounded by a number of countries), which

TABLE 1.4   India's Contribution to Processed Food in Comparison to the Top Processing Countries

Segment	India	Other Countries
Fruits and vegetables	2.2%	USA (65%)
		Philippines (78%)
		China (23%)
Milk	35%	60–75 % in developed countries
Marine	26%	60–70 % in developed countries
Poultry	6%	
Buffalo Meat	20%	

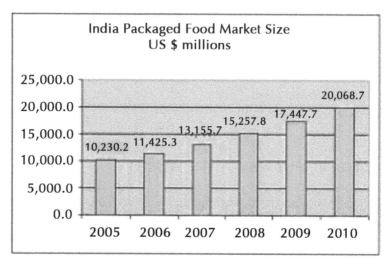

FIGURE 1.4   Trend of Processed foods in India (2006–2010).

have a demand for Indian products. The major importers of Indian processed food are countries in South Asia (34%), Middle East (29%), East Asia (17%), Western Europe (10%) Africa (1%), USA & Canada (1%) and rest of world (7%).

In 2013, IFI stood around US$ 39.03 billion and is expected to grow at the rate of 11% to reach US$ 64.3 billion by 2018. Looking at the share of packaged food products, the market size has increased from 10.23 billion dollars in (2006) to 20.06 billion dollars (2010) [5], thus recording an increase of 95% (Figure 1.4).

The export off processed foods has shown an upward trend from a value of 9.3 (billion $) to 22 (billion $) from 2011 to 2014 (Figure 1.5). Indian agricultural and processed food exports during April-May of 2014 stood at US $3813.63 million (APEDA). The numbers are increasing, indicating further growth in the sector.

1.5.1   PROCESSED FOOD SCENARIO WITH RESPECT TO SPECIFIC SECTORS

The industry structure and ongoing transformation offers opportunities for organized players to invest and grow. As the Indian market matures and

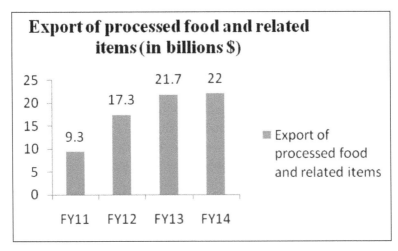

FIGURE 1.5 Trend of Exports recorded during FY (2011–2014).

consumers become more quality and brand conscious, the organized sector is poised to grow and gain prominence.

Food processing industry ranks third in terms of providing employment by employing approximately 18% of the total labor. This industry is amongst the horizon industries, which help in providing employment opportunities other than better food and nutrition. Grain, milk, fruits and vegetables, meat and poultry are the four major industries holding a major share in food processing sector in India.

Government is encouraging the entrepreneurs to invest in the agro-food processing sector by giving incentives. Various schemes listed below have been launched in the 11th five-year plan for strengthening the infrastructure in the processing sector.

- Food Parks Scheme;
- Scheme for Cold Chain, value addition and preservation infrastructure;
- Scheme for Modernization of Abattoirs;
- Establishment/modernization of food processing industries;
- Quality management, assurance and codex standards;
- Human resource development.

The growth expected in India's food processing sector is based on sound fundamentals. On the supply side, India is among the leading producers, now the focus is on improving productivity, reducing losses due to poor storage facilities, strengthening supply chain, improving rural infra-structure and providing quality food to all at affordable prices. On the demand side, there has been a major shift in the food consumption basket of Indians as a result of changing demographics and socio-economic struc-ture of India's population along with urbanization and rising disposable incomes. Dairy, poultry products and non-vegetarian packaged food items are increasingly occupying a significant share of food basket of Indians. As organized retail makes inroads into the Indian economy, the food pro-cessing sector will get another boost.

To bring India's food processing industry at par with the best in the world and be the preferred choice for processed foods, the Indian gov-ernment has adopted a positive outlook towards foreign investment in this sector. Under the automatic route, 100% FDI is allowed in the sec-tor, while the country is forging partnerships with various countries to enhance trade in processed foods and has granted a considerable degree of autonomy to state governments to identify partners to spur growth in food processing industry. The opportunity for NRIs and PIOs is immense in this sector, as the country is looking at best practices; technology and people who can help develop the sector and build a trained workforce for the ris-ing sector. As the country adopts a decentralized model for developing the food processing industry, partnering with states will be the best way to do business in this sector.

1.6   CONCLUSIONS

India is primarily an agriculture-based country, where 80% of the popula-tion is involved in this occupation. Green Revolution has helped India to fulfill its national need and reach the targets of production and yield, but industry has always played a vital role in the development of any nation. India being the top producer in grains, milk, poultry and fruits and vegeta-bles shows a huge potential in the development of food processing indus-try. The production, growth rate and returns showcase a promising future for this industry. To take it further the Government has introduced various schemes in the 11th five-year plan to encourage the investment from the big houses.

1.7   SUMMARY

The food processing industry provides vital linkages and synergies between industry and agriculture. The Food Processing Industry sector in India is one of the largest in terms of production, consumption, export and growth prospects. It helps to store perishable and semi-perishable agricultural commodities, avoid glut in the market, check postharvest losses and make the produce available during off-season, generates employment, development of ready-to-eat products, saves time for cooking, helps in improving palatability and organoleptic quality of the produce by value addition, enables transportation of delicate perishable foods across long distances, makes foods safe for consumption by checking of pathogenic microorganisms, modern food processing also improves the quality of living by way of healthy foods.

The government has accorded it a high priority, with a number of fiscal reliefs and incentives, to encourage commercialization and value addition to agricultural produce, for minimizing pre/postharvest wastage, generating employment and export growth. The food-processing sector in India is geared to meet the international standards. FSSAI has the mandate to develop standards and also to harmonize the same with International Standards consistent with food hygiene and food safety requirement and to the conditions of India's food industry.

KEYWORDS

- **agri-exports**
- **agro commodities**
- **codex**
- **cold chain**
- **dairy processing**
- **establishment/modernization of food processing industries**
- **food parks scheme**

- food processing
- food processing industry
- fruits processing
- government policies
- grain milling
- growth rate
- human resource development
- incentives
- Indian food industry
- meat products
- national horticulture mission
- preservation infrastructure
- processing
- quality management, assurance and codex standards
- scheme for cold chain, value addition and preservation infra-structure
- scheme for modernization of abattoirs
- value addition
- vegetable processing

REFERENCES

1. ASA and Associates (2015). A Brief Report on Food processing Sector in India.
2. Dutta, S. (2013). Economic importance of traditional Indian dairy products. National training on advances in Production, functional, Rheological and Quality aspects of Traditional Indian Dairy Products. NDRI, Karnal. pp. 237–241.
3. GOI (2004). Annual Report. Ministry of Food Processing, India.
4. GOI (2006). Committee on Agriculture (2005–2006), 14th Lok Sabha. Ministry of Agriculture, Department of Agricultural Research and Education (DARE) August 2006.
5. http://agropedia.iitk.ac.in/content-india-largest-producer-and-exporter-cereal-product
6. http://www.oifc.in/business
7. http://mofpi.nic.in/ContentPage.aspx?CategoryId=148
8. http://nhb.gov.in/

9. Mahajan, B. (2015). Maintenance of postharvest quality of horticultural crops. In: *Food Quality and Safety: Recent Advances in Evaluation Techniques*. CIPHET, PAU, Ludhiana. pp. 37–42.

10. NAARM (2003). Export Potential of Dairy Products. *Policy Paper 23* of session held at National Academy of Agricultural Research Management (NAARM) Hyderabad.

11. Patil, G. R. (2013). Current scenario, scope and challenges of traditional Indian dairy products. National Training on Advances in Production, Functional, Rheological and Quality aspects of Traditional Indian Dairy Products. NDRI, Karnal. pp. 1–11.

CHAPTER 2

CHALLENGES IN LIPASE MEDIATED SYNTHESIS OF FOOD FLAVOR ESTER COMPOUNDS

ANSHU SINGH and RINTU BANERJEE

CONTENTS

2.1   INTRODUCTION

Flavor, a characteristic of foods and beverages, is a collective perception build up through the binding of active metabolites with receptor proteins. Flavors are not only composed of different volatile compounds, but they also contain nonvolatile compounds, which are responsible for taste sensations [18, 19]. Flavor is governed by two factors, taste and aroma. Odor-active volatile compounds responsible for aroma includes, esters, alcohols, acids, lactones, carbonyl compounds, acetals, phenols, sulfur containing volatiles, nitrogen-containing volatiles, etc. Flavor esters are the largest qualitative constituent highly demanded in the market. These ester groups include ethyl esters of chain fatty acids (ethyl butyrate, hexanoate, octano-ate, decanoate and dodecanoate), which give fruity and wine-like aroma. Tropical fruit and banana-like aromas are due to the acetate esters of higher alcohols like isobutanol, isoamyl alcohol, and hexanol.

This chapter discusses status, challenges of lipase-mediated synthesis of food flavor ester compounds in foods.

Hundreds of short chain flavor esters have been identified, which are mostly ethyl esters formed by condensation reaction between the carboxyl group of an organic acid and the hydroxyl group of an alcohol or phenol. In plants, ethyl esters such as methyl, propyl, butyl, isobutyl, amyl and isoamyl esters are mostly responsible for flavors since alcohol residues are common metabolites produced during amino acid degradation. Enzymes such as lipase or esterase are mainly responsible for biosynthesis of esters in plant.

Short chain esters are known for the pleasant aroma of fruits and there-fore play a significant role in flavoring of baked foods, wines, cheese, cream, butter, yogurt, jam, jellies, candies, etc. [15]. In recent years, there has been a strong demand for flavors either extracted from natural sources or chemically synthesized. Extraction from natural sources results in expensive final products. Feron et al. [10] reported that vanilla extracted from vanilla pod costed about $4000 kg^{-1} while chemically synthesized counterpart was only $12 kg^{-1}. Synthetic compounds overcome issue of low yield and cost but their degree of purity is of big concern [27]. In such scenario, esters obtained through biological mean would better serve the food industry's needs and make product palatable among the consumers. The two methods by which flavors synthesis can be done are described in the following sections [5, 30].

2.1.1   BIOSYNTHESIS USING MICROBIAL CELL CULTURE

Biosynthesis using microbial cell culture occurs through solid-state fer-mentation. Addition or presence of precursors or intermediates to the culture medium is highly required. During the microbial growth, these biomolecules are converted to the desired form by entering metabolic cycles. Major drawback associated with such type of biotransformation is low yield, which is mainly caused due to slow microbial growth and substrate/product toxicity (Table 2.1).

TABLE 2.1 Fruity Aroma Compounds Synthesis Via Solid-State Fermentation Using Following Substrate

Substrate	Microbes
Cassava bagasse or giant palm bran	*Kluyveromyces marxianus*
Cassava bagasse, apple pomace or Soybean	*Ceratocystis fimbriata*
Coffee husk	*Ceratocystis fimbriata*
Pre-gelatinized rice	*Neurospora sp.*

2.1.2 BIOCONVERSION OF PRECURSORS *VIA* ENZYMATIC ROUTES

Enzymatic biotransformation has been well studied for the production of flavors. Esterases and lipases (triacylglycerol hydrolases; EC 3.1.1.3) are mainly used for the production. Among these two enzymes, lipases are usually highly specific and well tested enzyme for synthesis [11]. Lipase – from *Candida cylindracea, Candida antarctica Pseudomonas fluorescens, Mucor miehei, Aspergillus niger, Aspergillus javanicus Rhizopus arrhizus, Rhizopus oryzae* and *Candida rugosa* – possess the capability of catalyzing the formation of ester. Langrand et al. [19] have tested the ability of thirteen commercial lipase preparations among them four microbial lipases namely lipases from *Mucor miehei, Aspergillus, Candida rugosa* and *Rhizopus arrhizus* were found to be most suitable for the synthesis of esters containing short chain fatty acids. Therefore, present chapter provides the overview of the role of lipase in food fragrance sector.

2.2 FLAVOR ESTER PREPARATION

2.2.1 EXTRACTION FROM NATURAL SOURCES

The traditional method for obtaining the flavors component was by isolating them from natural resources. The mode of extraction includes: (i) organic solvents extraction, (ii) steam distillation, and (iii) supercritical fluid extraction.

Major drawbacks of these methods include, low selectivity, substantial loss of fragile flavors during extraction, high-energy costs, and the low yield. As plant growths are much prone to weather risk and pest attack therefore these methods are not confiable.

2.2.2 CHEMICAL SYNTHESIS

Esters are important compounds of the aroma can be directly extracted from natural sources. Unfortunately, due to the high recovery cost and low yields of the product during extraction, chemical synthesis provides an alternative method [15, 16]. After elucidation of flavor components, structure/synthetic flavors were produced by chemical synthesis. Flavor ester can be synthesized chemically by undergoing any of the reactions solvolytic, condensation or free radical processes. Commonly ester formation occurs under acidic conditions, by reaction of a carboxylic acid with an alcohol in the presence of a catalyst such as concentrated sulfuric acid. The chemically synthesized products are not accepted by today's market, as an ever-growing demand of consumer for food and health oriented products from natural sources. Chemical synthesis of food aromas possess several drawbacks such as, poor reaction selectivity leading to undesirable side reactions, low yields, pollution, high manufacturing costs, and impossibility of labeling the resulting products "natural" [5]. Although these flavor esters are manufactured mostly by chemical methods that include use of aggressive chemical catalysts, yet their application in food is highly questioned.

2.2.3 ENZYMATIC METHODS

A new prospect to produce natural esters was opened by use of biotechnological routes to produce natural flavors [3]. Direct biosynthesis by

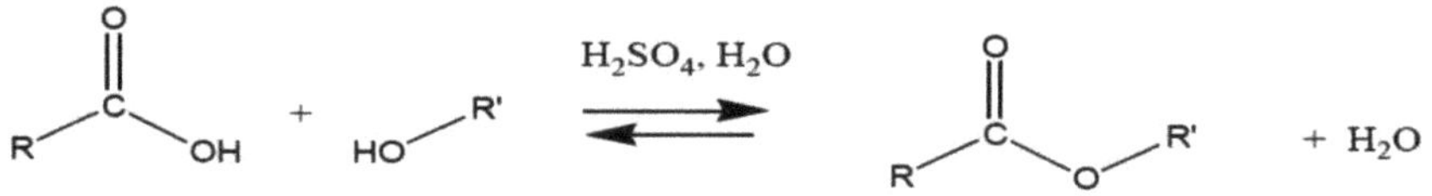

FIGURE 2.1   Chemical synthesis of esters under acidic condition.

fermentation, enzymatic esterification processes are techniques, which have provided large number of flavor esters. Lower productivity, a lesser ester concentrations, high purification cost are some of the problems common in fermentation process. For these reasons, enzymatic conversion can be employed to produce the particular short chain esters. Lipase, esterase, and oxidoreductase are some of the major enzymes of interest in this area. Among all enzymes, lipase is preferred for ester synthesis, as it acts on wide range of substrates, highly stable under extreme temperature, and carries out biotransformation in both aqueous phase and non-aqueous system.

2.3   LIPASE MEDIATED SYNTHESIS

Ester synthesis by means of either soluble or immobilized lipases constitutes an interesting application for the food additives industry [3, 6, 7]. Lipase-mediated ester synthesis is the alternative technique to chemical method as esters produced by organic phase biocatalysis comply with the US Food and Drug Administration's definition of "natural" [20]. Lipases play an important role in short chain ester synthesis by direct esterification (alcohol and acid) and transesterification (ester and alcohol) processes (Table 2.2).

TABLE 2.2   Short Chain Ester Formation in Organic Solvent System by Direct Esterification

Alcohol	Acid	Solvent	Ester	Sources of lipase	Ref.
Butanol	Acetic acid	Heptane	Butyl acetate	*Mucor miehei*	[22]
Citronellol	Buytric acid	Hexane	Citronellyl butyrate	*Rhizopus* sp.	[22]
Ethanol	Acetic acid	Heptane	Ethyl Acetate	*Candida antarctica*	[8, 9]
Geraniol; Citronellol	Acetic acid	Hexane Hexane	Geranyl acetate; citronellyl acetate	*Mucor miehei; Candida antarctica*	[25]
Hexanol	Acetic acid	Hexane	Hexyl acetate	*Staphylococcus simulans*	[22]

2.3.1 DIRECT ESTERIFICATION

Direct esterification between carboxylic acids and alcohols is highly dependent on alcohol and/or acid chain length [23, 30]. The use of lipolytic enzymes to catalyze the direct esterification reaction for producing flavor esters has been well investigated (Table 2.2). There are several reports [23, 24] indicating that direct esterification with acetic acid leads to inhibition of biocatalyst, which is disadvantage to this system. Free lipase from *M. miehie* was used for direct esterification of citronellol and geraniol with short-chain fatty acids in n-hexane [17]. It was observed that yield enhancement can be increased by varying the amount of acid.

The esterification of alcohols and organic acids catalyzed by lipolytic enzymes can be performed in hydrophobic organic solvent or in aqueous-organic two-phase systems [12, 26]. Immobilized lipase has also been used to increase the efficiency of the process. Covalent binded *M. miehei*lipase on the magnetic polysiloxane – polyvinyl alcohol particles was used for the synthesis using heptane as solvent [4].

2.3.2 TRANSESTERIFICATION

The transesterification is a reaction which occurs between ester and acid, ester and alcohol or ester and ester. This system is most favored and studied reaction system in flavor ester synthesis [28, 29, 31]. Hexyl butyrate has been synthesized by the immobilized lipase (Lipozyme IM-77) from *Mucor miehei* through mild trans-esterification of hexanol and tributrin. It has found enormous interest as the natural flavoring compound rather artificial or synthetic.

Two-flavor esters *n*-butyl acetate and *n*-propyl acetate occur naturally in various fruits like apple, strawberry, pear, etc. Mahapatra et al. [23] synthesized these flavor esters enzymatically by transesterification of vinyl acetate with alcohols namely *n*-butanol and *n*-propanol, respectively, in solvent-free systems using the lipase from *Rhizopus oligosporus* NRRL 5905 immobilized onto cross-linked silica gel. High yields of geranyl acetate and citronellyl acetate were achieved by transesterification of the respective terpenols with vinyl acetate using immobilized *Candida Antarctica* lipase [1]. Among the various solvents studied, high conversion of geranyl acetate was achieved in cyclohexane, hexane, heptane and iso-octane.

2.3.3 FACTORS AFFECTING LIPASE CATALYZED FLAVOR ESTER SYNTHESIS

2.3.3.1 Nature of Substrate

Lipases show degrees of selectivity towards the substrates [2]. During esterification reactions, lipases have high preference for long and medium chain fatty acids rather than short chain or branched ones due to stearic hindrance and electronic effects of the substrates. Substrate molar ratio also plays an important role in the esterification reaction.

2.3.3.2 Nature of Solvents

Lipase mediated synthesis are carried out in low-water content media by using organic solvent such as n-hexane, n-heptane, cyclohexane or isooctane. The choice of solvent has been found to be significant as it easily modulates the reaction rate, maximum velocity (V_{max}), specific activity (K_{cat}), substrate affinity (K_m) and specificity constants (K_{cat}/K_m). Extremely hydrophobic solvents are preferred as they favor better enzyme and substrate configuration. Recently major concern has been shown towards use of organic solvent in synthesis, therefore enzymatic synthesis of esters has also been attempted in solvent-free systems., i.e., where the reaction medium involves a reactant itself (i.e., an alcohol) as a solvent. High yield in solvent-free systems for lipase-catalyzed ester synthesis reactions yields have been well documented [20, 21]. Use of Supercritical carbon dioxide as suitable solvent increased conversions at very low enzyme loadings and substrate concentrations. Supercritical carbon dioxide opens up a new avenue for efficient conversion because they have low cost and possess nontoxicity.

2.3.3.3 Reaction Conditions

Reaction condition governs the progress of the catalysis [9, 32]. Selection of optimum reaction conditions helps to determine the minimum time necessary for obtaining good yield and directs the reaction towards cost-effectiveness. Reaction conditions, which influence the process, are time, temperature, pH, water addition, enzyme loading and reusability of enzyme. Sometime, agitation speed also plays critical role by overcoming external mass transfer limitations.

2.3.4   ADVANTAGES OF LIPASE MEDIATED ESTER SYNTHESIS

Esterification by lipases is an attractive alternative to bulk chemical routes because of the following facts:

- Lipases are substrate specific and stereoselective. Selectivity of lipase can be increased by using additives such as chiral or achiral amines, ethers and metal ions.
- Lipase acts on hydrophilic-hydrophobic interface and are tolerant to various organic solvents.
- Ester synthesis can be performed at room temperature, pressure and neutral pH in reaction vessels operated either batch wise or continuously, a feature that protects sensitive products or reactants from thermal degradation.
- Products obtained therefore are qualitatively more pure than the ones obtained by alternative chemical means because chemical catalysis tends to be unspecific and consequently generate several by-products.
- Handling of lipases is safe for the operator and the environment. Enzymatic processes are considered "environmentally friendly" as enzymes are biodegradable and there are fewer associated waste disposal problems.

2.3.5   DISADVANTAGES OF LIPASE MEDIATED ESTER SYNTHESIS

- The utilization of the enzyme is not extensive because of its high production and purification cost.
- The separation, toxicity and inflammability of organic solvents which are used in transesterification procedure remain as major problem.
- Use of specific instrument for enzymatic bioconversion with super-critical fluids hinders extensive commercial application.

2.4 CONCLUSIONS

Esters play a key role in sensory qualities of food products. The production of volatile ester by lipase is of major industrial interest, as the presence of natural flavor will decide the market value and acceptance of product among consumers. Lipases being a versatile enzyme are exploited in various sections of food industries, especially for flavor ester synthesis due to its high stability, specificity, catalytic activity both in aqueous media and non-aqueous media. Although lipase catalyzed esterification process has commercial significance, yet the utilization of the enzyme is not extensive because of its high cost. The future developments in low cost production and purification technologies would lower the cost of these enzymes for the increased applications. Various novel reactor designs with this immobilized enzyme would offer effective and efficient bioconversion using lipase. The development of lipases with novel properties by the directed evolution and molecular technologies, strain improvement by the site directed mutagenesis and medium optimization for lipase overproduction hold a major area of research in the future.

2.5 SUMMARY

Flavor esters are important compounds of the food industry. During the product development, the raw material undergoes through various processing steps which normally causes a loss of aroma therefore supplementation with flavor ester is of prime importance. Traditional method of extraction from natural sources or synthesizes via chemical means are well adapted by the industries. The high processing cost, the non-specific interactions, by-product formation and the most importantly the safety issues concern with chemical ways has increased the demand for the generation of "natural" flavors. For these reasons, biotransformation is becoming the best alternative to produce the short chain esters or fragrances instead of adopting the traditional methods. Flavor ester synthesis by means of either soluble or immobilized lipases constitutes an interesting green technology approach for food industries as esters produced by this method is considered to be safe and even FDA has defined such process as "natural."

Hence, the objective of the present chapter is to elaborate the greener route of lipase mediated synthesis of natural flavor ester compounds.

KEYWORDS

- aroma
- enzyme
- esterification
- fatty acid ester
- flavor esters
- lipase
- lipase mediated synthesis
- mediated synthesis
- natural flavor
- trans-esterification

REFERENCES

1. Akoh, C., & Yee, L. (1998). Lipase catalyzed transesterification of primary terpene alcohols with vinyl esters in organic media. *J Mol Catal B Enzym.*, 4, 149–153.
2. Alhir, S., Markajis, S., & Chandan, R. (1990). Lipase of *Penicillium caseicolum. J Agric Food Chem.*, 38, 598–601.
3. Berger, R. G. (2009). Biotechnology of flavors-the next generation. *Biotechnol Lett.*, 31, 1651–1659.
4. Bruno, L. M., de Lima Filho, J. L., Melo, E. H. M., & de Castro, H. (2004). Ester synthesis by *Mucor miehei*. Lipase immobilized on magnetic polysiloxane-polyvinyl alcohol particles. *Appl. Biochem. Biotech.*, 113(I-3), 189–200.
5. Brault, G., Shareck F., Hurtubise, Y., Lépine, F., & Doucet, N. (2014). Short-Chain flavor ester synthesis in organic media by an *E. coli* whole-cell biocatalyst expressing a newly characterized heterologous lipase. PLoS One, 9.
6. Carta, G., Gainer, J. L., & Benton, A. H. (1991). Enzymatic synthesis of esters using an immobilized lipase. *Biotechnol Bioeng.*, 37, 1004–1009.
7. Chiang, W., Chang, S., & Shieh, C. (2003). Studies on the optimized lipase-catalyzed biosynthesis of cis-3-hexen-1-yl acetate in n-hexane. *Proc Biochem.*, 38, 1193–1199.

8. Claon, P. A., & Akoh, C. C. (1994a). Enzymatic synthesis of geranyl acetate in n-hexane with *Candida antarctica* lipases. *J Am Oil Chem Soc.,* 71, 575–578.

9. Claon, P. A., & Akoh, C. C. (1994b). Effect of reaction parameters on SP435 lipase-catalyzed synthesis of citronellyl acetate in organic solvent. *Enz Microb Technol.,* 16, 835–838.

10. Feron, G., Bonnarme, P., & Durand, A. (1996). Prospects for the microbial production of food flavors. *Trends Food Sci Technol.,* 7, 285–293

11. Gillies, B., Yamazaki, H., & Armstrong, D. W. (1987). Production of flavor esters by immobilized lipases. *Biotechnol Letters*, 9709–9719.

12. Golberg, M., & Legoy, M. D. (1990) Water activity as a key parameter of synthesis reactions, the example of lipase in biphasic (liquid/solid) media. *Enz Microb Technol.,* 12, 976–981.

13. Guo, Z. W., & Sih, C. J. (1989). Enantio selective inhibition, a strategy for improving the enantioselectivity of biocatalytic systems. *J Am Oil Chem Soc.,* 111, 6836–6841.

14. Huang, S. Y., Chang, H. L., & Goto, M. (1998) Preparation of surfactant-coated lipase for the esterification of geraniol and acetic acid in organic solvents. *Enz Microb Technol.,* 22, 552–557.

15. Kim, J., Altreuter, D. H., Clark, D. S., & Dordick, J. S. (1998) Rapid Synthesis of fatty acid esters for use as potential food flavors. *J Am Oil Chem Soc.,* 75, 1109–1113.

16. Kumari, A., Mahapatra, P., Kumar, G. V., & Banerjee, R. (2008). Comparative study of thermostabilty and ester synthesis ability of free and immobilized lipases on cross-linked silica gel. *Bioprocess Biosyst Eng.,* 31, 291–298.

17. Laboret, F., & Perraud, R. (1999). Lipase catalyzed production of short-chain acids terpenyl esters of interest to the food industry.*Appl. Biochem. Biotechnol.,* 82(3), 185–198.

18. Langrand, G., Triantaphylides, C., & Baratti, J. (1988). Lipase catalyzed formation of flavor esters. *Biotechnol Lett.,* 10, 549–554.

19. Langrand, G., Rondot, N., Triantaphylides, C., & Baratti, J. (1990). Short chain flavor esters synthesis by microbial lipases. *Biotechnol Lett.,* 12, 581–586.

20. Liaquat, M., & Apenten, R. K. W. (2000). Synthesis of low molecular weight flavor esters using plant seedling lipases in organic media. *Food Chem Toxicol.,* 65, 295–299.

21. Lima, F. V., Pyle, D. L., & Asenjo, J. A. (1995). Factors affecting the esterification of lauric acid using an immobilized biocatalyst, enzyme characterization and studies in a well-mixed reactor. *Biotechnol Bioeng.,* 46, 69–79.

22. Longo, M. A., & Sanroman M. A. (2006).Production of food aroma compounds, microbial and enzymatic methodologies.*Food Technol Biotechnol.,* 44, 335–353.

23. Mahapatra, P., Kumari, A., Garlapati, V. K., Banerjee, R., & Nag, A. (2009). Enzymatic synthesis of fruit flavor esters by immobilized lipase from *Rhizopus oligosporus* optimized with response surface methodology. *J Mol Catal B Enzym.,* 60, 57–63.

24. Mahapatra, P., Kumari, A., Garlapati, V. K., Banerjee, R., & Nag, A. (2010). Optimization of process variables for lipase biosynthesis from *Rhizopus oligosporus* NRRL 5905 using evolutionary operation factorial design technique. *Ind J Microbiol.,* 50, 396–403.

25. Melo, L. L. M. M., Pastore, G. M., & Macedo, G. A. (2005). Optimized synthesis of citronellyl flavor esters using free and immobilized lipase from *Rhizopus* sp. *Proc Biochem.*, 40, 3181–3185.

26. Monot, F., Borzeix, F., Bardin, M., & Vandecasteele, J. P. (1991). Enzymatic esterification in organic media, role of water and organic solvent in kinetics and yield of butyl butyrate synthesis. *Appl Microbiol Biotechnol.*, 35, 759–765.

27. Park, Y. C., Shaffer, C. E. H., & Bennett, G. N. (2009). Microbial formation of esters. *Appl Microbiol Biotechnol.*, 85, 13–25.

28. Romero, M. D., Calvo, L., Alba, C., Daneshfar, A., & Ghaziaskar, H. S. (2005). Enzymatic synthesis of isoamyl acetate with immobilized *Candida antarctica* lipase in n-hexane. *Enz Microb Technol.*, 37, 42–48.

29. Shaw, J. F., Wang, D. L., & Wang, Y. J. (1991). Lipase-catalyzed ethanolysis and 2-propanolysis of triglycerides with long-chain fatty acids. *Enz Microb Technol.*, 13, 544–546.

30. Vandamme, E. J., & Soetaert, W. (2002).Bioflavors and fragrances via fermentation and biocatalysis.*J Chem Technol Biotechnol.*, 77, 1323–1332.

31. Ved, J. J., & Pai, J. S. (1996). Preparation of short esters by the lipase from *Mucor miehei* using heptane and silica gel. *Biotechnol Tech.*, 10, 855–856.

32. Yadav, G. D., & Lathi, P. S. (2003). Kinetics and mechanism of synthesis of butyl isobutyrate over immobilized lipases. *J Biochem Eng.*, 16, 245–252.

PART II

FOOD SCIENCE AND TECHNOLOGIES: EMERGING TRENDS AND ADVANCES

CHAPTER 3

ROLE OF HYDRATION IN GRAIN PROCESSING: A REVIEW

VINAY MANNAM

CONTENTS

3.1 INTRODUCTION

Cereals grains and legumes are one of the primary plant based food source for majority of the world population [46, 72]. The term "legumes is used to define dry beans (navy, black, pinto, lentil, lima, kidney, chickpea, faba, pigeon pea) and oilseeds (soybean, peanut, lupin seeds). The term 'cereal' is for seeds obtained from grass crops (rice, wheat, barley, maize, sorghum). Collectively, cereals and legumes are often referred to as pulses. Together, pulses constitute an important source of protein and energy uptake in the diet. According to FAO food supply data, total world consumption of cereals as food is 146.7 kg/capita/year, and legumes are 6.6 kg/capita/year [59]. The consumption in of cereals and legumes is higher in developing (152.3 and 11.9 kg/capita/year) and under-developed (113.8 and 12 kg/capita/year) countries, when compared to United States of America (108.2 and 4.5 kg/capita/year).

Nutritionally, cereals are abundant in carbohydrates and minerals, while legumes are protein rich [25]. Proteins are important nutrients obtained from legumes and are relatively economical to produce when compared to protein obtained from animal sources [38]. This prominence is more evident in under developed and developing countries, where lack of animal protein is compensated with a diet rich in legumes. In these regions, more than two thirds of protein intake is acquired from legumes (40.9 g/capita/day out of 65 g/capita/day). In United States, proteins from cereals and legumes have found new recognition due to their higher nutrient density and protein to calorie ratio with enhanced health benefits, resulting in 2010 dietary guideline recommendations (40 grams of beans and 110 grams of whole grains) [94, 101]. Colloquially used with rehydration, hydration defines the processing step for increasing the water content of dried edible grains.

In this chapter, different processing methods in preparation of functional foods using pulses are discussed with emphasis on soaking. Various physiological and chemical changes that occur during soaking are discussed for legumes and cereals with particular emphasis on barley. Current research involving factors effecting hydration of foods are reviewed in detail. Experimental and mathematical models are presented on hydration process of cereals and legumes. Important challenges are summarized and discussed.

3.2   IMPORTANCE OF HYDRATION IN PROCESSING OF SEEDS

Availability of proteins in cereal and pulses depends on the processing method [36]. Elimination of anti-nutrients from legumes and cereals is also an important function of processing [5]. While factors such as soil pH, organic matter content, available nutrients and soil-water relationships affect initial crop nutritional value, it is the post-harvest processing that will enable the bioavailability of all the nutrients in finished product

[35]. Refined and milled grains tend to possess reduced levels of valuable dietary fibers and important nutrients during processing. Traditionally, cereals and legumes are prepared in common households by soaking followed by cooking or thermal treatment. Different preparation methods for cereals and legumes and their effect for plant based diet are reviewed and discussed in depth elsewhere and are summarized in Table 3.1

TABLE 3.1 Different Processing Methods in Preparation of Cereals and Legumes

Method	Description	Effects	Seeds
Thermal	Boiling Cooking Retorting	Breaks structure by puncturing of cells walls to enable bioavailability of nutrients. Over exposure to thermal processing may result in loss of heat-liable and water soluble nutrients.	Processing dry beans by canning
	High pressure cooking	Thermal processing is also carried out to eliminate harmful bacteria such as *Clostridium botulinum.*	
Mechanical	Mechanical shear forces Milling Grinding	Shear forces are used to remove barn/germ from cereals, which contains phytates, thus enhancing availability of iron, zinc and calcium.	Rice, Wheat, Maize, Sorghum
		Cereal grains are also milled to make flour.	
		Grinding and milling reduce valuable nutrients.	
Soaking	Cereals and legumes soaked in water	Water-soluble phytates diffuse passively, and cell wall is exposed making important nutrients bio-available. Over soaking can cause textural damages and loss of nutrients into water. Soaking is often used as secondary process aid to most of the plant based-food preparation methods.	Dry beans, Rice, Barley, Wheat, Maize
Fermentation	Exposure of seeds to microbial enzymes in controlled environment	Phytates are hydrolyzed by microbial enzymes, which preserve protein content and increases digestibility of cereals. Fermentation is often preceded by soaking.	Barley, Wheat, Millet, Maize, Rice
Malting	Activating internal enzymes by allowing germination	Germination triggers internal enzymes in cereals and legumes which induce phytate hydrolysis. The rate of this activity can be controlled by temperature and Ph content.	Barley, Rye, Wheat

TABLE 3.2   Hydration Methods and Their Impact on Different Seed Varieties Downstream Processing [2, 50, 74, 102]

Cereal/ Grains	Type of Hydration	Effect	Product
Dry beans	Soaked in water prior to cooking, often at temperatures higher than 120° F for short time.	Softens seed, and removes phytic acid and enzyme inhibitors.	Prepared beans such as baked beans.
Wheat, Barley, Rye, Sorghum, dry beans	Steeped in water prior to germination.	Activate amylase enzymes that enable conversion of starch to sugars	Cereals are steeped for malt used for preparation of alcoholic beverages. Dry beans are steeping for sprouts.
Soybeans	Dry soybeans are hydrated for 3 -12 hours based on temperature of water.	Hydration of powdered soybeans help extract soy slurry through wet grinding.	Preparation of soy milk and tofu.
Rice	Rice is soaked prior to cooking for 3–4 hours at room temperature, or in some cases for longer periods of up to 12–14 hours.	Soaking prior to cooking softens seeds, and saves energy while cooking and longer soaking is preferred for better nutrient retention.	Cooked prepared rice.

[22, 35, 36, 47]. Simple hydration followed by cooking or germination of cereal and legumes has been considered more effective preparation technique then other intrusive methods such as milling, roasting, and grinding [90].

The type of hydration a seed receives in processing is tailored for specific cereal or legume and requires consideration of end product. Each hydration method has its own particular advantages and disadvantages as summarized in Table 3.2. Hydration of cereals and legumes can be broadly classified into simple soaking (continuous immersion) and steeping (intermittent submersion). This distinction is noteworthy because static water soaking conditions rapidly become anaerobic. As demonstrated in the Table 3.2, soaking is generally accompanied with temperature to improve hydration rates, and steeping is carried out to promote germination.

The common component in all the hydration methods is water and it primarily softens texture and ensures easier usage of one or more hydration methods is recommended and adopted in food industry to get the best nutritional retention and reduction of anti-nutrients of each preparation protocol [26, 71]. Factors such as time of exposure to water, properties of water and temperature impact the end product of the particular hydration method. Soaking or simple hydration is the initial step for most preparation methods and thus is the most studied and analyzed processing step for cereals and legumes. Great fundamental understanding of soaking and its effects on thermal processing, fermentation and germination/malting can lead to adoption of methods that improve processing times and yield higher bioavailability of nutrients [32, 36].

3.3 FACTORS AFFECTING HYDRATION

Cereal grains and legumes are naturally dried prior to harvest or immediately dried following harvest to control growth of bacteria and fungus during storage. Soaking reduces mineral complexing anti-nutrients from seeds by passive diffusion phytate salts of Na, Mg, and K into water [61]. The extent of reduction depends on the type of cereal or legume, water conditions and length of soaking. Rehydration is essential to solubilize essential nutrients and render palatability to dried beans and cereals. Hydration is referred to as 'amount of water absorbed by dry seed' and is frequently expressed on either a fresh weight or dry weight basis. Although there is no consistent nomenclature and standard definition for rehydration, the term varies in usage with different seed [47]. External variables play a prominent role in seed hydration [66, 75]. Studies have been carried out to analyze the effect of extrinsic parameters on hydration kinetics of seeds during soaking, and Table 3.3 summarizes some of the important work done [103].

3.3.1 INITIAL SEED CONDITIONS

Cereals and legumes are commonly stored in relatively low moisture conditions, immediately after harvesting. Postharvest conditions heavily

TABLE 3.3 Summary of Factors Affecting Soaking of Cereals and Legumes

Factor	Description	Effects	Sample
Initial conditions	Initial Storage conditions Pre-hydration treatments like de-hulling	Decrease in moisture content, and hard-to-cook phenomena *Seeds stored at high temperature and high humidity can become hard-to-cook, which is an important defect*	Dry beans Barley Chick-peas
Soaking time	Time of exposure of seeds to water	Water uptake, nutrients, anti-nutrients leaching and degradation *Longer soak time increase water uptake*	Rice Wheat Black beans
Tempera-ture	Temperature of hydration medium	Rate of water uptake [diffusivity] and final water content [water holding capacity], *Higher temperatures increase water diffusivity and holding capacity until equilibrium is achieved*	Amaranth White beans Rice Red- beans Chick-pea
Additives	Salts – effecting pH	Soak time, flavor *Presence of monovalent salts [Na+, K+] decreases cooking time, and slightly impacts flavor*	Dry beans Faba beans Pinto beans
Pressure	Pressure application in hydration container	water uptake, shorter soaking time, flavor, texture *Optimum pressure will increase water update, decreases solid loss, preserves flavor and texture*	Chick pea Green pea

impact hydration kinetics. Prior to and frequently during storage, moisture content of cereals and legumes is decreased to improve quality and reduce growth of microorganisms [35, 54]. Seeds that are stored for longer periods in drier environment tend to lose further moisture and are more difficult to hydrate than normally stored products. Initial moisture content before hydration in most cereals and legumes is between 12–15% [39, 66]. Lower moisture content in seeds possess a reduced rate of water uptake increasing the time required for seeds to achieve optimum water content. Whereas, higher moisture content induces textural damages and initiates germination while in storage [58, 96]. Also, temperature of seed before soaking effects the hydration rate, as lower seed temperature will hinder water diffusivity until equilibrium temperature is attained [96]. In cereal grains, de-hulling enables quicker rehydration rates [49]. Hard-to-cook seeds (seeds that resist hydration are often called steely seeds) are mainly associated with irregular moisture profiles and adverse storage conditions [67]. Treatment of seeds with enzymes and chemicals, such as protease and $NaHCO_3$, alters cell wall structure in seeds to improve water diffusivity in steely seeds. Optimized pretreatments with care can reduce soak and cooking times and effects of the enzyme and chemical pretreatments on legumes are studied elsewhere [85].

3.3.2   SOAKING TIME

In practice, preparation by soaking is often a slow and highly variable process with few control points. The rate of diffusion varies during soaking is based on type of grain or legume [36]. The relationship between rate of hydration and soak time has been extensively studied and is well understood, where water uptake increase as soak time is increases, while overall water diffusion rate decrease over time resulting in moisture content reaching a plateau [12, 75, 84]. Soak time is also dependent on other factors and is highly dependent on temperature [39]. Extended soaking of seeds in water can result in textural damage, leaching out of important nutrients, and may result in insanitary conditions with food safety and health implications.

3.3.3   SOAK TEMPERATURE

Temperature of soaking medium is one of the critical parameters affecting diffusivity of water into grain. Increases in soak temperature have been shown to increase overall water uptake and hydration rate [2, 66]. Controlled low temperatures are also important in soaking when germination of cereals or legumes has to be achieved for preservation of valuable nutrients [40]. Hydration of foods and the effect of temperature have been explained by diffusion kinetics such as Fick's law, which states that diffusion of water increases as the temperature increases, leading to faster inhibition rate [66, 103]. The dependence of diffusivity on temperature is described by an Arrhenius type equation [62, 93].

$$D_e = D_o e^{\left(E_a/R_g T\right)} \tag{1}$$

where, D_e is Diffusion co-efficient [m²/s], D_o is pre-exponential factor [m²/s], E_a is activation energy [KJ mol⁻¹], R_g is universal gas constant [8.314 KJ mol⁻¹K⁻¹], T is absolute temperature [K].

The effect of soak water temperature and its relationship with hydration rate and water uptake are inter-dependent. Thus, as soak temperature increases the water holding capacity will decrease following a critical mechanistic point (e.g., hydrolysis of polymers, starch gelatinization, or protein denaturation) while diffusivity continues to increase [93].

3.3.4   SOAK WATER ADDITIVES

Chemistry of soak water impacts hydration of cereals and legumes. It is known that as pH increases, water uptake rate and capacity also increases [31]. pH in soak water is altered by presence of monovalent ions such as Na⁺ and K⁺ ions in form of salt has shown to decrease soak time and increase protein retention [19]. Alkaline solutions are proven to improve water-holding capacity of both normal seeds and hard-to-cook seeds [28]. Although salts help seeds hydrate, high concentrations known to diminish germination properties of seeds, whereas low salt concentrations (up to certain level) have proven to viable germination, as uptake of monovalent ions increases, the chances of germination decreases [91, 92]. It is demonstrated that both monovalent ions such as Na^+, K^+ compete with divalent ions such as Ca^{2+}, Mg^{2+} in water and influence numerous relationships among macro and micro nutrients including the extent of cross-linked pectin complexes. Studies have shown that optimizing these two ions in water can reduce soaking time and further enhance softening of the seeds during cooking. Additives in form of salts like NaCl, $CaCl_2$, and $NaHCO_3$ are added to improve the ionic concentration in soak medium to optimum conditions. Excessive levels of salt adversely impacts one flavor without concomitant improvement in cooking rates of the final product [19]. Thus, additives in soak water can play an important role in changing the pH of water and in turn enhancing water holding capacity. The specific mechanisms of these ions and their interactions are reviewed elsewhere [12, 79].

3.3.5 PRESSURE

Hydrostatic and osmotic pressures of seeds increase during processing which impact water absorption rate in seed. This mechanism is utilized processing of seeds using pressure to increase water uptake with lower temperatures [63]. These pressure differentials have been applied as direct over-riding air/steam pressure or through vacuum impregnation during soaking.

It is evident that temperature can impact water uptake of seeds during soaking, and effect of pressure paired with temperature has been a keen interest. Pressure is extensively employed in processing of rice, which are hard to soak grains among cereals and legumes [4, 12]. Hydrostatic and osmotic pressures of seeds increase during processing which impact water absorption rate in seed. This mechanism is utilized processing of seeds using pressure to increase water uptake with lower temperatures [63]. Ultra High Pressure can be used to inactivate microorganisms at lower temperatures, which will help in preserving heat-liable protein in cereals and legumes [65]. It is widely reported that combination of thermal and pressure works best for foods [42]. Among cereals and legumes in industry, only rice and occasionally soybean is processed with pressure, and it is very rarely utilized commercially in processing of dry beans and other cereal grains. There have been plenty of studies carried out to analyze viability of pressure of dry beans processing, and factors such as complete inactivation of microorganisms hinder its application [24]. The energy saving advantages of pressure by replacing thermal treatments, combined with improved functional and nutritional retention of ingredients, with improved food quality parameters make it one of the exciting novel processing methods for cereals and legumes.

3.4 PHYSIOLOGICAL AND NUTRITIONAL CHANGES DURING HYDRATION

The primary objective of soaking in legumes and cereals is providing the seeds with sufficient water to enable required physical and chemical changes associated with product palatability. Based on desired end product, the extent of hydration and the supporting soak environment may be altered to achieve optimally soaked grain appropriate for the specific end-product. As discussed, hydration is primarily provided to cereals and legumes through simple soaking (continuous water exposure) or steeping (intermittent water exposure) processes. During both soaking and steeping, the physiological changes are initiated through mass transfer of water through seeds and leaching out of solids into the water medium. The process of dynamic mass transfer into and out of the seed continues until the soaking/steeping protocol is terminated. These protocols are frequently aided by adjustment of temperature or pressure, and medication of soak water chemistry. The primary nutritional and physiological changes that occur during soaking and steeping of cereals and legumes are presented.

3.4.1 SOAKING

3.4.1.1 Legumes

Legumes such as soybeans, navy beans, pinto beans, black beans, kidney beans, lima beans, faba beans, and lentils gain both weight and volume with water absorption during soaking. Typically, the moisture content of beans increases from 10–15% to 50–60% during soaking. Weight increase is proportional to moisture, whereas volume changes may vary among bean types and varieties [9]. Soaking in dry beans is performed to facilitate fast cooking and to facilitate quicker gelatinization of starch, along with enabling higher protein quality due to shorter thermal processing. Oligosaccharides such as raffinose, and sachyose are non-digestible, and 30% of them leach out during soaking process [81]. Anti-nutrient components such as phytic acid, tanins and trypsin inhibitors are decreased by 28.75%, 7.02% and 28.75% respectively during soaking [20]. Vitamins and minerals in dry beans are lost primarily through leaching out into water. Reduction in levels of vitamins [E, B_{12}] and minerals [Ca, Mg] during soaking is highest at 50–60°C [80]. Further, during soaking of soybeans, a 12% loss of iso-flavones is observed [99].

3.4.1.2 Cereals

Among cereals, rice is soaked prior to thermal processing to achieve adequate starch gelatinization and to decrease time of exposure at higher temperatures and reduce energy inputs [12]. Moisture content of rice changes from 10–15% to 55–65% during soaking depending on the soak water temperature. Soaking proved to be effective in reducing phytate content in rice, maize, millet, rice and sorghum up to 28%, 21%, 4%, and 17% respectively. Minerals such as Fe and Zn are lost by 33% & 3% respectively in millet, 7% & 11% respectively in maize, 41% &1.3% respectively in sorghum and 60% & 30 % respectively in rice [45].

3.4.2 STEEPING

3.4.2.1 Legumes

Steeping of dry beans is carried out for production of sprouts that are consumed either whole-fresh or pre-treated with minimal processing procedures. Moth beans, soybeans, alfalfa, lentils, pea, chickpeas and mung beans are legumes that are commonly steeped for sprouting. During steeping, enzymes are activated with optimum soak water temperatures, which initiate biochemical changes in seed associated with germination.

Germination through steeping is considered to be effective technology to retain maximum nutrients in legumes. Bioavailability of nutrients in legumes improved considerably after sprouting. Steeping reduced enzyme inhibitors such as trypsin inhibitor by 50% in kidney beans and improves their digestibility [23]. Starch content of beans is also reduced by 55.6% during steeping process of kidney beans. Also, vitamin content (thiamin, riboflavin, niacin) of beans such as soy increased (20%, 50% and 30%, respectively) during steeping. Further, ascorbic acid content dynamically increased during sprouting [1].

3.4.2.2   Cereals

Wheat, rye, sorghum and barley are mainly used for human consumption by steeping to obtain malt (although barley is most commonly malted grain). Barley and other cereals are converted to malt for beer preparation, salads, soups or consumed as a powdered nutritional supplement [48]. During steeping the bulk volume of cereals increases about 25%, with increase in width and depth of seeds, and no appreciable change in length [15]. Steeping of barley enables release of enzymes, which hydrolyze complex starches to maltodextrins and simple sugars (glucose and maltose). Also proteins and nutrients are released through disruption of the protein matrix [7]. Absorption of water during steeping facilitates release of gibberellins [GAs] from the seed embryo. GAs help trigger release of hydrolytic enzymes such as amylase and convert endosperm starch to soluble sugars. The sugars, which are converted from starch during steeping, are subsequently utilized by yeast during fermentation. The germination of other cereals, such as wheat and rye are similar to barley during steeping and malting [15].

3.5   RESEARCH STUDIES ON HYDRATION

Hydration of cereals and legumes has been studied to understand the impact of extrinsic and intrinsic factors on water uptake [2, 30, 34, 41, 63, 87]. Extensive studies focused on cereals and legumes have been carried out in preparation and processing as well as the expansive discipline of plant physiology to understand water uptake required for germination. The primary factors that are different among both cases are extrinsic. Although, most of cereals and legumes are not germinated but instead used directly for processing foods, there has been recent interest in developing functional foods using germinated seeds [16]. Studies are either experimental or utilize mathematical modeling to assess hydration phenomena. The following section reviews both these experimental and mathematical studies of seed hydration.

3.5.1 EXPERIMENTAL STUDIES

Experimental studies on cereals and legume are carried out to understand hydration rate as affected by changing extrinsic factors such as soak parameters, processing method, and seed conditions. Most of the primary research and subsequent reviews published of cereals and legumes focus on nutritional changes occurring during processing. While the role of hydration in seeds is well understood, there is no standard parameter or set of parameters that are consistently measured. While, moisture content and perhaps seed volume are the primary indicators of water uptake in seeds, recent advance in specialized technologies has enabled direct visualized water diffusion into seeds through Nuclear Magnetic Resonance [NMR] and scanning electron microscopy [SEM] [56, 88, 95, 97]. Description of different techniques used to determine hydration kinetics of cereals and legumes are presented:

3.5.1.1 Moisture Content

Most legume and cereal moisture content during storage ranges from 12 to 15%, while hydration, this is brought up to a range of 40–60%, depending on the desired end-product. The conventional method used to measure moisture content through complete drying of the hydrated sample and performing gravimetric analysis for their ease of application, as described in AOAC method 925.10 [38]. The type of drying method and accuracy varies by standard method employed [13]. During soaking of cereals and legumes, moisture content is measured to determine the time required to reach equilibrium moisture content, which varies from 50–65% [66]. During steeping, moisture content of cereals and legumes are monitored to assess the extent of hydration and correlate the data to germination times of seeds during micro-malting studies [53]. Seeds with rapid water uptake or initial moisture increase are considered superior (in yield and improved germination and malt quality) to those seeds that hydrate slowly [15].

3.5.1.2   Volume

Measurement of volume during soaking of cereals and legumes is an essential when further processing is dependent up on the size of the seed, for example in thermal processing. Volume change varies across the spectrum of legumes and cereals, however in general, legumes tend to increase in volume to a greater extent than cereals [86]. Volume change also correlates to moisture content of the seed based on seed properties, where legumes gain more than 100% of their volume, cereals gain about 25% during hydration [89]. Methods employed to measure seed volume employ both direct and indirect techniques. Volume of seeds measured by Archimedes principle of displacement [55]. Further, weight and density may be measured to derive volume using a pycnometer filled with toluene [57]. Numerous manual volume measurement techniques have been employed, by approximating seed particle shape (ellipsoid or spheroid), however, these fail to reflect volume of bulk particles with large sample. Even with assistance of imaging technology, the manual measurement has very little accuracy associated with to its small sample size [37, 77, 78].

3.5.1.3   Imaging Techniques

Modern imaging techniques have been used to understand diffusion of water into seeds [76–78]. Proton Nuclear Magnetic Resonance paired with Magnetic Resonance Imaging provides a non-invasive technique to study water uptake in seeds [88, 95]. These techniques provide adequate understanding of water migration and distribution in seeds, however it is difficult to quantify the rate and amount of water uptake. Further, these methods are widely employed to understand HTC phenomena in seeds [3, 44].

3.5.2 MATHEMATICAL MODELS FOR SEED HYDRATION

The difficulty and complexity of experimental methods have led to the usage of mathematical models to describe hydration kinetics of seeds. Modeling of seed hydration has become a de facto standard to emphasize the rate and extent of hydration and to determine the impact of temperature, and soak water parameters. Many models are proposed to describe hydration of seeds, both in food processing and plant sciences [14, 30]. Research groups have routinely used two approaches to describe hydration of porous biomaterials. One approach is based on Fick's law of diffusion, which constituted the earliest attempts to model water uptake in seeds [98], while other approaches are based on empirical and semi-empirical methods [30, 60]. Empirical models like Peleg and Weibull are more widely employed to simulate hydration in foods, due to their simplicity, and having to work with fewer parameters [100]. Of the models that are described below, the empirical models have proven to be best suited for hydration of whole intact foods, whereas, physical models are best utilized to describe the properties (e.g., diffusivity) and process conditions (e.g., temperature) of both the hydration medium and the rehydrating food particles.

3.5.2.1 Diffusion Model

The diffusion model is based on following assumptions using Fick's second law of diffusion: (i) The water diffusion is constant and primary method of transfer; (ii) Shape of the seed is spherical; (iii) No external resistance to heat and mass transfer exists; and (iv) Surface concentration of water reaches saturation immediately after immersion [11]. The model equation is given as:

$$\frac{m - m_s}{m_0 - m_s} = \frac{6}{\pi^2} exp^{\left(\frac{-\pi^2 D_{ef} t}{R^2}\right)} \tag{2}$$

where, m_s and m_0 are saturation and initial moisture contents [kg/kg, dry basis], t is the time [s], R the equivalent radius (radius of the sphere having the same volume as the grain) [m] and D_{ef} is the effective diffusivity [m²/s].

The moisture absorption curve is fitted by adjusting the value of D_{ef}. In the study reported above, and other work done on hydration analysis using Fick's law, the limitation of determining saturation moisture content is reported for rice. That is due to variation in moisture content of seeds with gelatinization depending on starch content. Bello et al. [11] studied the impact of temperature on diffusivity rate of rice while soaking for 24 hours using the diffusion model derived from Fick's second law. Figure 3.1 shows the equilibrium moisture content of rice remained relatively constant after 45°C, with slight increase from 25°C to 35°C, while the diffusivity increased with increase in temperature. The moisture content was determined using the Fick's law with less than 3% error.

3.5.2.2   Capillary Flow Model

Studies are carried out to look at water uptake through foods as a capillary flow in the porous media driven by energy potential, rather than diffusion, accounting for anisotropy effects [100]. One of the main advantages proposed for capillary model is that it can be derived through capillary relationships from physical-based water activity curves commonly used in

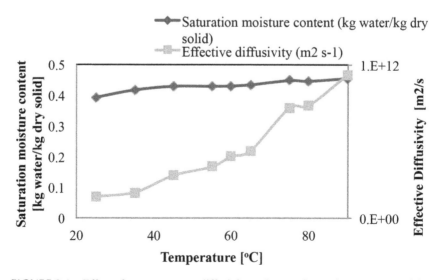

FIGURE 3.1   Effect of temperature on diffusivity and saturation moisture content of rice.

food science and food engineering disciplines. Different models for rehydration of dry food particles based on porous media are summarized elsewhere [70]. The porous media models are derived from Lucas-Washburn equation with following main assumptions: (i) food is in form of microscopically uniform porous media; and (ii) pore size is average throughout. The equation for equilibrium liquid rise in porous foods is given below:

$$y(t) = \frac{k_1}{k_2}\left[1 - e^{\left(-\frac{k_2 y[t]}{k_2}\right)}e^{\left(-\frac{k_2^2 t}{k_1}\right)}\right] ; k_1 = \frac{r\gamma \cos[\delta]}{2\mu} \text{ and } k_2 = \frac{r^2 g\rho}{8\mu} \quad (3)$$

where, k_1 and k_2 are permeability parameters; y is distance traveled by the liquid at time t [m]; r is pore radius [m]; t is time [s]; γ is surface tension [N/m]; δ is contact angle [°]; μ is viscosity [N s/m^2]; g is gravity [m/s^2].

Goula et al. [29, 30] analyzed the application of capillary flow approach to describe water absorption through dried tomatoes, and found that the capillary model adequately describes rehydration behavior of dried foods [29]. Figure 3.2 shows the capillary model predicted and measured moisture content values of dried tomato with no differences among ($P < 0.05$) slope and intercept. The greatest advantage of capillary model among

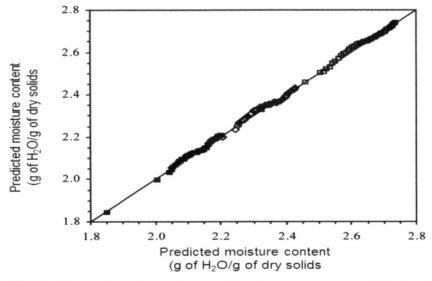

FIGURE 3.2 Experimental and capillary model predicted moisture content of rehydrated dry tomatoes. (From Goula, A. M., & Adamopoulos, K. G. (2009). Modeling the rehydration process of dried tomato. Drying Technology, 27, 1078–1088. Reprinted by permission of Taylor & Francis Ltd, http://www.tandfonline.com.)

both physical models is due its consideration of gravity, which is lacking in diffusion model and it is comprehensive in its constitutive relationships required for capillary flow, its complexity might be a hindrance to practical utilization in cereal and legume hydration.

3.5.2.3   Becker Model

Becker's model is derived from Fick's law of diffusion considering solids with arbitrary shape [10]. The primary drawback of the equation is its basis using law of diffusion and difficulty in calculation of ΔM_0, which is initial moisture gain, caused by capillary forces and error in determining surface moisture content. Becker's model is given as:

$$M_t = M_0 + \Delta M_0 + \frac{2}{\sqrt{\pi}}(M_e - M_0)\left(\frac{S}{V}\right)\sqrt{(D_{eff})(t)} \qquad (4)$$

where, M_t is moisture content [% dry basis] at time, t [s]; M_e is equilibrium moisture content [% dry basis]; M_0 is initial moisture content [% dry basis]; ΔM_0 is initial moisture gain [% dry basis]; M_s is surface moisture content [% dry basis]; S is surface area [m²]; V is volume [m³]; D_{eff} is effective diffusion coefficient [m²/s].

The Becker model was used to predict hydration of four different rice cultivars with a correlation of r = 0.98 as shown in Figure 3.3 [8].

3.5.2.4   Exponential Model

The simplest empirical model described in literature is exponential model, which describes relationship of moisture content for particular food over time [21].

$$\frac{M_t - M_e}{M_0 - M_e} = a e^{[-bt]} \qquad (5)$$

where, M_t is moisture content [% dry basis] at time, t [s]; M_e is equilibrium moisture content [% dry basis]; M_0 is initial moisture content [% dry basis]; a and b are empirical constants.

A modified version of above exponential equation is proposed by Maskan to model water uptake of wheat kernels [52] during soaking considering temperature, T:

$$\frac{M_t - M_e}{M_0 - M_e} = 7259 \times exp^{[0.031t - 1470/T]} \tag{6}$$

Maskan reported that model was correlated with experimental value with r = 0.925.

3.5.2.5   Peleg Model

The Peleg model is first introduced to assess moisture sorption curves for both granular foods and seeds [60]. It is one of the first non-exponential

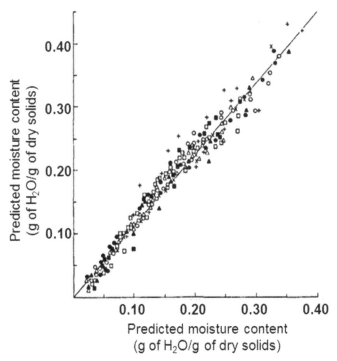

FIGURE 3.3   Correlation between predicted and measured moisture content of different rice varieties using Becker model [8]. (From Bandyopadhyay, S., & Roy, N. (1978). A semi-empirical correlation for prediction of hydration characteristics of paddy during parboiling. *International Journal of Food Science & Technology*, 13, 91–98. © 1978. Reprinted with permission from John Wiley and Sons.)

models to be used for understanding hydration. Since its introduction, the model has been widely used to model hydration processes of many food materials [18, 30, 83, 84, 93]. The model is described as:

$$M_t = M_0 + \frac{t}{[k_1 + k_2 t]} \tag{7}$$

where, M_t is moisture content [% dry basis] at time, t [s]; M_0 is initial moisture content [% dry basis]; k_1 is constant related to the initial rate of sorption [hour per % weight of dry solids to water]; k_2 is constant related to equilibrium moisture content [% weight of dry solids to water].

Peleg model has been extensively used to study cereal and legume hydration. The model was also used as basis for testing new empirical and mechanistic models [17]. Solomon used Peleg model to understand hydration kinetics of lupin seeds at temperatures of 20°C, 30°C, 40°C and 50°C for 9 to 12 hours. Figure 3.4 shows experimental moisture content of lupin seeds along with Peleg model curves. Solomon reported that Peleg model predicted the experimental results with co-efficient of determination, $R^2 = 0.96$ to 0.99. Sopade et al. [1990] used the Peleg equation to model water absorption of maize, millet and sorghum at temperatures of 10°C, 30°C and 50°C. Figure 3.5 shows the Peleg model predicted and experimental values of maize, millet and sorghum. The correlation between predicted and

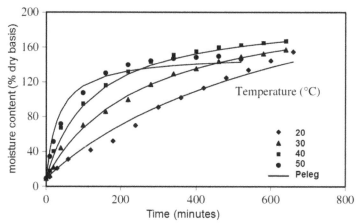

FIGURE 3.4   Measured and predicted (Peleg model) moisture content data for lupin seeds at different temperatures [82]. (Reprinted from Sopade, P., & Obekpa, J. (1990). Modelling water absorption in soybean, cowpea and peanuts at three temperatures using Peleg's equation. *Journal of Food Science*, 55, 1084–1087. © 1990. With permission from John Wiley.)

experimental values is reported as, $r^2= 0.98$ to 0.99. Sapode et al. [1990] also reported that Peleg constant k_1 was temperature dependent, while k_2 was unaffected. Similar high correlation (>0.95) was reported for Peleg model used to study the hydration kinetics of cereals and legumes [18, 27, 30, 93].

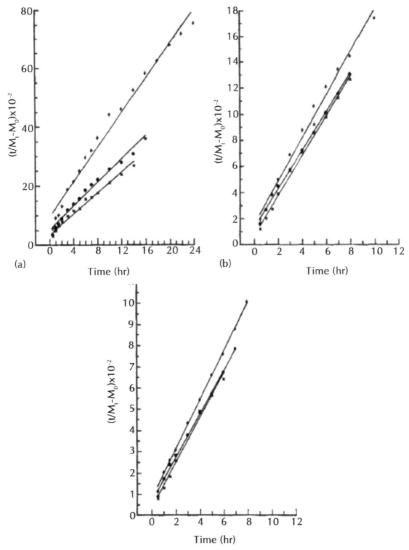

FIGURE 3.5 Application of Peleg model to experimental data on maize [a]; millet [b]; and sorghum [c]. (Reprinted from Sopade, P., & Obekpa, J. (1990). Modelling water absorption in soybean, cowpea and peanuts at three temperatures using Peleg's equation. *Journal of Food Science*, 55, 1084–1087. © 1990. With permission from John Wiley.)

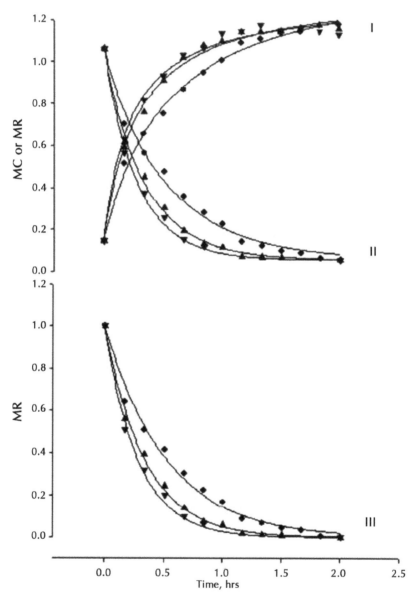

FIGURE 3.6   Fitting of hydration models for chick peas: Peleg [I], Weibull [II] and Exponential [III] models at 40°C, 50°C, and 60°C hydration temperatures [62]. (Reprinted with permission from Prasad, K., Vairagar, P., & Bera, M. (2010). Temperature dependent hydration kinetics of Cicer arietinum splits. *Food Research International*, 43, 483–488. © 2010. With permission from Elsevier.)

3.5.2.6 Weibull Model

An improved version of probabilistic Weibull distribution function has been utilized to describe rehydration of dried foods [51]. The equation is:

$$\frac{M_t}{M_e} = 1 - e^{\left[-\left(\frac{t}{\alpha}\right)^{\beta}\right]} \tag{8}$$

where, M_t is moisture content [% dry basis] at time, t [s]; M_e is equilibrium moisture content [% dry basis]; The model utilizes two empirical parameters: α, which is reciprocal of the process rate constant and is related rapid hydration; and β is shape parameter, related to the shape of initial lag phase [68].

The equation has been modified numerous times to fit different foods [18, 27, 30, 69]. Prasad et al. [62] studied hydration kinetics of chick peas using different empirical and mechanistic models at temperatures of 40°C, 50°C and 60°C. Figure 3.6 shows fitting of Peleg, Weibull and exponential models of measured moisture ratio (MR) values with time for chick peas.

3.6 CURRENT CHALLENGES IN HYDRATION PROCESS

The hydration processes of cereals and legumes have been extensively studied and the general phenomena are well understood. However, as individual treatments hydration rates of seeds vary based on their properties, it is difficult to generalize hydration process in dry foods. When considering hydration as a processing step in preparation of functional foods, it is important to understand the factors affecting the process. Past and current research indicates a very good fundamental understanding of science associated with hydration [6, 26, 64, 67, 73, 97]. Changes in nutritional value and ability to preserve important nutrients during hydration have been important foci [26, 35, 43]. Extensive mathematical modeling of seed hydration has been conducted; however, these are limited by the lack of readily obtained and accurate experimental data [69, 97]. New technologies in imaging and microstructural analyzes are used to support the established mathematical theories [3].

The main challenge very seldom addressed when working with hydration of seeds is the lack of standard experimental techniques with a broad enough scope to study a wide range of factors impacting both soaking and steeping processes. Current modeling techniques fall short of usage in commercial environment due to the time required to obtain enough measurements to obtain the data required for rigorous statistical analyzes. Improved analytical tools to measure and record various seed parameters during hydration will both improve the quality of hydration studies, and also provide an approach toward enhanced commercialization of improved practices [33].

3.7   SUMMARY

Cereals and legumes are important sources of vegetable-based human nutrition. Together they account for 48.6% of protein and 8.7% carbohydrate consumption around the world. During preparation, majority of these agricultural staples are re-hydrated to aid in their digestibility, palatability and the bioavailability of the nutrients. Different processing methods in preparation of functional foods using cereals legumes are discussed with emphasis on soaking. Various physiological and chemical changes that occur during soaking are discussed for legumes cereals with particular emphasis on barley. Current research involving factors effecting hydration of foods are reviewed in detail. Experimental and mathematical models are presented on cereals and legume hydration. Important challenges were summarized and discussed. Hydration: Colloquially used with rehydration, is the term used to define the processing step for increasing the water content of dried edible grains

KEYWORDS

- **Becker model**
- **capillary flow**
- **cereal nutrition**
- **cereals**
- **cooking methods**

- dry beans
- edible seeds
- experimental model
- Fick's second law
- food processing
- hydration
- imaging techniques
- legumes
- mathematical models
- moisture content
- navy beans
- nutrition
- Peleg model
- pinto beans
- rice
- soak temperature
- soak water additives
- soaking
- soaking time
- soybeans
- steeping
- volume
- water chemistry
- Weibull model
- weight

REFERENCES

1. Abdullah, A., & Baldwin, R. E. (1984). Mineral and vitamin contents of seeds and sprouts of newly available small-seeded soybeans and market samples of mung beans. *Journal of Food Science,* 49, 656–657.
2. Abu-Ghannam, N. (1998). Modeling textural changes during the hydration process of red beans. *Journal of Food Engineering,* 38(3), 341–352.

3. Aguilera, J. M., & Lillford, P. J. (2000). Microstructural and imaging analyzes as related to food engineering. *Food Engineering,* 23–38, 1997.

4. Ahromrit, A., Ledward, D., & Niranjan, K. (2006). High pressure induced water uptake characteristics of Thai glutinous rice. *Journal of Food Engineering,* 72, 225–233.

5. Ayyagari, R., Rao, B. S. N., & Roy, D. N. (1989). Lectins, trypsin inhibitors, BOAA and tannins in legumes and cereals and the effects of processing. *Food Chemistry,* 34, 229–238.

6. Bamforth, C. W. (2003). Barley and malt starch in brewing: a general review. *Master Brewers Association of the Americas TQ,* 40(2), 89–97.

7. Bamforth, C. W. (2006). *Scientific Principles of Malting and Brewing.* American Society of Brewing Chemists.

8. Bandyopadhyay, S., & Roy, N. (1978). A semi-empirical correlation for prediction of hydration characteristics of paddy during parboiling. *International Journal of Food Science & Technology,* 13, 91–98.

9. Barampama, Z., & Simard, R. E. (1995). Effects of soaking, cooking and fermentation on composition, in-vitro starch digestibility and nutritive value of common beans. *Plant Foods for Human Nutrition (Formerly Qualitas Plantarum),* 48, 349–365.

10. Becker, H. (1960). On the absorption of liquid water by the wheat kernel. *Cereal Chem.,* 37, 309–323.

11. Bello, M., Tolaba, M. P., & Suarez, C. (2004). Factors affecting water uptake of rice grain during soaking. *LWT-Food Science and Technology,* 37(8), 811–816.

12. Bello, M., Tolaba, M. P., & Suárez, C. (2008). Hydration kinetics of rice kernels under vacuum and pressure. *International Journal of Food Engineering,* 4(4), ISSN (Online) 1556–3758

13. Bouraoui, M., Richard, P., & Fichtali, J. (1993). A review of moisture content determination in foods using microwave oven drying. *Food Research International,* 26, 49–57.

14. Bradford, K. J. (2002). Applications of hydrothermal time to quantifying and modeling seed germination and dormancy. *Weed Science,* 50, 248–260.

15. Briggs, D. E. (1998). *Malts and Malting.* Springer.

16. Charalampopoulos, D., Wang, R., Pandiella, S. S., & Webb, C. (2002). Application of cereals and cereal components in functional foods: a review. *International Journal of Food Microbiology,* 79, 131–141.

17. Coutinho, M. R, Omoto, E. S, dos Santos Conceição, W. A, Andrade, C. M. G., & Jorge, L. M. M. (2010). Evaluation of Two Mathematical Models Applied to Soybean Hydration. *International Journal of Food Engineering,* 6(6), 1556–3758.

18. Cunningham, S. E., McMinn, W. A. M, Magee, T. R. A., & Richardson, P. S. (2007). Modeling water absorption of pasta during soaking. *Journal of Food Engineering,* 82, 600–607.

19. De León, L., Elias, L., & Bressani, R. (1992). Effect of salt solutions on the cooking time, nutritional and sensory characteristics of common beans (*Phaseolus vulgaris*). *Food Research International,* 25, 131–136.

20. Deshpande, S., & Cheryan, M. (1983). Changes in phytic acid, tannins, and trypsin inhibitory activity on soaking of dry beans (*Phaseolus vulgaris*, L.). *Nutrition Reports International,* 27.

21. Diamante, L., & Munro, P. (1991). Mathematical modeling of hot air drying of sweet potato slices. *International Journal of Food Science & Technology,* 26, 99–109.

22. Egli, I., Davidsson, L., Juillerat, M., Barclay, D., & Hurrell, R. (2002). The influence of soaking and germination on the phytase activity and phytic acid content of grains and seeds potentially useful for complementary feeding. *Journal of Food Science,* 67, 3484–3488.

23. El-Hag, N., Haard, N., & Morse, R. (1978). Influence of sprouting on the digestibility coefficient, trypsin inhibitor and globulin proteins of red kidney beans. *Journal of Food Science,* 43, 1874–1875.

24. Estrada-Giron, Y., Swanson, B., & Barbosa-Cánovas, G. (2005). Advances in the use of high hydrostatic pressure for processing cereal grains and legumes. *Trends in Food Science & Technology,* 16, 194–203.

25. Ford, J. E., & Hewitt, D. (1979). Protein quality in cereals and pulses. *British Journal of Nutrition,* 41, 341–352.

26. Friedman, M. (1996). Nutritional Value of Proteins from Different Food Sources: A Review. *J. Agric. Food Chem., 44,* 6.

27. García-Pascual, P., Sanjuán, N., Melis, R., & Mulet, A. (2006). Morchella esculenta [morel] rehydration process modeling. *J Food Eng.,* 72(4), 346–353.

28. García-Vela, L., & Stanley, D. (1989). Water-Holding Capacity in Hard-to-Cook Beans (*Phaseolus vulgaris*): Effect of pH and Ionic Strength. *Journal of Food Science,* 54, 1080–1089.

29. Goula, A. M., & Adamopoulos, K. G. (2009). Modeling the rehydration process of dried tomato. *Drying Technology,* 27, 1078–1088.

30. Goula, A. M., & Adamopoulos, K. G. (2009). A new technique for spray drying orange juice concentrate. *Innovative Food Science and Emerging Technologies,* 11(2), 342–351.

31. Haladjian, N., Fayad, R., Toufeili, I., Shadarevian, S., Sidahmed, M., Baydoun, E., & Karwe, M. (2003). The pH, temperature and hydration kinetics of faba beans (*Vicia faba,* L.). *Journal of Food Processing and Preservation,* 27, 9–20.

32. Han, I. H., & Baik, B. K. (2006). Oligosaccharide content and composition of legumes and their reduction by soaking, cooking, ultrasound, and high hydrostatic pressure. *Cereal Chemistry,* 83, 428–433.

33. Harte, F., Mannam, V., Worley, S., Wilkerson, J., & Smith, D. (2014). Soak Chamber and System to Measure the Seed Density Hydration Profile of Seeds. *Google Patents.*

34. Hegarty, T. W. (2012). The physiology of seed hydration and dehydration, and the relation between water stress and the control of germination: a review. *Plant, Cell & Environment,* 1, 101–119.

35. Hornick, S. B. (1992). Factors affecting the nutritional quality of crops. *American Journal of Alternative Agriculture,* 7, 63–68.

36. Hotz, C., & Gibson, R. S. (2007). Traditional food-processing and preparation practices to enhance the bioavailability of micronutrients in plant-based diets. *The Journal of Nutrition,* 137, 1097–1100.

37. Igathinathane, C., Pordesimo, L., Columbus, E., Batchelor, W., & Methuku, S. (2008). Shape identification and particles size distribution from basic shape parameters using Image. *J. Computers and Electronics in Agriculture,* 63, 168–182.

38. Iqbal, A., Khalil, I. A., Ateeq, N., & Sayyar, Khan, M. (2006). Nutritional quality of important food legumes. *Food Chemistry, 97,* 331–335.
39. Kashaninejad, M., Dehghani, A. A., & Kashiri, M. (2009). Modeling of wheat soaking using two artificial neural networks [MLP and RBF]. 91, 602–607.
40. Khan, M. A., Gul, B., & Weber, D. J. (2000). Germination responses of Salicornia rubra to temperature and salinity. *Journal of Arid Environments, 45,* 207–214.
41. Khazaei, J., & Mohammadi, N. (2009). Effect of temperature on hydration kinetics of sesame seeds (*Sesamum indicum,* L.). 91, 542–552.
42. Knorr, D. (1999). Process assessment of high pressure processing of foods: an overview. *Process Optimization and Minimal Processing of Foods, 1.*
43. Kon, S. (1979). Effect of soaking temperature on cooking and nutritional quality of beans. *Journal of Food Science, 44,* 1329–1335.
44. Laurent, B., Ousman, B., Dzudie, T., Carl, M. F. M., & Emmanuel, T. (2010). Digital camera images processing of hard-to-cook beans. *Journal of Engineering and Technology Research, 2,* 177–188.
45. Lestienne, I., Icard-Vernière, C., Mouquet, C., Picq, C., & Trèche, S. (2005). Effects of soaking whole cereal and legume seeds on iron, zinc and phytate contents. *Food Chemistry, 89,* 421–425.
46. Leterme, P. (2002). Recommendations by health organizations for pulse consumption. *British Journal of Nutrition, 88,* 239–242.
47. Lewicki, P. P. (1998). Some remarks on rehydration of driedfoods. *Journal of Food Engineering, 36*(1), 81–87.
48. Lorenz, K., & D'Appolonia, B. (1980). Cereal sprouts: composition, nutritive value, food applications. *Critical Reviews in Food Science & Nutrition, 13,* 353–385.
49. Lucas, D. A. (2010). *Effect of Storage Pre-Treatments and Conditions on the Dehulling Efficiency and Cooking Quality of Red Lentils. M.Sc. Thesis,* University of Manitoba.
50. Mahatma, M., Bhatnagar, R., Solanki, R., & Mittal, G. (2009). Effect of seed soaking treatments on salinity induced antioxidant enzymes activity, lipid peroxidation and free amino acid content in wheat (*Triticum aestivum,* L.) leaves. *Indian Journal of Agricultural Biochemistry, 22,* 108–112.
51. Marabi, A., & Saguy, I. S. (2009). Rehydration and reconstitution of foods. *Advances in Food Dehydration,* 237–284.
52. Maskan, M. (2001). Effect of maturation and processing on water uptake characteristics of wheat. *Journal of Food Engineering, 47,* 51–57.
53. Mayolle, J., Lullien-Pellerin, V., Corbineau, F., Boivin, P., & Guillard, V. (2011). Water diffusion and enzyme activities during malting of barley grains: A relationship assessment. *Journal of Food Engineering.*
54. Molina, M., Fuente, G., & Bressani, R. (1975). Interrelationships between storage, soaking time, cooking time, nutritive value and other characteristics of the black bean (*Phaseolus vulgaris*). *Journal of Food Science, 40,* 587–591.
55. Moreira, R., Chenlo, F., Chaguri, L., & Fernandes, C. (2008). Water absorption, texture, and color kinetics of air-dried chestnuts during rehydration. *Journal of Food Engineering, 86,* 584–594.
56. Muñoz, L., Cobos, A., Diaz, O., & Aguilera, J. (2012). Chia seeds: Microstructure, mucilage extraction and hydration. *Journal of Food Engineering, 108,* 216–224.

57. Muramatsu, Y., Tagawa, A., Sakaguchi, E., & Kasai, T. (2006). Water absorption characteristics and volume changes of milled and brown rice during soaking. *Cereal Chemistry*, 83, 624–631.

58. Nasar-Abbas, S., Siddique, K., Plummer, J., White, P., Harris, D., Dods, K., & D'Antuono, M. (2009). Faba bean (Vicia faba, L.) seeds darken rapidly and phenolic content falls when stored at higher temperature, moisture and light intensity. *LWT-Food Science and Technology*, 42, 1703–1711.

59. Organization FAA. (2013). *Food Supply*. United Nations.

60. Peleg, M. (1998). An empirical model for the description of moisture sorption curves. *Journal of Food Science*, 53, 1216–1217.

61. Perlas, L. A., & Gibson, R. S. (2002). Use of soaking to enhance the bioavailability of iron and zinc from rice-based complementary foods used in the Philippines. *Journal of the Science of Food and Agriculture*, 82, 1115–1121.

62. Prasad, K., Vairagar, P., & Bera, M. (2010). Temperature dependent hydration kinetics of Cicer arietinum splits. *Food Research International*, 43, 483–488.

63. Ramaswamy, R., Balasubramaniam, V., & Sastry, S. (2005). Effect of high pressure and irradiation treatments on hydration characteristics of navy beans. *International Journal of Food Engineering*, 1, 1–15.

64. Rao, M., & Lund, D. (1986). Kinetics of thermal softening of foods–a review. *Journal of Food Processing and Preservation*, 10, 311–329.

65. Rastogi, N., Raghavarao, K., Balasubramaniam, V., Niranjan, K., & Knorr, D. (2007). Opportunities and challenges in high pressure processing of foods. *Critical Reviews in Food Science and Nutrition*, 47, 69–112.

66. Resio, A. N. C., Aguerre, R. J., & Suárez, C. (2006). Hydration kinetics of amaranth grain. *Journal of Food Engineering*, 72(3), 247–253.

67. Reyes-Moreno, C., Paredes-López, O., & Gonzalez, D. E. (1993). Hard-to-cook phenomenon in common beans: A review. *Critical Reviews in Food Science & Nutrition*, 33, 227–286.

68. Saguy, I. S., & Marabi, A. (2010). Rehydration **of** dried **food** particulates. In: *Encyclopedia of Agricultural, Food, and Biological Engineering*, Dennis, R. Heldman (Ed.). Taylor & Francis.

69. Saguy, I. S., Marabi, A., & Wallach, R. (2005). New approach to model rehydration of dry food particulates utilizing principles of liquid transport in porous media. *Trends in Food Science & Technology*, 16, 495–506.

70. Saguy, I. S., Troygot, O., Marabi, A., & Wallach, R. (2010). Rehydration Modeling of Food Particulates by Using Principles of Water Transport in Porous Media. Chapter 17, pages 219–236), In: *Water Properties in Food, Health, Pharmaceutical and Biological Systems*, Eds. David, S. Reid, Tanaboon Sajjaanantakul, Peter, J. Lillford, Sanguansri Charoenrein. John Wiley Online.

71. Sandberg, A. S. (2002). Bioavailability of minerals in legumes. *British Journal of Nutrition*, 88, 281–285.

72. Sathe, S. K. (2012). Chemistry and Implications of Antinutritional Factors in Dry Beans and Pulses. *Dry Beans and Pulses: Production, Processing and Nutrition*, 359.

73. Sathe, S. K., Deshpande, S. S., Salunkhe, D. K., & Rackis, J. J. (1984). Dry beans of phaseolus. A review. Part 1. Chemical composition: Proteins. *Critical Reviews in Food Science and Nutrition*, 20(1), 1–46.

74. Sayar, S., Turhan, M., & Gunasekaran, S. (2001). Analysis of chickpea soaking by simultaneous water transfer and water–starch reaction. *Journal of Food Engineering,* 50, 91–98.

75. Sefa-Dedeh, S., Stanley, D., & Voisey, P. (1978). Effects of soaking time and cooking conditions on texture and microstructure of cowpeas [*Vigna unguiculata*]. *Journal of Food Science,* 43, 1832–1838.

76. Shahin, M., Symond, S., Schepdael, L., & Tahir, A. (2006). Three dimensional seed shape and size measurement with orthogonal cameras. *Annual International Meeting of ASABE.*

77. Shahin, M., & Symons, S. (2005). Seed sizing from images of non-singulated grain samples. *Canadian Biosystems Engineering,* 47, 49–55.

78. Shahin, M. A., Symons, S. J., & Poysa, V. W. (2006). Determining soya bean seed size uniformity with image analysis. *Biosystems Engineering,* 94, 191–198.

79. Siddiq, M., Butt, M. S., & Sultan, M. T. (2011). Dry Beans: Production, Processing, and Nutrition. *Handbook of Vegetables and Vegetable Processing,* p. 545–564.

80. Siddiq, M., & Uebersax, M. A. (2012). *Dry Beans and Pulses: Production, Processing and Nutrition.* Wiley-Blackwell.

81. Silva, H. C., & Braga, G. L. (2006). Effect of soaking and cooking on the oligosaccharide content of dry beans (*Phaseolus vulgaris,* L.). *Journal of Food Science,* 47, 924–925.

82. Solomon, W. (2007). Hydration kinetics of lupin [Lupinus albus] seeds. *Journal of food process engineering,* 30, 119–130.

83. Sopade, P., & Obekpa, J. (1990). Modelling water absorption in soybean, cowpea and peanuts at three temperatures using Peleg's equation. *Journal of Food Science,* 55, 1084–1087.

84. Sopade, P. A., Ajisegiri, E. S., & Badau, M. H. (1992). The use of Peleg's equation to model water absorption in some cereal grains during soaking. *Journal of Food Engineering,* 15, 269–283.

85. Sreerama, Y. N., Sashikala, V. B., & Pratape, V. M. (2009). Expansion properties and ultrastructure of legumes: Effect of chemical and enzyme pre-treatments. *LWT-Food Science and Technology,* 42, 44–49.

86. Taylor, A. G., & Kenny, T. J. (1985). Improvement of germinated seed quality by density separation. *J. Am. Soc. Hort. Sci.,* 110, 347–349.

87. Taylor, A., Prusinski, J., Hill, H., & Dickson, M. (1992). Influence of seed hydration on seedling performance. *HortTechnology,* 2, 336–344.

88. Terskikh, V., Müller, K., Kermode, A. R., & Leubner-Metzger, G. (2011). In Vivo 1H-NMR Microimaging During Seed Imbibition, Germination, and Early Growth. *Methods in Molecular Biology,* 773, 319.

89. Thakor, N., Sokhansanj, S., Patil, R., & Deshpande, S. (1995). Moisture sorption and volumetric changes of canola during hydrothermal processing. *Journal of Food Process Engineering,* 18, 233–242.

90. Tharanathan, R., & Mahadevamma, S. (2003). Grain legumes: a boon to human nutrition. *Trends in Food Science & Technology,* 14, 507–518.

91. Tobe, K., Li, X., & Omasa, K. (2004). Effects of five different salts on seed germination and seedling growth of Haloxylon ammodendron [Chenopodiaceae]. *Seed Science Research,* 14, 345–354.

92. Tobe, K., Li, X., & Omasa, K. (2002). Effects of sodium, magnesium and calcium salts on seed germination and radicle survival of a halophyte, Kalidium caspicum [Chenopodiaceae]. *Australian Journal of Botany,* 50, 163–169.

93. Turhan, M., Sayar, S., & Gunasekaran, S. (2002). Application of Peleg Model to Study Water Absorption in Chickpea During Soaking. 53, 153–159.

94. USDHSS, (2010). *Dietary Guidelines for Americans.* Washington, DC: U. S. Government Printing Office.

95. Vashisth, A., Joshi, D., & Singh, R. (3012). Characterization of water uptake and distribution in chickpea (*Cicer arietinum,* L.) seeds during germination by NMR spectroscopy. *African Journal of Biotechnology,* 11, 12286–12297.

96. Vorwald, J., & Nienhuis, J. (2009). Effects of seed moisture content, cooking time, and chamber temperature on nuña bean (*Phaseolus vulgaris* L.) popping. *HortScience,* 44, 135–137.

97. W., J. B., Rene, V. A., & Paul, R. B. (2012). Review: Microsite characteristics influencing weed seedling recruitment and implications for recruitment modeling. *Canadian Journal of Plant Science.*

98. Waggoner, P. E., & Parlange, J. Y. (1976). Water uptake and water diffusivity of seeds. *Plant Physiology,* 57, 153.

99. Wang, H. J., & Murphy, P. A. (1996). Mass balance study of isoflavones during soybean processing. *Journal of Agricultural and Food Chemistry,* 44, 2377–2383.

100. Weerts, A., Lian, G., & Martin, D. (2003). Modeling rehydration of porous biomaterials: Anisotropy effects. *Journal of Food Science,* 68, 937–942.

101. Williams, P. G., Grafenauer, S. J., & O'Shea, J. E. (2008). Cereal grains, legumes, and weight management: a comprehensive review of the scientific evidence. *Nutr Rev. United States,* pp. 171–182.

102. Xu, B., & Chang, S. K. C. (2008). Effect of soaking, boiling, and steaming on total phenolic content and antioxidant activities of cool season food legumes. *Food Chemistry,* 110, 1–13.

103. Xu, S. (2010). *Development and Application of an Automatic System for Determining Seed Volume Kinetics During Soaking. MSc Thesis,* University of Tennessee.

CHAPTER 4

APPLICATIONS OF NUCLEAR MAGNETIC RESONANCE IN FOOD PROCESSING AND PACKAGING MANAGEMENT

DIPENDRA KUMAR MAHATO, DEEPAK KUMAR VERMA, SUDHANSHI BILLORIA, MANDIRA KOPARI, P. K. PRABHAKAR, V. AJESH KUMAR, SUNIL MANOHAR BEHERA, and PREM PRAKASH SRIVASTAV

CONTENTS

4.1 INTRODUCTION

Food is a source of energy for all living beings. But due to lack of proper packaging and management about half of the food intended for human consumption in developed countries is wasted, for example, 4.2 million tons of food buried in Australia every year [38]. In order to contain, preserve and store for a longer duration, it must be packaged in an appropriate material. The packaging material can be classified into three groups. The packaging material, which is in direct contact with the food, is termed as primary packaging material. This serves to contain and protect the food as well as satisfy the consumers' need with respect to convenience and safety. Secondary packaging is a physical protection of the product, which can be either a wooden box or a corrugated box containing primary packaged food inside it. Secondary packaging has information like lot number, date of manufacture, expiry date, etc. on it. Tertiary packaging includes both primary and secondary packaging in a transportation package system [21].

With the passage of time and emergence of new technologies, the life style and living standards of people have changed, so is the case with the processing and packaging of food materials. The design and choice of packaging materials depend on various factors such as type of food, processing conditions, environmental impact on the packaging, consumer demands, logistics, economy, marketing as well as the distribution [12, 78, 93, 144]. The same food materials may need different packaging materials before and after processing. Nowadays, a combination of processes is utilized for processing and preserving the food materials. This technology is known as Hurdle Technology and is used to design different food products for consumers' need as well as for safety concern [2, 128]. Not all the commodities are packaged after processing but some might require processing after packaging. The interaction between the food and the packaging materials both during and after processing assists in selection and optimization of packaging materials [28, 37, 101].

The researchers at Stanford and Massachusetts Institute of Technology (MIT), USA developed Nuclear Magnetic Resonance (NMR) in 1946, which was initially used by chemists to analyze the structure of different chemical compounds. Now NMR is widely used in food processing industries. Different food processing and reactions can be monitored by the use of low-field NMR spectroscopy [33]. The *in situ* monitoring of reactions using probes and magnet have been studied by different NMR techniques [58]. The molecular structure of the packaging materials can be characterized with the help of NMR technique. Different foods and package interactions can be better revealed through NMR. NMR is a fingerprinting technique and non-destructive or non-invasive methods for assay of structural information taking the advantage of magnetic properties of certain nuclei. It can provide us with the complete structural information of any compounds while chromatographic methods like gas-chromatography mass spectroscopy (GC-MS), high performance liquid chromatography (HPLC), thin layer chromatography (TLC), infrared spectroscopy (IRS) and others can give clue to functional groups only. The different molecules that diffuse through the edible films or coatings can be found out by NMR spectroscopy [62, 83].

This chapter focuses on principles of NMR and its applications in food processing and packaging management.

4.2   PRINCIPLES OF NUCLEAR MAGNETIC RESONANCE (NMR)

4.2.1   CLASSICAL MODEL

According to this model, the positively charged nuclei are treated as microscopic magnets and as they spin, a magnetic field is generated. When these spinning nuclei were placed in an externally applied magnetic field (B_0), the atomic magnetic field opposes the externally applied magnetic field (B_0) thereby decreasing the effective field (B) at the nucleus by a fraction as in Eq. (1).

$$B_{eff} = B_0(1 - \sigma) \tag{1}$$

$$\sigma = \sigma_p + \sigma_d \tag{2}$$

where, σ = screening/shielding constant ($\sim10^{-6}$); σ_p = paramagnetic contribution; and σ_d = diamagnetic contribution.

The resultant external field exerts a torque on the spinning nuclei resulting in their precessional motion. The rate of this precession (also called Larmor frequency, υ) is proportional to the external magnetic field (B_0) and to the strength of the nuclear magnet Υ (called 'gyromagnetic ratio' which is a constant for a given nucleus) as in Eq. (3).

$$\upsilon = \frac{\Upsilon B_0 (1-\sigma)}{2\pi} \qquad (3)$$

From Eq. (3), the frequency is directly proportional to the external magnetic field (B_0), and makes it difficult to compare NMR spectra taken on spectrometers operating at different field strengths. To overcome this problem, the term chemical shift (δ) was introduced. It is the difference between the resonance frequency of the nucleus and a standard, relative to the standard and is reported in ppm [4].

$$\delta = \frac{\upsilon - \upsilon_{ref}}{\upsilon_{ref}} \times 10^6 \qquad (4)$$

The chemical shift for different compounds is taken with reference to tetramethylsilane (TMS) which is set to 0 ppm. TMS is suitable for reference because of its solubility in most organic solvents, inertness, and volatile nature with 12 equivalent ^{1}H and 4 equivalent ^{13}C [88].

4.2.2 QUANTUM MODEL

The nucleons (protons and neutrons) of atoms are spin particles. In some atoms, they cancel each-other producing zero overall spin while in others they possess an overall spin, i.e., nonzero spin angular number and are NMR active [88]. The nuclei of atoms have magnetic properties which are being utilized to obtain the chemical information. The overall spin of nucleus can be predicted by following rules:

Rule 1: No. of neutrons, no. of protons = even

Such nuclei are NMR inactive as they have zero spin quantum number ($I = 0$), e.g., ^{12}C, ^{16}O, ^{32}S.

Rule 2: No. of neutrons + no. of protons = odd

Such nuclei have one or more unpaired nucleon. These nuclei have a half-integer spin ($I = \frac{1}{2}, \frac{3}{2}, \frac{5}{2}$), e.g., ^{1}H, ^{13}C, ^{19}F, ^{15}N, ^{31}P, etc.

Rule 3: No. of neutrons, no. of protons = odd

Then such nuclei have an integer spin (i.e., 1, 2, 3).

The magnetic field generated by spinning charge has a magnetic moment (μ), which is proportional to spin quantum number (I) as in Eq. (5).

$$\mu = YP = Y\sqrt{I(I + 1)}\,\frac{h}{2\pi} \tag{5}$$

where, $p = mvr$ is the spin angular momentum.

The most common nuclei analyzed by NMR are the proton (H), ^{13}C isotope of carbon, ^{19}F and ^{31}P, all having a spin of half [114].

4.2.3   NMR INSTRUMENTATION

NMR uses superconducting magnets surrounded by two layers of jackets. The outer jacket is of liquid nitrogen while inner one of liquid helium to cool the magnets. The probe inside the magnet bore having small magnetic coils receive power from the NMR instrument. It also has coils to receive and transmit RF energy. The sample is inserted at the top of the bore and lowered down to the magnetic probe. The probe is connected to an electronic chamber consisting of transmitter, receiver and other systems to control NMR instrument (Figure 4.1). The transmitter has to

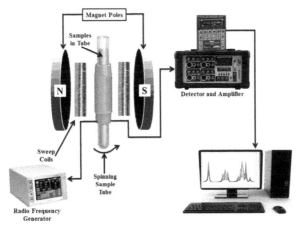

FIGURE 4.1   Diagrammatic presentation of NMR spectrometer [40]. (Modified from Ellis, D. I., Brewster, V. L., Dunn, W. B., Allwood, J. W., Golovanov, A. P., & Goo¬dacre, R. (2012). Fingerprinting food: current technologies for the detection of food adulteration and contamination. Chemical Society Reviews, 41, 5706–5727.)

produce a pulse at correct frequency in order to observe a nucleus. Proton (^{1}H) nucleus analysis requires a 500 MHz NMR spectrometer to produce 500 MHz and a second transmitter system at 125 MHz for ^{13}C-NMR analysis [114].

4.3   FOOD ANALYSIS

The produce obtained after post-harvest is ultimately undergone different processing to make the food edible and easily digestible along with enhancement in its quality and nutritional parameters. NMR spectroscopy is used for structural and compositional aspects in food science [49, 147], food analysis [30], authentication [44] and quality control of foodstuff [119]. Most of the chemical compounds of interest in food science are in the form of phosphate and the substituent on the oxygen atoms of phosphate influence the electron density on the phosphorus nucleus, and ultimately ^{31}P chemical shifts. Thus, ^{1}H along with ^{31}P-NMR (Phosphorus-NMR) spectroscopy is used for the analysis of food.

Among the four "omics" technologies, i.e., genomics (gene analysis), metabolomics (investigation of metabolites present in biological system), proteomics (study of the entire protein complement) and transcriptomics (gene expression analysis) of foodstuff, NMR gives complete view of foodstuffs metabolites [83]. Solid-state ^{13}C Cross Polarisation Magic Angle Spinning-NMR (CP-MAS NMR) has been used for monitoring the chemical composition of intact food samples such as mushrooms [108] or intact seeds [54] and for characterizing important insoluble components of food including carbohydrate [42], cellulose [154], cutin [59], and proteins [35]. ^{13}C-NMR studies mainly on the lipid fraction of vegetable oils [53, 120], fish [8], meat [14] and milk [4, 5] along with evaluation of glucose and fructose isoforms in Modena balsamic vinegar [31] while minor components like mono-/di-glycerides, free fatty acids, glycerol, total free sterols and polyphenols containing carboxyl and hydroxyl groups have been identified by ^{31}P-NMR. On the other hand, the effect of heat on saturated and unsaturated seed oils has been studied by ^{31}P-tagging NMR. NMR metabolomics on fish is focused on fatty acid profiling using ^{13}C-NMR spectroscopy. Salting has been used traditionally for centuries to preserve the food, e.g., light salting is used to increase the moisture content and to reduce drip in fish [140].

The use of LF-NMR to check the combined effect of salt and modified atmosphere packaging (MAP) on cod (*Gadus morhua*) during super-chilled storage revealed longer relaxation times with increasing salt concentration while super-chilling storage and MAP enhanced the shelf life of the cod loins but brine injection adversely effected it [51]. ^{31}P-NMR spectra of raw meat reveal different levels of phosphorylated metabolites and their changes after post-mortem. ^{31}P-NMR chemical shifts of ATP (adenosine triphosphate), PME (pectin methylesterase) and Pi (pigments) are key to intracellular pH of meat muscle. In case of milk ^{1}H-NMR is not used directly because of the hindrance of signals from water (major component of milk) on the signals from other components. ^{31}P-NMR is used to determine phosphoglycerides in milk at different storage conditions [15] and ^{13}C-NMR distinguishes milk of different animal species as they vary in their fatty acid composition. ^{1}H-NMR is used to quantify the key markers of beer quality viz., acetic, citric, lactic, malic, pyruvic and succinic acids [116].

Tea is a popular drink consumed widely in the world in which about 31 compounds have been identified in the ^{1}H-NMR spectrum of green tea extracts [76]. Spraul et al. [132] proposed a fully automated NMR screening approach for quality control and authentication of different fruit juices. In case of orange juice, NMR metabolomics focuses on the examination of adulteration [77]. A combination of NMR metabolomic data and data from all other "omics" finally authenticate foodstuff properties [83].

4.3.1 NON-DESTRUCTIVE ANALYSIS OF FOOD BY MAGNETIC RESONANCE IMAGING

Magnetic Resonance Imaging (MRI) is also based on the principle of NMR, which has capability of the on-line detection of food quality as well as internal properties and structural composition of the food after passing various stages of post harvesting (effects of various external physiological factors) as NMR can only judge testing chemical composition and internal properties [23]. Within the food industries we are witnessing the transition of classical process of quality analysis of food to the non-destructive or non-invasive on-line assessment of food. Non-destructive techniques are more rapid and precise. The microstructure is a target in case of semi-solid foods like sauces, dressings, yogurt, spreads and ice cream. MRI is extensively used for the microbial analysis of fruits and vegetables [111] and also for meat [51]. MRI is generally use to evaluate the internal fruit tissues for certain defects like bruising in avocado [87], chilling injury in micro-tom tomato [137] and insect infestation in mango [27] and heat treatment causing injuries to mesocarp tissues [60]. MRI was used to detect seeds in orange [55], internal browning in apple [23] and pears [56] and translucency in overripe pineapple [24].

MRI could even distinguish red tomato from green [24] and provide an indication of fruit maturity by measuring the oil/water ratio in intact avocado [25]. Joyce et al., [60] conducted a limited study of MRI on 'Kensington Pride' mangoes during ripening which indicated an outward-moving flux of water activity (a_w). In the case of internal browning in pears, alteration of the fruit membrane leads to the tissue disintegration. The disintegrated tissue undergoes the higher transverse relaxation rates compare to the sound tissues under the high magnetic field strength and long pulse spacing, i.e., NMR relaxometry [56]. Browning caused by CO_2 may also lower the antioxidant amount of the pears like *conference* and *rocha* pears [146].

4.4 PROCESS MONITORING AND QUALITY CONTROL

On-line analytical techniques are classified as at-line, on-line, in-line, and non-invasive techniques [115]. Reaction monitoring inside the probe and the magnet by Magic Angle Spinning-NMR (MAS-NMR) techniques was used to examine polymerization [58, 74]. The on-line NMR method takes an advantage of a bypass system for the sample detection, which has lowered the typical delay time with ^{1}H-NMR spectroscopy to less than one minute compared to 5–10 minutes using conventional NMR tubes [33].

The NMR-MOUSE for *in-situ* measurements in industries as for *in-situ* monitoring of dairy product quality has proved to be the simplest and most robust device [102]. LF-NMR uses T_1 and T_2 relaxation time for characterization and quality control studies. It is used to determine the water holding capacity (WHC) in meat and studies stability and its performance after recovery in an enzyme treated poultry meat using T_1 and T_2 relaxation time [94]; texture and quality of frozen fish using T_2 relaxation time analysis [121, 135]; albumen quality in hen eggs by T_1 relaxation time [72] and NMR spin relaxation for crystal formation in ice cream [81].

A device using s three-electrode electrochemical cell (i.e., an ECNMR cell) is used for *in situ* quantitative structural and mechanical characterization of a chemical system [153]. The oxidation of caffeic acid is responsible for the browning effect in food and beverages, especially in wine production [79]. Electrochemical studies have been performed in different types of electrodes to understand the behavior of caffeic acid [48, 57, 95, 125].

The oxidation products of caffeic acid have been characterized by an ECNMR cell. MRI is used to study moisture and fat content of cheese [118], NMR to investigate water behavior change in Mozzarella cheese [71] while MRI and NMR to study ice formation and its effect in Mozzarella cheese during freezing [70]; monitoring changes in feta cheese dur-

TABLE 4.1 List of Antioxidants Used As Stabilizer in Polymer [75]

Antioxidant	Chemical name
Chimasorb 81	2-Hydroxy-4-*n*-octyloxybenzophenone
Irgafos 168	Tris-(2,4-di-*tert.*-butylphenyl) phosphate
Irgafos P-EPQ	Tetrakis(2,4-di-*tert.*-butylphenyl)-4,49-biphenylenediphosphonite
Irganox 1010	Pentaerythrityl-tetrakis-3-(3,5-di-*tert.*-butyl-4-hydroxyphenyl) propionate
Irganox 1076	Octadecyl-3-(3,5-di-*tert.*-butyl-4-hydroxyphenyl) propionate
Irganox 1330	(1,3,5-Trimethyl-2,4,6-tris(3,5-di-*tert.*-butyl-4-hydroxyphenyl) propionate
Tinuvin 234	2-[2-Hydroxy-3,5-bis(1-methyl phenyl)phenyl]benzotriazole
Tinuvin 326	2-(5-Chloro-2H-benzotriazole-2-yl)-6-(1,1-dimethylethyl)-4-methylphenol
Tinuvin 770 DF	Bis(2,2,6,6-tetramethyl-4-piperidyl)sebacate
Tinuvin P	2-(2'-Hydroxy-5'-methyl-phenyl) benzotriazole

ing brining [1] and a shrinkage study at high brine concentration on cheese surface [91]. A higher salt concentration lowers final cheese moisture by increasing syneresis [90, 110] while a lower salt concentration results in more hydrated paracasein with weaker interaction between proteins [67].

4.5   POLYMERS AND PAPER AS PACKAGING MATERIAL

Polymers are used for the manufacture of different packaging materials. Polymers constitute of monomers, oligomers and additives like antioxidants (Table 4.1). Polymers may undergo degradation by oxidation under exposure to UV light. The antioxidants are added to stabilize the polymers by degrading themselves [75]. Polyethylene (PE) and polypropylene (PP) are highly consumed thermoplastics in the world. Copolymerization process is used for creating new materials with novel and enhanced properties [134]. The properties of copolymers depend on many factors, such as molecular structure (composition, molecular weight, molecular weight distribution, monomer distribution, etc.) and supra molecular structure (crystallinity, entanglements, morphology, etc.) [151].

Copolymer composition is determined by ^{13}C-NMR spectroscopy, which reveals the influence the co-monomer size on polymer conformation. Conformation of polymer chains in solution is analyzed according to the scaling law between radius of gyration and the number of repeating units [136]. Paper and paperboard are replacing these in food and drink packages. About 48% of all packaging is being occupied by paper packaging [131].

The EU's paper production data shows that almost half of the paperboard production is linked with the food contact standards [41]. The plastic packaging material was implemented within the European Communities to prevent transfer of harmful substances from packaging to foodstuff. The European Committee norms ensure protection of consumers' health from adulteration of the foodstuff by unacceptable transfer of components from packaging. Mainly two types of migration limits for plastic materials, i.e., overall migration limit (OML) and specific migration limit (SML) have been established. OML of 60 mg (of substances)/Kg (of foodstuff or food simulants) apply to the entire foodstuff while SML is fixed on the basis

TABLE 4.2   Threshold Limit of Elements for Packaging Material and Whole Packaging [9]

Element	Dry substance (mg/kg)
As	5
Cd	0.5
Cr	50
Cu	50
F	100
Hg	0.5
Mo	1
Ni	25
Pb	50
Se	0.75
Zn	150

of the toxicological evaluation of the substance. The SIVIL is established based on the acceptable daily intake (ADI) or the tolerable daily intake (TDI) set by the Scientific Committee on Food (SCF). The threshold limit of elements for packaging material and whole packaging is shown in Table 4.2.

4.5.1 GAMMA IRRADIATION TREATMENT OF POLYMERS

Most foods are irradiated after packaging. A radiation treatment up to 10 K Gy is found safe for foods [29]. Irradiation has marked effect on the paper and paperboard. The effects mainly include discoloration and loss of strength [20, 103]. The radiation-induced changes in polymers are studied by the use of NMR [6]. ^1H and ^{13}C-NMR spectra in solution were observed for oligomers and additives [22]. In an experiment conducted by Pentimalli et al. [104], polymers were taken from EniChem, residual monomers and oligomers were purified to remove commercial additives by solvent/no-solvent precipitation methods followed by irradiation in glass ampoules with ^{60}Co Y-cell at 6 KGy/hr. The pulsed low-resolution ^1H-NMR and solid-state ^{13}C CPMAS-NMR spectra were taken at 56 MHz 50.33 MHz respectively. The irradiation of packaging materials during sterilization may degrade the additives.

The effect of irradiation and migration of degraded additives' by-products are shown in Table 4.3. The radiation effects on polymers are either scission or cross-linking among the polymer chains leading to a decrease and/or increase in its molecular weight; production of small volatile compounds by chemical reactions and structural modifications of polymer chains [103]. Aromatic polymers have greater resistance to irradiation than aliphatic polymers due to presence of phenyl rings [50]. The glass packaging shows an unpleasant esthetic effect of brown tinting after radiation but plastic material becomes ineffective against microbial contaminants and small particles released into food deteriorating its quality [92].

Paper and boards are mainly used for packaging of disposable products. These can be recycled for use as packaging for economy purpose. Recycled papers are used for packaging dry foods like flour, grain, pasta, rice, salt and sugar [18]. These recycled paper and boards interact with the foods being packaged and there is transfer of components between the two. Modified polyphenylene oxide (MPPO) called tenax has been found by NMR and being quantified by GC-MS as a contaminant transferred from recycled paper packages to the food [43, 143]. The migration of compounds from the paperboard to the food samples depends on the nature of

TABLE 4.3   Effect of Gamma Radiation on Polymers: An NMR Study [103]

Polymers	Effect of Gamma Radiation	Remarks
Acrylonitrile butadiene styrene (ABS)	• Presence of additives as in HIPS pre-vents both cross-linking and degradation	HIPS and ABS do not show detectable effects
High-impact poly-styrene (HIPS)	• Quite unaffected by irradiation	Most widely used material
Polybutadiene (PB)	• Easily cross-linked and degraded even after addition of stabilizers	Not very suitable material for packaging of irradiated foods
Polystyrene (PS)	• No significant effects can be observed under 100 kGy • Mechanical properties are slightly affected	Most suitable material for packaging of irradiated foods
Styrene-acrylonitrile (SAN)	• No significant effects observed	Not as resistant as poly-styrene

the compounds, their molecular structure, concentration, etc. The migration is also temperature dependent, i.e., with the increase of temperature at/or during processing or packaging, there is increase in the percentage of migrants. It also depends on the type and the nature of the foods being packaged.

In foods with high fat content, there is prominent increase in migrants with increase in temperature. Dry foods have no free fats but at high temperature fatty compounds melt and can penetrate into paper material. It is also being noted that Benzophenone, a photo-initiator for inks and varnishes, is highly fat soluble and get mixed into the milk-powder being packaged in paper packaging [3]. In crystalline phase, the cellulose fibrils are tightly packed compared to the amorphous phase (cellulose, lignin and hemi-cellulose) in which water filled cavities with hydrogen bonds are responsible for tight pack. The longer T_2 value from large pore diameter indicates the γ-irradiation enlargement of the pore size.

4.6 STRUCTURAL CHARACTERIZATION OF POLYMERS

Plastic packaging materials are non-biodegradable and ultimately pollute our environment. In addition, they are made from valuable and limited non-renewable resource, i.e., petroleum [138]. The packaging materials used are usually made from plastic, glass, paper or metal. These materials contribute about two-third of total packaging waste by volume [84]. The biodegradable aliphatic polyesters, such as poly-butylenes adipate terephthalate (PBAT), polycaprolactone (PCL), poly-3-hydroxybutyrate (PHB) and polylactic acid (PLA) are expensive and not feasible for all possible applications [10, 126, 149].

The solid-state ^{13}C-NMR spectroscopy was applied for structural characterization of starches of various botanical origins like acid-modified tapioca starch [7]; xylans in the cell wall of *Palmaria palmate* [73] and for the molecular mobility of enzymatically modified potato starch [59]. Traditional plastic materials are easily manufactured and are resistant to corrosion and microbial attack. Their intense use increases municipal waste and environmental pollution since they are undegradable [127]. Naturally occurring polymers as Polyhydroxyalkanoate (PHA) [16], synthetic aliphatic polyester such as poly(ϵ-caprolactone) [141] and aliphatic-aromatic co-polyester [137] are being used by micro-organisms as a source of carbon and thus minimizes environmental pollution [141].

$$C_{polymer} + O_2 \longrightarrow CO_2 + H_2O + C_{residue} + C_{biomass} \qquad (6)$$

The water vapor permeability (WVP) for biodegradable is greater than synthetic polymers as shown in Table 4.4.

4.6.1   STARCH/CLAY NANO-COMPOSITE FILM

The nano-composites are used to improve the mechanical and barrier properties of the films. The starch matrices (like neat potato starch and a mixture of potato starch with biodegradable polyester) are dispersed with

TABLE 4.4   Water Vapor Permeability (WVP) for Biodegradable and Synthetic Films

Film type	Film formulation	WVP $(g\ m^{-1}\ s^{-1} Pa^{-1})$	Ref.
1. Biodegradable films	Amylose	3.8×10^{-10}	[46]
	Chitosan 2%	$3.66\text{–}4.80 \times 10^{-10}$	[148]
	Corn zein	5.35×10^{-10}	[47]
	Corn zein plasticized with glycerol	8.90×10^{-10}	[46]
	Fish skin gelatine	2.59×10^{-10}	[12]
	Gelatin	1.6×10^{-10}	[126]
	Whey protein plasticized With sorbitol	7.17×10^{-10}	[89]
2. Synthetic films	Cellophane	8.4×10^{-11}	[123]
	HDPE	2.31×10^{-13}	[129]
	LPDE	9.14×10^{-13}	[129]
	PVDC	2.22×10^{-13}	[123]

PVDC: polyvinyl dichloride; LDPE: low density polyethylene; HDPE: high density polyethylene.

functionalized layered silicates (clay minerals) in thermoplastic starch by polymer melt processing techniques to obtain nano-composite films [9]. The organization of different nano-composite constituents may not result in a well-defined 3-dimensional periodicity, thus X-ray diffraction failed to resolve the structure of aluminum-containing pillars in aluminum-pillared clays. One-phase ^{27}Al-NMR (Aluminum-NMR) reveals the pillars to have tridecamers of $[Al_{13}O_4(OH)_{24}(H_2O)_{12}]^{+7}$ having the so called "ε-Kegging" structure [145]. ^{27}Al-NMR spectra especially give an idea of the first coordination sphere of the Al (III) ion.

For mechanical characterization, the Young's modulus, strength and strain at break was calculated from the experimental stress–strain curves. It was found that the Young's modulus decreased when bio-degradable polyester was added to the mixture of PS and PE while strain at break increased with respect to other samples. Similarly, the clay nano-particles increased the mechanical properties of starch (PS sample), while the addition of nano-powder to the PS/PE matrix decreased both modulus and stress. Water facilitates gelatinization and hence affects the modulus of starch blends [9].

4.6.2 BIODEGRADATION TEST

Biodegradation process involves the characterization by Gel Permeation Chromatography (GPC) and different NMR techniques followed by chemical analysis of extracted matter from the soil [124, 137]. Samples were then analyzed by ^{1}H- and ^{31}P-NMR to determine different phenolic and other lignin functionalities [68], study lipid metabolites [36, 99] and/or to quantify fatty acids and glycerides for quality control purposes [32, 133]. The mineral content in the soil is calculated by the following formula:

$$\text{Min}\ (\%) = \left[\frac{\text{mg CO}_{2\ \text{sample}} - \text{mg CO}_{2\ \text{blank}}}{\text{mg sample} \times \%\text{C} \times 3.6667} \right] \times 100 \qquad (7)$$

where, mg CO_2 sample is the amount of CO_2 measured in the test jars; mg CO_2 blank is the amount of CO_2 produced in the blank jars; and 3.6667 is the ratio between CO_2 molecular weight and C atomic weight.

The moist soil samples (amended with polyester and blanks incubated for different time durations) were oven-dried at 105°C to remove water followed by Soxhlet extraction and analysis for mineral content by GPC and NMR. ^{31}P-NMR data reveals the hydrolysis of polymer and decrease in the molecular weight also confirmed by mass loss of extracted material and increase in mineral content [127]. In fact, the degradation of the poly-butylene sebacate started with the terminal ester bonds' hydrolysis and the release of monomer or oligomers in the soil. Only a fraction of the polymer remained in the soil in the form of oligomers after degradation.

4.7   BIO-PLASTIC

Many efforts have been done to develop degradable, renewable and recyclable green materials [17, 45, 95, 109, 113]. Composites from biodegradable polymers degrade completely in the soil without producing any toxic compounds [66]. Biodegradable polymers like PCL, PHB, PLA and poly-butylene succinate adipate (PBSA) unlike conventional plastics PS and PE can degrade in soil within few years [80]. PBSA has wide application in food industry and agricultural packaging [61, 122]. The production of PBSA is complex and expensive which can be overcome by blending PBSA with cost-effective agricultural residues such as rice husk (RH), wheat straw, and corn stover [39, 98]. Blending PBSA with agricultural residues reduces cost and improves the mechanical properties of the blend [19]. The physical properties and biodegradability of rice husk (RH) composites with PBSA and acrylic acid rich PBSA is characterized using Fourier transform infrared (FTIR) spectroscopy and ^{13}C-NMR [150].

The solid-state ^{13}C-NMR spectra of PBSA/RH (20 wt %), PBSA-g-AA/RH (20 wt %), and RH reveals the formation of ester groups by the reactions between hydroxyl groups in RH and carboxyl groups of PBSA-g-AA. This interaction significantly affects the mechanical and biodegradation properties PBSA-g-AA/RH. The tensile strength of the blend is enhanced as shown in Figure 4.2.

4.8   EDIBLE-FILMS AND COATINGS

Edible films and coatings are generally applied to limit the mass transfer and increase the shelf-life of the food product [64]. Edible films consist of

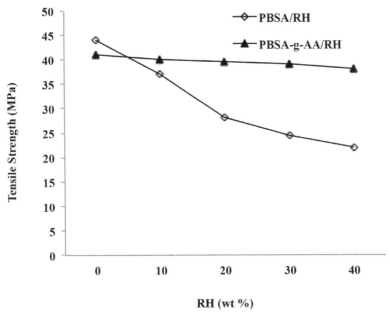

FIGURE 4.2 Effect of RH content on tensile strength for PBSA/RH and PBSA-g-AA/RH composites [150]. (Reprinted from Polymer Degradation and Stability, Volume 97, Issue 1, January 2012, Pages 64–71, Characterization and biodegradability of polyester bioplastic-based green renewable composites from agricultural residues, Chin-San Wu. © 2012, with permission from Elsevier.)

polysaccharides, proteins, lipids, and waxes. Polysaccharides and proteins provide good matrix cohesion while lipids and waxes act as barrier substance because they are impermeable to water [62]. The emulsified films could be used to combine the water barrier efficiency of the lipid phase with the plasticity of the continuous matrix as plastic films are more efficient than lipid films in terms of water barrier efficiency [97]. The barrier properties of the edible films and coatings depend on different factors, such as on the drying conditions [105, 106], on the lipid concentration [69], on the repartition of the lipid phase [85], on the solid fat content [34] and on the type of emulsifier used [107]. The plasticizers used for improving the film deformability have been found to influence mass transfer [130]. The permeability of edible films to aroma compounds, or the diffusion of small molecules in these films, such as antimicrobial agents is the major areas of focus nowadays [100, 112].

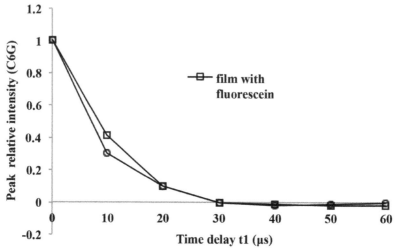

FIGURE 4.3   Intensity decay of the C6G peak of i-carrageenan film with and without fluorescein [65]. (Reprinted from Food Chemistry, Volume 106, Issue 4, 15 February 2008, Pages 1340–1349, Diffusion of small molecules in edible films: Effect of water and interactions between diffusant and biopolymer, T. Karbowiak, RD Gougeon, S Rigolet, L Delmotte, F Debeaufort, A Voilley. © 2008, with permission from Elsevier.)

4.8.1   MASS TRANSFER THROUGH EDIBLE FILMS

The mass transfer of small molecules like water, solutes, coloring agents, and aroma compounds can be prevented between the foodstuff and the surrounding medium or among different phases of a composite food by the application of an edible barrier. The iota-carrageenans on cooling below the critical temperature forms thermo-reversible gels with the formation of double-helices of carrageenan chains [152]. The i-carrageenan films and coatings are used for the encapsulation [13], prevention of mass transfers [63] or support for active molecules [26].

Diffusion in porous system is controlled by voids and channels while in dense system, diffusion follows Fick's law. Fick and Henry laws describe permeability coefficient in terms of solubility and diffusion coefficients. The transport of small molecules through the film is described by sorption, diffusion and desorption processes [117]. The diffusion of a small reference molecule (fluorescein) in i-carrageenan films samples with or without fluorescein is observed through ^{1}H- and ^{13}C-NMR. The ^{1}H–^{13}C CP-MAS (cross-polarization- magic angle of spinning) spectra were obtained by cross-polarization, with proton dipolar decoupling [65]. The $^1H-^{13}C$ experiment of Wideline Separation (WISE) and raw NMR were processed using MATLAB and the Win-NMR software (Bruker Biospin) and the WinFit program [86]. The $^1H-^{13}C$ CP-MAS spectra intensity decay of the C6G peak of i-carrageenan film, with and without fluorescein (Figure 4.3) shows the greatest broadening of C6G carbon. It suggests the interaction of fluorescein with the carrageenan network and i-carrageenan helix diameter is enough for fluorescein diffusion [65]. The diffusion process in solid polymers is more complex than in simple liquids or gels.

4.9 FUTURE PROSPECTUS

The modes of processing and packaging are changing day by day. The processing industries need to focus on on-line monitoring of all bioprocesses and reactions for the purpose of convenience and better quality. Even all sorts of packaging materials from polymer to recent edible, anti-microbial and nano-packaging needs more research to determine their stability, durability, bio-safety and bio-hazards. The packaging materials need to be assessed for the possible contaminants and toxic materials being transferred through the biodegradable and edible films which ultimately determine the health of the consumers. Another area, which needs more focus, is the bio-plastic and other forms of degradable and recyclable green materials to lead the future into green packaging.

4.10 CONCLUSIONS

The raw foods are first being processed to edible and digestible form. The processing conditions and factors are set to meet the Hazard Analysis & Critical Control Points (HACCP) principles for the safety food as well as consumers. Nowadays different on-line NMR techniques have been applied to monitor the food processing and its quality. Food must be packaged in an appropriate packaging material after or before processing to preserve its quality parameters and increase the shelf-life. Among polymers PS has been the most suitable material for the packaging of irradiated food.

High Impact Polystyrene (HIPS) is quite unaffected by irradiation, hence mostly widely used for packaging. Paper packaging materials are disposable as well as recyclable but present threats to foods being packaged due to migration of some compounds like MPPO (tenax) and Benzophenone. Biodegradable packaging materials are recommended for use as Green Packaging. These are decomposed and used by micro-organisms as a source of carbon-dioxide. Recently different nano-composites and bioplastics have been designed and analyzed by NMR spectroscopy for the mass transfer & bio-safety of consumers as well as the environment. Edible films and coatings are applied to limit the mass transfer and increase the shelf-life of the product. Edible films are used to wrap food products while coatings are applied directly on foods. The diffusion of solutes and aroma-compounds through the films and coatings are assessed by ^{1}H- and ^{13}C-NMR for the quality parameter and overall safety of consumers.

4.11 SUMMARY

Food processing and packaging determine the ultimate safety of food being packaged in terms of shelf-life, microbial infestation, contaminants and toxic materials from within or outside environment. NMR-techniques have been used by the researchers to characterize the different types of processes and packaging materials from plastic to recent edible and biodegradable packaging materials. NMR non-destructively provides structural information of any compound and enables understanding of the interactions between the binding sites of diffusant and the polymer along with their effects on polymer organization.

Owing to safety and durability, the packaging materials are structurally, chemically, and mechanically characterized by different NMR-techniques and also looked for the mass transfer through the packaging films and between packaging materials and food products. NMR is used for in-situ and online monitoring of process and chemical reactions going on in food industries for quality control purposes. The chapter also focused on nano-composite films and biodegradability assay of packaging materials and different on-line techniques for the quality analysis and reaction monitoring in food industries. The Green Packaging is preferred nowadays for the consumer as well as environmental safety.

KEYWORDS

- analytical technique
- antioxidant
- authentication
- biodegradability
- biodegradable
- carbohydrate
- cellulose
- chemical composition
- chemical compound
- consumer
- contaminants
- edible coatings
- edible film
- electrochemical
- energy
- environmental safety
- food
- food analysis
- food industry
- food processing
- food product
- food science
- foodstuff
- genomics
- green packaging
- magnetic
- magnetic resonance imaging
- mass transfer
- material
- metabolite

- metabolomics
- microbial
- molecular
- molecular structure
- monitoring
- nano-composite
- nano-composite film
- nmr spectroscopy
- oligomer
- Omics technology
- packaging
- packaging management
- packaging material
- plastic material
- plastic packaging
- polyethylene
- polymer
- polypropylene
- primary packaging
- primary packaging material
- protein
- proteomics
- quality
- quality analysis
- quality control
- quantum
- safety
- secondary packaging
- shelf-life
- spectrometer
- spectroscopy
- stabilizer

- **structural**
- **structural information**
- **tertiary packaging**
- **toxic**
- **transcriptomics**

REFERENCES

1. Altan, A., Oztop, M. H., McCarthy, K. L., & McCarthy, M. J. (2011). Monitoring changes in feta cheese during brining by magnetic resonance imaging and NMR relaxometry. *Journal of Food Engineering,* 107, 200–220.
2. Alvarez, I. (2007). Hurdle technology and the preservation of food by pulsed electric fields. In: *Food Preservation by Pulsed Electric Fields* by Lelieveld, H. L. M., Notermans, S., & de Haan, S. W. H. (Eds.). CRC Press, pp. 165–178.
3. Anderson, W. A. C., & Castle, L. (2003). Benzophenone in cartonboard packaging materials and the factors that influence its migration into food. *Food Additives and Contaminats*, 20, 607–618.
4. Andreotti, G., Lamanna, R., Trivellone, E., & Motta, A. (2002). ^{13}C NMR spectra of TAG: an easy way to distinguish milks from different animal species. *Journal of American Oil Chemist's Society*, 79, 123–127.
5. Andreotti, G., Trivellone, E., Lamanna, R., Di Luccia, A., & Motta, A. (2000). Milk identification of different species: ^{13}C NMR spectroscopy of triacylglycerols from cows and buffaloes' milks. *Journal of Dairy Science*, 83, 2432–2437.
6. Anelli, P., Baccaro, S., & Casadio, C. (1998). Gamma radiation effects on an Amine antioxidant added in an ethylene-propylene copolymer. *Radiation Physics and Chemistry*, 52(16), 183–186.
7. Atichokudomchaia, N., Varavinita, S., & Chinachotic, P. (2004). A study of ordered structure in acid-modified tapioca starch by ^{13}C CP/MAS solid-state NMR. *Carbohydrate Polymers*, 58, 383–389.
8. Aursand, M., & Grasdalen, H. (1992). Interpretation of the ^{13}C NMR spectra of omega-3 fatty acids in lipid extracted from white muscle of Atlantic salmon (Salmo salar). *Chemistry and Physics of Lipids*, 62, 239–251.
9. Avella, M., De Vlieger, J. J., Errico, E. M., Fischer, S., Vacca, P., & Volpe, G. M. (2005). Biodegradable starch/clay nanocomposite films for food packaging applications. *Food Chemistry,* 93, 467–474.
10. Avella, M., Erico, M. E., Laurienzo, P., Martuscelli, E., Raimo, M., & Rimedio, R. (2000). Preparation and characterization of compatibilised polycaprolactone/starch composites. *Polymer*, 41, 3875–3881.

11. Avena-Bustillos, R. J., & Krochta, J. M. (1993). Water vapor permeability of caseinate-based edible films as affected by pH, calcium cross-linking and lipid content. *Journal of Food Science*, 58, 904–907.
12. Banati, D. (2005). Adulteration of foodstuffs: from misleading to poisoning. Diet Diversification and Health Promotion. *Forum of Nutrition Basel*, 57, 135–146.
13. Bartkowiak, A., & Hunkeler, D. (2001). Carrageenan–oligochitosan microcapsules: Optimization of the formation process. *Colloids and Surfaces B*, 21 (4), 285–298.
14. Beauvallet, C., & Renou. J. P. (1992). Applications of NMR spectroscopy in meat research. *Trends in Food Science and Technology*, 3, 241–246.
15. Belloque, J., Carrascosa, A. V., & Lopez-Fandino, R. (2001). Changes in phosphoglyceride composition during storage of ultrahigh-temperature milk, as assessed by 31P-nuclear magnetic resonance. Possible involvement of thermoresistant microbial enzymes. *Journal of Food Protection*, 64, 850–855.
16. Bertini, F., Canetti, M., Cacciamani, A., Elegir, G., Orlandi, M., & Zoia, L. (2012). Effect of ligno-derivatives on thermal properties and degradation behavior of poly(3-hydroxybutyrate)-based biocomposites. *Polymer Degradation and Stability*, 97, 1979–1987.
17. Bhardwaj, R., Mohanty, A. K., Drzal, L. T., Pourboghrat, F., & Misra, M. (2006). Renewable resource-based green composites from recycled cellulose fiber and poly (3-hydroxybutyrate-co-3-hydroxyvalerate) bioplastic. *Biomacromolecules*, 7, 2044–2051.
18. Binderup, M. L., Pedersen, G. A., Vinggaard, A. M., Rasmussen, E. S., Rosenquist, H., & Cederberg, T. (2002). Toxicity testing and chemical analyzes of recycled fiber-based paper for food contact. *Food Additives & Contaminants*, 19,13–28.
19. Bledzki, A. K., Mamun, A. A., & Volk, J. (2010). Physical, chemical and surface properties of wheat husk, rye husk and soft wood and their polypropylene composites. *Composites Part A*, 41, 480–488.
20. Brody, A. L., & Marsh, K. S. (1997). The Wiley Encyclopedia of Packaging Technology. Wiley, New York.
21. Brown, W. E. (1992). Plastics in Food Packaging. Properties, Design and Fabrication. Marcel Dekker, New York.
22. Capitani, D., & Segre, A. L. (1996). NMR spectroscopy in solid polymers. *Trends in Polymer Science*, 4 (1), 4–5.
23. Chayaprasert, W., & Stroshine, R. (2005). Rapid sensing of internal browning in whole apples using a low-cost, low-field proton magnetic resonance sensor. *Postharvest Biology and Technology*, 36 (3), 291–301.
24. Chen, P., McCarthy, M. J., & Kauten, R. (1989). NMR for internal quality evaluation of fruits and vegetables. *Transactions of the ASAE*, 32, 1747–1753.
25. Chen, P., McCarthy, M. J., Kauten, R., Sarig, Y., & Han, S. (1993). Maturity evaluation of avocados by NMR methods. *Journal of Agricultural Engineering Research*, 55(3), 177–187.
26. Choi, J. H., Choi, W. Y., Cha, D. S., Chinnan, M. J., Park, H. J., & Lee, D. S. (2005). Diffusivity of potassium sorbate in kappa-carrageenan based antimicrobial film. *Lebens Wissens Technology*, 38 (4), 417–423.
27. Clark, C. J., Hockings, P. D., Joyce, D. C., & Mazucco, R. A. (1997). Application of magnetic resonance imaging to pre- and post-harvest studies of fruits and vegetables. *Postharvest Biology and Technology*, I (1), 1–21.

28. Clough, R. L. (2001). High-energy radiation and polymers: a review of commercial processes and emerging applications. *Nuclear Instruments and Methods in Physics Research Section B*, 185, 8–33.

29. Codex Standard 106, (1983). Norma Generale del Codex per glialimenti irradiati. In: *The Wiley Encyclopedia of Packaging Technology* by Brody, A. L., & Marsh, K. S. (Eds.). Wiley & Sons, New York.

30. Colquhoun, I. J., & Goodfellow, B. J. (1994). NMR spectroscopy. In: *Spectroscopic Techniques in Food Analysis* by Wilson, R. H. (Ed.). VCH Publishing, New York, pp. 87–145.

31. Consonni, R., Cagliani, L. R., Rinaldini, S., & Incerti, A. (2008). Analytical method for authentication of traditional balsamic vinegar of Modena. *Talanta*, 75, 765–769.

32. Dais, P., & Spyros, A. (2007). ^{31}P- NMR spectroscopy in the quality control and authentication of extra-virgin olive oil: a review of recent progress. *Magnetic Resonance in Chemistry*, 45, 367–377.

33. Dalitz, F., Cudaj, M., Maiwald, M., & Guthausen, G. (2012). Process and reaction monitoring by low-field NMR spectroscopy. *Progress in Nuclear Magnetic Resonance Spectroscopy,* 60, 52–70.

34. Debeaufort, F., Quezada-Gallo, J. A., Delporte, B., & Voilley, A. (2000). Lipid hydrophobicity and physical state effects on the properties of bilayer edible films. *Journal of Membrane Science*, 180 (1), 47–55.

35. Demuth, D., Haase, N., Malzacher, D., & Vogel, M. (2015). Effects of solvent concentration and composition on protein dynamics: ^{13}C MAS NMR studies of elastin in glycerol–water mixtures. *Biochimica et Biophysica Acta,* 1854, 995–1000.

36. DeSilva, M. A., Shanaiah, N., Gowda, G. A. N., Rosa-Perez, K., Hanson, B., & Raftery, D. (2009). Application of ^{31}P NMR spectroscopy and chemical derivatization for metabolite profiling of lipophilic compounds in human serum. *Magnetic Resonance in Chemistry,* 47, S74–S80.

37. Devlieghere, F., Vermeiren, L., & Debevere, J. (2004). New preservation technologies: possibilities and limitations. *International Dairy Journal*, 14(4), 273–285.

38. DEWHA, (2009). National waste policy: less waste, more resources, W. Department of the Environment, Heritage and the Arts, Editor, Commonwealth of Australia: Canberra.

39. Dias, A. B., Muller, C. M. O., Larotonda, F. D. S., & Laurindo, J. B. (2010). Biodegradable films based on rice starch and rice flour. *Journal of Cereal Science*, 51, 213–219.

40. Ellis, D. I., Brewster, V. L., Dunn, W. B., Allwood, J. W., Golovanov, A. P., & Goodacre, R. (2012). Fingerprinting food: current technologies for the detection of food adulteration and contamination. *Chemical Society Reviews*, 41, 5706–5727.

41. Escabasse, J. Y., & Ottenio, D. (2002). Food-contact paper and board based on recycled fibers: regulatory aspects-new rules and guidelines. *Food Additives & Contaminants*, 19, 79–92.

42. Fransen, C. T. M., van Laar, H., Johannis P. Kamerling, J. P., Johannes F.G., & Vliegenthart, J. F. G. (2000). CP-MAS NMR analysis of carbohydrate fractions of soybean hulls and endosperm. *Carbohydrate Research,* 328, 549–559.

43. Franz, R. (2002). Program on the recyclability of food-packaging materials with respect to food safety considerations: polyethylene terephthalate (PET), paper and

board, and plastics covered by functional barriers. *Food Additives & Contaminants,* 19, 93–110.

44. Gall, G. L., & Colquhoun, I. J. (2003). NMR spectroscopy in food authentication. In: *Food Authenticity and Traceability* by Lees, M. (Ed.). Woodhead Publishing, Cambridge, pp. 131–155.

45. Gandini, A. (2008). Polymers from renewable resources: a challenge for the future of macromolecular materials. *Macromolecules,* 41, 9491–9504.

46. Gennadios, A., Weller. C. L., & Gooding, C. H. (1994). Measurement errors in water vapor permeability of highly permeable, hydrophilic edible films. *Journal of Food Engineering,* 21, 395–409.

47. Ghanbarzadeh, B., Musavi, M., Oromiehie, A. R., Rezayi, K., Razmi, E., & Milani, J. (2007). Effect of plasticizing sugars on water vapor permeability, surface energy and microstructure properties of zein films. *LWT – Food Science and Technology,* 40, 1191–1197.

48. Giacomelli, C., Ckless, K., Galato, D., Miranda, F. S., & Spinelli, A. (2002). Electrochemistry of caffeic acid aqueous solutions with pH 2.0 to 8.5. *Journal of Brazilian Chemical Society,* 13, 332–338.

49. Gil, A. M. (2003). Spectroscopy: nuclear magnetic resonance. In: *Encyclopedia of Food Science and Nutrition* by Caballero, B. (Ed.). Elsevier, pp. 5447–5454.

50. Grassie, N., & Scott, G. (1985). Polymer degradation and stabilization. Cambridge University Press, Cambridge.

51. Gudjonsdottir, M., Jonsson, A., Bergsson, A. B., Arason, S., & Rustad, T. (2011a). Shrimp processing assessed by low field nuclear magnetic resonance, near infrared spectroscopy, and physicochemical measurements—The effect of polyphosphate content and length of pre-brining on shrimp muscle. *Journal of Food Science,* 76(4), E357–E367.

52. Gudjonsdottir, M., Lauzon, H. L., Magnusson, H., Sveinsdottir, K., Arason, S., Martinsdottir, E., & Rustad, T. (2011b). Low field Nuclear Magnetic Resonance on the effect of salt and modified atmosphere packaging on cod (*Gadus morhua*) during superchilled storage. *Food Research International,* 44, 241–249.

53. Gunstone, F. D. (1993). Information on the composition of fats from their high resolution ^{13}C nuclear magnetic resonance spectra. *Journal of the American Oil Chemist's Society,* 70, 361–366.

54. Gussoni, M., Greco, F., Pegna, M., Bianchi, G., & Zetta, L. (1994). Solid state and microscopy: NMR study of the chemical constituents of Afzelia cuanzensis seeds. *Magnetic Resonance Imaging,* 12, 477–486.

55. Hernandez-Sanchez, N., Barreiro, P. J., & Ruiz-Cabello, J. (2006). On-line identification of seeds in mandarins with magnetic resonance imaging. *Biosystems Engineering,* 95 (4), 529–536.

56. Hernandez-Sanchez, N., Hills, B. P., Barreiro, P., & Marigheto, N. (2007). An NMR study on internal browning in pears. *Postharvest Biology and Technology,* 44, 260–270.

57. Hotta, H., Ueda, M., Nagano, S., Tsujino, Y., Koyama, J., & Osakai, T. (2002). Mechanistic study of the oxidation of caffeic acid by digital simulation of cyclic voltammograms. *Analytical Biochemistry,* 303, 66–72.

58. Hunger, M., & Weitkamp, J. (2001). In situ IR, NMR, EPR, and UV/Vis spectroscopy: tools for new insight into the mechanisms of heterogeneous catalysis. *Angewandte Chemie International*, 40, 2954–2971.

59. Jarvinen, R., Silvestre, A. J. D., Ana M. Gil, A. M., & Kallio, H. (2011). Solid state ^{13}C CP-MAS NMR and FT-IR spectroscopic analysis of cuticular fractions of berries and suberized membranes of potato. *Journal of Food Composition and Analysis*, 24, 334–345.

60. Joyce, D. C., Hockings, P. D., Mazucco, R. A., Shorter, A. J., & Brereton, I. M. (1993). Heat treatment injury of mango fruit revealed by nondestructive magnetic resonance imaging. *Postharvest Biology and Technology*, 3, 305–311.

61. Kale, G., Kijchavengkul, T., Auras, R., Rubino, M., Selke, S. E., & Singh, S. P. (2007). Compostability of bioplastic packaging materials: an overview. *Macromolecular Bioscience*, 7, 255–277.

62. Karbowiak, T., Debeaufort, F., & Voilley, A. (2007). Influence of thermal process on structure and functional properties of emulsion-based edible films. *Food Hydrocolloids*, 21, 879–888.

63. Karbowiak, T., Debeaufort, F., Champion, D., & Voilley, A. (2006). Wetting properties at the surface of iota-carrageenan-based edible films. *Journal of Colloid and Interface Science*, 294(2), 400–410.

64. Karbowiak, T., Debeaufort, F., Voilley, A., & Trystram, G. (2009). From macroscopic to molecular scale investigations of mass transfer of small molecules through edible packaging applied at interfaces of multiphase food products. *Innovative Food Science and Emerging Technology*, 10, 116–127.

65. Karbowiak, T., Goegeon, D. R., Rigolet, S., Delmotte, L., Debeaufort, F., & Voilley, A. (2008). Diffusion of small molecules in edible films: Effect of water and interactions between diffusant and biopolymer. *Food Chemistry*, 106, 1340–1349.

66. Kim, H. S., Kim, H. J., Lee, J. W., & Choi, I. G. (2006). Biodegradability of bio-flour filled biodegradable poly (butylene succinate) bio-composites in natural and compost soil. *Polymer Degradation and Stability*, 91, 1117–1127.

67. Kindstedt, P. S. 2007. Low-moisture Mozzarella cheese. In: *Cheese Problems Solved* by McSweeney, P. L. H. (Ed.). CRC Press, pp. 299–329.

68. King, G. A. W. T., Zoia, L., Filpponen, I., Olszewska, A., Xie, H., Kilpelainen, I., & Argyropoulos, D. S. (2009). In situ determination of lignin phenolics and wood solubility in imidazolium chlorides using ^{31}P-NMR. *Journal of Agricultural and Food Chemistry*, 57, 8236–8243.

69. Koelsch, C. M., Labuza, T. P. (1992). Functional, physical and morphological properties of methyl cellulose and fatty acid-based edible barriers. *Lebens Wissens Technology*, 25, 404–411.

70. Kuo, M. I., Anderson, M. E., Gunasekaran, S. (2003). Determining effects of freezing on pasta filata and non-pasta filata Mozzarella cheeses by nuclear magnetic resonance imaging. *Journal of Dairy Science*, 86, 2525–2536.

71. Kuo, M. I., Gunasekaran, S., Johnson, M., & Chen, C. (2001). Nuclear magnetic resonance study of water mobility in pasta filata and non-pasta filata Mozzarella. *Journal of Dairy Science*, 84, 1950–1958.

72. Laghi, L., Cremonini, M. A., Placucci, G., Sykora, S., Wright, K., & Hills, B. (2005). A proton NMR relaxation study of hen egg quality. *Magnetic Resonance Imaging*, 23, 501–510.

73. Lahaye, M., Rondeau-Mouro, C., Deniaud, E., & Buleon, A. (2003). Solid-state ^{13}C NMR spectroscopy studies of xylans in the cell wall of *Palmaria palmata* (L. Kuntze, Rhodophyta). *Carbohydrate Research*, 338, 1559–1569.

74. Landfester, K., Spiegel, S., Born, R., & Spiess, H. W. (1998). On line detection of emulsion polymerization by solid state NMR spectroscopy. *Colloid and Polymer Science*, 276, 356–361.

75. Lau, W. O., & Wong, K. S. (2000). Contamination in food from packaging material. *Journal of Chromatography A*, 882, 255–270.

76. Le Gall, G., Colquhoun, I. J., & Defernez, M. (2004). Metabolite profiling using ^{1}H NMR spectroscopy for quality assessment of green tea, *Camellia sinensis* (L.). *Journal of Agricultural and Food Chemistry*, 52, 692–700.

77. Le Gall, G., Puaud, M., & Colquhoun, I. J. (2001). Discrimination between orange juice and pulp wash by ^{1}H nuclear magnetic resonance spectroscopy: identification of marker compounds. *Journal of Agricultural and Food Chemistry*, 49, 580–588.

78. Lelieveld, H. L. M. (2007). Pitfalls of pulsed electric filed processing. In: *Food Preservation by Pulsed Electric Fields* by Lelieveld, H. L. M., Notermans, S., & de Haan, S. W. H. (Eds.). CRC Press, pp. 294–300.

79. Li, H., Guo, A., & Wang, H. (2008). Mechanisms of oxidative browning of wine. *Food Chemistry*, 108, 1–13.

80. Liu, W., Mohanty, A. K., Drzal, L. T., & Misra, M. (2005). Novel biocomposites from native grass and soy based bioplastic: processing and properties evaluation. *Industrial and Engineering Chemistry Research*, 44, 7105–7112.

81. Lucas, T., Wagener, M., Barey, P., & Mariette, F. (2005). NMR assessment of mix and ice cream. Effect of formulation on liquid water and ice. *International Dairy Journal*, 15, 1064–1073.

82. Macomber, R. S. (1998). A complete introduction to modern NMR spectroscopy. A Wiley-Inter Science Publication, pp. 56–103.

83. Mannina, L., Sobolev, A. P., & Viel, S. (2012). Liquid state ^{1}H high field NMR in food analysis. *Progress in Nuclear Magnetic Resonance Spectroscopy* 66: 1.

84. Marsh, K., & Bugusu, B. 2007. Food packaging: Roles, materials, and environmental issues. *Journal of Food Science*, 72 (3), 39–55.

85. Martin-Polo, M., Mauguin, C., & Voilley, A. (1992). Hydrophobic films and their efficiency against moisture transfer. 1. Influence of the film preparation technique. *Journal of Agricultural and Food Chemistry*, 40, 407–412.

86. Massiot, D., Fayon, F., Capron, M., King, I., Le Calve, S., & Alonso, B. (2002). Modelling one- and two-dimensional solid-state NMR spectra. *Magnetic Resonance in Chemistry*, 40 (1), 70–76.

87. Mazhar, M., Joyce, D., Cowin, G., Brereton, I., Hofman, P., Collins, R., & Gupta, M. (2015). Non-destructive 1H-MRI assessment of flesh bruising in avocado (*Persea americana* M.) cv. Hass. *Postharvest Biology and Technology*, 100, 33–40.

88. Mazumder, A., & Dubey, D. K. (2013). Nuclear Magnetic Resonance (NMR) Spectroscopy. Vertox Laboratory, Defence Research and Development Establishment. Gwalior, India. Elsevier Inc.

89. McHugh, T. H., Krochta, J. M. (1994). Sorbitol- vs glycerol-plasticized whey protein edible films: integrated oxygen permeability and tensile property evaluation. *Journal of Agricultural and Food Chemistry*, 42, 841–845.

90. McMahon, D. J., Motawee, M. M., & McManus, W. R. (2009). Influence of brine concentration and temperature on composition, microstructure, and yield of feta cheese. *Journal of Dairy Science, 92,* 4169–4179.

91. Melilli, C., Carco, D., Barbano, D. M., Tumino, G., Carpino, S., & Licitra, G. (2005). Composition, microstructure, and surface barrier layer development during brine salting. *Journal of Dairy Science,* 88, 2329–2340.

92. Milz, J. 1987. Food product-package compatibility. In: *Proceedings* by Gray, J. I., Harte, B. R., & Miltz, J. (Eds.). PA: Technomic Publishing, Lancaster, pp. 30–43.

93. Min, S., & Zhang, Q. H. (2005). Packaging for non-thermal food processing. In: *Innovations in food packaging* by Han, J. H. (Ed.). Elsevier Academic Press, Oxford, pp. 482–500.

94. Mitchell, J., Gladden, L. F., Chandrasekera, T. C., & Fordham, E. F. (2014). Low-field permanent magnets for industrial process and quality control. *Progress in Nuclear Magnetic Resonance Spectroscopy* 76, 1–60.

95. Moghaddam, A. B., Ganjali, M. R., Dinarvand, R., Norouzi, P., Saboury, A. A., & Moosavi-Movahedi, A. A. (2007). Electrochemical behavior of caffeic acid at single-walled carbon nanotube: graphite-based electrode. *Biophysical Chemistry*, 128, 30–37.

96. Mohanty, A. K., Misra, M., & Drzal, L. T. (2002). Sustainable bio-composites from renewable resources: opportunities and challenges in the green materials world. *Journal of Polymers and the Environment,* 10, 19–26.

97. Morillon, V., Debeaufort, F., Bond, G., Capelle, M., & Voilley, A. (2002). Factors affecting the moisture permeability of lipid-based edible films: A review. *Critical Reviews in Food Science and Nutrition*, 42 (1), 67–89.

98. Nyambo, C., Mohanty, A. K., & Misra, M. (2010). Polylactide-based renewable green composites from agricultural residues and their hybrids. *Biomacromolecules*, 11, 1654–1660.

99. Oostendorp, M., Engelke, U. F. H., Willemsen, M. A. A. P., & Wevers, R. A. (2006). Diagnosing inborn errors of lipid metabolism with proton nuclear magnetic resonance spectroscopy. *Clinical Chemistry*, 52, 1395–1405.

100. Ozdemir, M., & Floros, J. D. (2001). Analysis and modeling of potassium sorbate diffusion through edible whey protein films. *Journal of Food Engineering*, 47 (2), 149–155.

101. Ozen, B. F., & Floros, J. D. (2001). Effects of emerging food processing techniques on the packaging materials. *Trends in Food Science and Technology*, 12 (2), 60–67.

102. Pedersen, H. T., Ablett, S., Martin, D. R., Mallett, M. J. D., & Engelsen, S. B. (2003). Application of the NMR-MOUSE to food emulsions. *Journal of Magnetic Resonance*, 165, 49–58.

103. Pentimalli, M., Capitani, D., Ferrando, A., Ferri, D., Ragni, P., & Segre, A. L. (2000a). Gamma irradiation of food packaging materials: an NMR study. *Polymer*, 41, 2871–2881.

104. Pentimalli, M., Ragni, P., Righini, G., & Capitani, D. (2000b). Polymers and paper as packaging materials of irradiated food: An NMR study. *Radiation Physics and Chemistry*, 57, 385–388.

105. Perez-Gago, M. B., & Krochta, J. M. (2000). Drying temperature effect on water vapor permeability and mechanical properties of whey protein-lipid emulsion films. *Journal of Agricultural and Food Chemistry*, 48 (7), 2687–2692.

106. Phan The, D., Debeaufort, F., Peroval, C., Despre, D., Courthaudon, J. L., & Voilley, A. (2002a). Arabinoxylan-lipid-based edible films and coatings. 3. Influence of drying temperature on film structure and functional properties. *Journal of Agricultural and Food Chemistry*, 50 (8), 2423–2428.

107. Phan The, D., Peroval, C., Debeaufort, F., Despre, D., Courthaudon, J. L., & Voilley, A. (2002b). Arabinoxylan-lipids-based edible films and coatings. 2. Influence of sucroester nature on the emulsion structure and film properties. *Journal of Agricultural and Food Chemistry*, 50 (2), 266–272.

108. Pizzoferrato, L., Manzi, P., Bertocchi, F., Fanelli, C., Rotilio, G., & Paci, M. (2000). Solid-state ^{13}C CP MAS NMR spectroscopy of mushrooms gives directly the ratio between proteins and polysaccharides. *Journal of Agricultural and Food Chemistry*, 48, 5484–5488.

109. Pommet, M., Juntaro, J., Heng, J. Y. Y., Mantalaris, A., Lee, A. F., & Wilson, K. (2008). Surface modification of natural fibers using bacteria: depositing bacterial cellulose onto natural fibers to create hierarchical fiber reinforced nanocomposites. *Biomacromolecules*, 9, 1643–1651.

110. Prasad, N., & Alvarez, V. B. (1999). Effect of salt and chymosin on the physico-chemical properties of feta cheese during ripening. *Journal of Dairy Science*, 82, 1061–1067.

111. Pykett, I. L. (2000). NMR-A powerful tool for industrial process control and quality assurance. *IEEE Transactions on Applied Superconductivity*, 10 (1), 721–723.

112. Quezada Gallo, J. A., Debeaufort, F., & Voilley, A. (1999). Interactions between aroma and edible films. 1. Permeability of methylcellulose and low density polyethylene films to methylketones. *Journal of Agricultural and Food Chemistry*, 47 (1), 108–113.

113. Raquez, J. M., Deleglise, M., Lacrampe, M. F., & Krawczak, P. (2010). Thermosetting (bio) materials derived from renewable resources: a critical review. *Progress in Polymer Science*, 35, 487–509.

114. Reuhs, L. B., & Simsek, S. (2010). Nuclear Magnetic Resonance. In: *Food Analysis* by Nielsen, S. S. (Ed.). Springer, New York, pp. 443–456.

115. Riebe, M. T., & Eustace, D. J. (1990). Process analytical-chemistry – An industrial perspective. *Analytical Chemistry*, 62, 65–71.

116. Rodrigues, J. E. A., Erny, G. L., Barros, A. S., Esteves, V. I., Brandao, T., Ferreira, A. A., Cabrita, E., & Gil, A. M. (2010). Quantification of organic acids in beer by nuclear magnetic resonance (NMR)-based methods. *Analytica Chimica Acta*, 674, 166–175.

117. Rogers, C. E. (1985). Permeation of gases and vapors in polymers. In: *Polymer permeability* by Comyn, J. (Ed.). Elsevier Applied Science Publishers, New York, pp. 11–73.

118. Ruan, R., Chang, K., Chen, P. L., Fulcher, R. G., & Bastian, E. D. (1998). A magnetic resonance imaging technique for quantitative mapping of moisture and fat in a cheese block. *Journal of Dairy Science*, 80, 9–15.

119. Sacchi, R., & Paolillo, L. (2007). NMR for food quality and traceability. In: *Advances in Food Diagnostics* by Nollet, L. M. L., & Toldra, F. (Eds.). Blackwell Science, pp. 101–118.

120. Sacchi, R., Addeo, F., Giudicianni, I., & Paolillo, L. (1992). Analysis of the positional distribution of fatty acids in olive oil triacylglycerols by high-resolution ^{13}C NMR of the carbonyl region. *Italian Journal of Food Science*, 2, 117–122.

121. Sanchez-Alonso, I., Martinez, I., Sanchez-Valencia, J., & Careche, M. (2012). Estimation of freezing storage time and quality changes in hake (*Merluccius merluccius*, L.) by low field NMR. *Food Chemistry*, 135, 1626–1634.

122. Shah, A. A., Hasan, F., Hameed, A., & Ahmed, S. (2008). Biological degradation of plastics: a comprehensive review. *Biotechnology Advances*, 26, 246–265.

123. Shellhammer, T. H., & Krochta, J. M. (1997). Water vapor barrier and rheological properties of simulated and industrial milk fat fractions. Trans. *ASAE*, 40, 1119–1127.

124. Shi, G., Cooper, D. G., & Maric, M. (2011). Poly (3-caprolactone)-based 'green' plasticizers for poly (vinyl chloride). *Polymer Degradation and Stability*, 96, 1639–1647.

125. Silva, L. F., Stradiotto, N., & Oliveira, H. P. (2008). Determination of caffeic acid in red wine by voltametric method. *Electroanalysis*, 20, 1252–1258.

126. Singh, R. P., Pandey, J. K., Rutot, D., Degee, P., & Dubois, P. (2003). Biodegradation of poly (ε-caprolactone)/starch blends and composites in composting and culture environments: the effect of compatibilization on the inherent biodegradability of the host polymer. *Carbohydrate Research*, 338, 1759–1769.

127. Siotto, M., Zoia, L., Tosin, M., Innocenti, D. F., Orlandi, M., & Mezzanotte, V. (2013). Monitoring biodegradation of poly(butylene sebacate) by Gel Permeation Chromatography, ^{1}H-NMR and ^{31}P-NMR techniques. *Journal of Environmental Management*, 116, 27–35.

128. Siripatrawan, U., & Jantawat, P. (2008). A novel method for shelf life prediction of a packaged moisture sensitive snack using multilayer perception neural network. *Expert Systems with Applications*, 34 (2), 1562–1567.

129. Smith, M. A. (1986). Polyethylene, high density. In: *The Wiley Encyclopedia of Packaging Technology* by Bakker, M. (Ed.). John Wiley & Sons, New York, NY. pp. 214–223.

130. Sobral, P. J. A., Menegalli, F. C., Hubinger, M. D., & Roques, M. A. (2001). Mechanical, water vapor barrier and thermal properties of gelatin based edible films. *Food Hydrocolloids*, 15 (4–6), 423–432.

131. Song, Y. S., Park, H. J., & Komolprasert, V. (2000). Analytical procedure for quantifying five compounds suspected as possible contaminants in recycled paper/paperboard for food packaging. *Journal of Agricultural and Food Chemistry*, 48, 5856–5859.

132. Spraul, M., Schutz, B., Humpfer, E., Mortter, M., Schafer, H., Koswig, S., & Rinke, P. (2009). Mixture analysis by NMR as applied to fruit juice quality control. *Magnetic Resonance in Chemistry*, 47, S130–S137.

133. Spyros, A., & Dais, P. (2000). Application of ^{31}P-NMR spectroscopy in food analysis. I. Quantitative determination of mono- and diglycerides in virgin olive oils. *Journal of Agricultural and Food Chemistry*, 48, 802–805.

134. Stephens, C. H., Poon, B. C., Ansems, P., Chum, S. P., Hiltner, A., & Baer, E. (2006). Comparison of propylene/ethylene copolymers prepared with different catalysts. *Journal of Applied Polymer Science,* 100, 1651–1658.

135. Streen, C., & Lambelet, P. (1997). Texture changes in frozen cod mince measured by low field nuclear magnetic resonance spectroscopy. *Journal of the Science of Food and Agriculture,* 75, 268–272.

136. Suarez, I., Losio, S., & Coto, B. (2013). Polymer chain conformation of copolymers with different monomer size: ^{13}C NMR spectroscopy and MALS study. *European Polymer Journal,* 49, 3402–3409.

137. Tan, F. T., Cooper, D. G., Maric, M., & Nicell, J. A. (2008). Biodegradation of a synthetic copolyester by aerobic mesophilic microorganisms. *Polymer Degradation and Stability,* 93, 1479–1485.

138. Tang, X. Z., Kumar, P., Alavi, S., & Sandeep, K. P. (2012). Recent Advances in Biopolymers and Biopolymer-Based Nanocomposites for Food Packaging Materials. *Critical Reviews in Food Science and Nutrition,* 52, 426–442.

139. Tao, F., Zhang, L., McCarthy, M. J., Beckle, D. M., & Saltveit, M. (2014). Magnetic resonance imaging provides spatial resolution of Chilling Injury in Micro-Tom tomato (Solanum lycopersicum L.) fruit. *Postharvest Biology and Technology,* 97, 62–67.

140. Thorarinsdottir, K. A., Arason, S., & Thorkelsson, G. (2002). The effects of light salting on physiochemical characteristics of frozen cod (*Gadus morhua*) filets. *Journal of Aquatic Food Product Technology,* 11, 287–301.

141. Tokiwa, Y., & Calabia, B. P. (2007). Biodegradability and biodegradation of polyesters. *Journal of Polymers and the Environment,* 15, 259–267.

142. Tokiwa, Y., Calabia, B. P., Ugwu, C. U., & Aiba, S. (2009). Biodegradability of plastics. *International Journal of Molecular Sciences,* 10, 3722–3742.

143. Triantafyllou, V. I., Karamani, A. G., Akrida-Demertzi, K., & Demertzis, P. G. (2002). Studies on the usability of recycled PET for food packaging applications. *European Food Research and Technology,* 215, 243–248.

144. Trienekens, J. H., Hagen, J. M., Beulensc, A. J. M., & Omta, S. W. F. (2003). Innovation through (international) food supply chain development: a research agenda. *International Food and Agribusiness Management Review,* 6 (3), 1–15.

145. Tucker, N., & Johnson, M. (2004). Low environmental impact polymers. Rapra Technology Shawbury.

146. Veltman, R. H., Kho, R. M., van Schaik, A. C. R., Sanders, M. G., & Oosterhaven, J. (2000). Ascorbic acid and tissue browning in pears (*Pyrus communis* L. cvs. Rocha and Conference) under controlled atmosphere conditions. *Postharvest Biology and Technology,* 19, 129–137.

147. Webb, G. A. (2006). Applications in Materials Food, and Marine Sciences. Modern Magnetic Resonance. Vol. 3, Springer.

148. Wong, D. W. S., Gastineau, F. A., Gregorski, K. S., Tillin, S. J., Pavlath, A. E. (1992). *Journal of Agricultural and Food Chemistry,* 40, 540.

149. Wu, C. S. (2003). Physical properties and biodegradability of maleated- polycaprolactone/starch composite. *Polymer Degradation and Stability,* 80, 127–134.

150. Wu, S. C. (2012). Characterization and biodegradability of polyester bioplastic-based green renewable composites from agricultural residues. *Polymer Degradation and Stability,* 97, 64–71.

151. Xu, X., Xu, J., Feng, L., & Chen, W. (2000). Effect of short chain-branching distribution on crystallinity and modulus of metallocene-based ethylene–butene copolymers. *Journal of Applied Polymer Science,* 77, 1709–1715.

152. Yuguchi, Y., Thu Thuy, T. T., Urakawa, H., & Kajiwara, K. (2002). Structural characteristics of carrageenan gels: Temperature and concentration dependence. *Food Hydrocolloids,* 16(6), 515–522.

153. Zhang, X., & Zwanziger, J. W. (2011). Design and applications of an *in situ* electrochemical NMR cell. *Journal of Magnetic Resonance,* 208, 136–147.

154. Zuckerstattera, G., Terinte, N., Sixtac, H., & Schuster, K. C. (2013). Novel insight into cellulose supramolecular structure through ^{13}C CP-MAS NMR spectroscopy and paramagnetic relaxation enhancement. *Carbohydrate Polymers,* 93, 122–128.

CHAPTER 5

A REVIEW ON PRINCIPLES AND APPLICATIONS OF ULTRASOUND PROCESSING

GABRIELA JOHN SWAMY, SANGAMITHRA ASOKAPANDIAN, and CHANDRASEKAR VEERAPANDIAN

CONTENTS

5.1   INTRODUCTION

Traditional thermal treatment processes are keystone of the food industry to provide the required safety profile and addendums to the existing shelf-life period. However, such processes may result in the losses of desired organoleptic properties and may destruct the thermo-labile nutrients. As a result, novel thermal and non-thermal technologies have emerged to convene the requisite food safety demands whilst reducing the effects on the nutritional and sensory attributes of the food product. Novel technologies are of benefit to both food processors and consumers. However depending on the complexity of food material and the variety of foods produced, the validation process is a challenge for the food industry. The factors that drive the necessity of the validation process include: extension of shelf-life, nutritional and sensory aspects, organoleptic properties, consumer acceptance and impact on the environment [1]. Ultrasound is one of the important technologies among the non-thermal methods that have gained widespread acceptance from the industry and consumers.

Ultrasound is known to have a major influence on different processes in the food and bioprocessing industry. Most processes can be completed in few seconds or minutes with high reproducibility using ultrasound, thereby reducing the processing cost, with higher purity of the final product and consuming only a small portion of the energy normally needed for conventional processes [2]. Several ultrasound-assisted processes such as extraction, freezing and thawing, cutting and drying have been carried out efficiently in the food industry [3]. Food processes executed by the action of ultrasound are influenced by cavitation phenomena and mass transfer enhancement [4].

There are number of novel processing technologies currently available: high hydrostatic pressure, irradiation, oscillating magnetic field, high intensity pulsed light, high intensity pulsed electric field, ultrasound and ozonation [5].

This chapter presents a review on applications of ultrasound in food and bioprocessing technology. It presents the essential theoretical background and the applications for various industrial processes. The factors, which make the possible combination of food processing and ultrasound one of the most assuring research areas in the field of modern food engineering, are also discussed.

5.2   PRINCIPLES OF ULTRASOUND PROCESSING

Ultrasound technology is based on mechanical sound waves at a frequency above the threshold of human hearing, approximately >16 kHz. The ultrasound waves have two modes of travel. One is through the bulk of a material and the other one is on the surface of the material at a speed which is characteristic of the nature of the wave and the material through which it is propagating. Based on the frequency and power ranges, ultrasound can be classified into following two types [6]:

a. High frequency (100 kHz – 1 MHz); Low power (typically <1 Wcm^{-2});

b. Low frequency (16–100 kHz); High power (10–1000 Wcm^{-2}).

Most applications of ultrasound in the food industry involve a non-destructive analysis such as estimation of firmness, ripeness, sugar content, and acidity of food products [7]. These analyzes use high frequency and low power ultrasound. Low frequency and high power ultrasound is used to alter the food properties, either physically or chemically [8]. Also it can be used for the extraction of oil for cooking and biodiesel production [9].

Ultrasound is propagated through the medium as a series of compression and rarefaction waves induced on the molecules of the medium. At high power ranges, the rarefaction cycle surpasses the attractive forces of the liquid molecules resulting in the formation of cavitation bubbles from gas nuclei present within the fluid [10]. Over a period of time, the bubbles reach the critical size and become unstable leading to immediate collapse. The implosion of the bubbles leads to energy accumulations in hot spots, causing extreme temperatures ~5000 K and pressures ~1000 atm [11]. The temperature and pressure changes resulting from the implosions can kill some microbes, but they are localized and do not affect a large area.

5.3 APPLICATIONS OF ULTRASOUND IN FOOD INDUSTRY

5.3.1 MICROBIAL INACTIVATION

Pasteurization and ultra-high temperature are few of the conventionally used thermal treatments to inactivate microbes. Generally, these treatments result in undesirable flavor formation and nutrient loss. However in ultrasonic treatment, the cavitation as a result of changes in temperature and pressure created by the waves are responsible for bacterial elimination. The mechanism of microbial inactivation is mainly because of the thinning of cell membranes, localized heating and production of free radicals [12]. The number of factors such as nature of the dissolved gas, hydrostatic pressure, specific heat of the liquid and gas in the bubble and the tensile strength of the liquid are necessary to determine the minimum oscillation pressure that is required to produce cavitation (cavitation threshold) of a medium [13]. Another important variable is temperature, which is inversely proportional to cavitation threshold. For cavitation to occur, the suitable ultrasonic frequency used must be <2.5 MHz [14].

Microbial spores are quite resistant to the effects and therefore require extended periods of ultrasonication to render a product safe. For practical purposes, ultrasound is used in conjunction with pressure (manosonication), temperature (thermosonication) or both (manothermosonication) to facilitate enhanced mechanical disruption of cells. Inactivation experiments with ultrasound-resistant strain of *Listeria monocytogenes* in apple cider was conducted at sub lethal (20, 30, and 40°C) and lethal (50, 55, and 60°C) temperatures with and without application of ultrasound (20 kHz, 457 Wm^{-1}) [15]. The survival tests showed a possibility of using a mild treatment condition in combination with ultrasound to achieve a 5-log reduction. The effects of ultrasound treatment at various amplitudes (50, 75, and 100%) and times (0, 6, 12, 18, 24, and 30 min) on *Escherichia coli* and *Saccharomyces cerevisiae* and physicochemical characteristics were determined in pomegranate juice [16]. More than 5-log inactivation of *E. coli* and 1.36-log inactivation of *S. cerevisiae* were achieved at 100% ultrasound amplitude after 30 min of treatment. The log-linear and Weibull models were effectively utilized to estimate the microbial inactivation as a function of ultrasound treatment time ($R^2 > 0.97$). The organisms such as *Escherichia coli* O157:H7, *Salmonella enterica, Listeria monocytogenes, Debaryomyceshansenii, Clavisporalusitaniae, Torulasporadelbrueckii, Pichiafermentans,* and *Saccharomyces cerevisiae* in orange juice were exposed to multi-frequency power ultrasound treatment and the inactivation kinetic parameters [17].

The time required to reduce initial inoculated populations by 5 log cycles (99.99%), T_5D values, was significantly increased with acid adaptation. After acid exposure, the T_5D of *E. coli, S. enterica,* and *L. monocytogenes* increased from 37.64, 36.87, and 34.59 to 54.72, 40.38, and 37.83 min, respectively. Pineapple, grape and cranberry juice were thermos sonicated (24 kHz, 400 W, 120 µm) at 40, 50 and 60[18]. Inactivation of Saccharomyces cerevisiae was tested from 0 to 10 min and color and pH were measured. Survivor's curves were fitted using Weibull distribution, four-parameter model and modified Gompertz equation. Grape juice showed total inactivation (7-log) after 10 min. The modified Gompertz equation showed the best fit. Energy analysis showed that pineapple juice (4287.02 mW/mL) required a higher amount of energy, whereas grape juice showed the lowest value (3112.13 mW/mL).

The usage of sonication and thermosonication ($53 \pm 1°C$) was investigated as potential methods for decreasing the numbers of *Campylobacter, Enterobacteriaceae* and total viable counts (TVC) on raw poultry [19]. *Campylobacter jejuni* was more susceptible to thermosonication compared to thermal or sonication treatment with mean inactivations of about 4.72, 1.45 and 3.17 $\log_{10}$ CFU/mL, respectively. No viable *Campylobacter* or *Enterobacteriaceae* were detected in broiler skin pieces in HI (high-intensity) unit treated with 16 min thermal, sonication and thermosonication treatments and also TVC were decreased by 1.93, 1.34 and 2.49 $\log_{10}$ CFU/g, respectively. Thermosonication treatment in the LI (low-intensity) unit reduced *Enterobacteriaceae* and TVC populations by 2.74 and 1.69 $\log_{10}$ CFU/g, respectively.

The effect of simultaneous sonication and thermosonication on inactivation of *C. perfringens* spore in beef slurry was also investigated [20]. Less than 1.5 log reduction was obtained for both *C. perfringens* NZRM 898 and NZRM 2621 spores exposed to 60 min thermosonication process (24 kHz, 0.33 W/g) at 75°C. Also, thermal inactivation of *C. perfringens* spores in beef slurry was assessed for the two strains, which followed the first order kinetics. The D value at 105°C and z-values were 2.5 min and 10.6°C for NZRM 898 and 1.8 min and 10.9°C for NZRM 2621, respectively. The thermal inactivation rate of spore in beef slurry was doubled by the combined treatment. At 95°C, D-value of 20.2 min decreased to 9.8 min, revealing that spore exposure to heat shock followed by ultrasonication enhanced its thermal inactivation.

The influence of ultrasound and temperature on *Escherichia coli* and *Staphylococcus aureus* in milk were studied and it was noticed that Gram-negative bacteria (*E. coli*) are more susceptible to the ultrasonic treatment than the Gram-positive ones (*S. aureus*). The maximum inactivation of *E. coli* and *S. aureus* were achieved an optimal conditions such as temperature of 59.99*E. coli* and 117.27 mm for *S. aureus* [21]. Inactivation of *E. aerogenes* in skim milk with different protein concentrations using low-frequency (20 kHz) and high-frequency (850 kHz) ultrasonication were also studied [22]. The log reductions were −3.64 (), −2.73 (), −2.31 () and −2.21 () for the bacteria in water, 5% milk, 10% milk and 15% milk, respectively when ultrasonicated for 60 min. In the case of high-frequency ultrasound treatment of *E. aerogenes* in milk (5 and 10 %), there was no change for the viable cells after ultrasonication up to 60 min at 50 W. A central composite response surface model was used to analyze the influence of temperature (20–52), sound intensity (60–120 W/cm^2) and treatment time (40–240 s) at a constant pressure (225 kPa) by manothermosonication processing on microbial inactivation in milk [23]. Reductions up to 1.6 log CFU/mL were achieved for *E. coli* and *Pseudomonas fluorescens*. Lower inactivation values were reported for *Staphylococcus aureus* (1.05 log CFU/mL) and such values were achieved using conditions of 36°C, 90 W/cm^2 and 240 s. Inactivation of *E. coli* by manothermosonication (R^2=0.90) was described by an exponential curve, whereas inactivation of *S. aureus and P. fluorescens* by manothermosonication (R^2 ≥ 0.73) were described by a linear trend.

5.3.2   EXTRACTION OF BIOACTIVES

Applying ultrasound as a technique for extraction of bioactive compounds such as polyphenols, aromatic compounds, anthocyanins, polysaccharides and functional components from plant sources and waste materials have been widely studied. However, limited research publications include continuous ultrasonic process as an aid to assist extraction.

Desired plant components are localized in the surface glands of living tissues. These components can be stimulated to release by comparatively mild ultrasonic stressing [2]. However, when the desired material is located internally within cells, a pre-ultrasound treatment is required to increase the surface area to achieve quick and complete extraction. Ultrasonic treatment can also effectively accelerate the hydration process, when pre-hydration is required to maximize extraction. All these mentioned extraction enhancements by ultrasound have been accredited to the spread of sonic pressure waves that result in the cavitation phenomena. The high shearing force amplifies the mass transfer process and the implosion of cavitation bubbles produces turbulence and high velocity particle collisions thereby accelerating eddy and internal diffusion. This phenomenon can be confirmed by carrying out a scanning electron micrography.

The ultrasound extraction variables such as time, temperature, power and solvent to sample ratio were studied and their effect on stability, antioxidant activity and yield of papaya seed oil were also investigated [24]. Optimum extraction was attained at temperature of 62.5°C; time of 38.5 min; ultrasound power of 700 W; solvent to sample ratio about ~7:1 (v/w). This optimum extraction process provided the desirable yield (23.3 ± 0.6%) and reliable antioxidant activity (88.1 ± 0.8%) and oxidative stability (Peroxide value, 0.18 ± 0.02 (meq/g); p-anisidine value, 0.07 ± 0.009; Totox value, 0.43 ± 0.02). The extracted oil had very reliable antioxidant activity (88.1 ± 0.8%) and desirable stability (Peroxide value, 0.18 ± 0.02 (meq/g); p-anisidine value, 0.07 ± 0.009; Totox value, 0.43 ± 0.02).

The influence of three amplitudes of pulsed ultrasound-assisted solvent extraction (PUASE) (0, 25, and 50%; 100 W, 30 kHz; the 0% treatment as control) on kinetics, yield and quality of extracted oil from *Pistaciakhinjuk* kernel at different temperatures (30, 40, and 50) was evaluated [25]. A maximum oil yield of 77.5% (w/w) was obtained for samples treated with PUASE at 50% amplitude and 50°C. The influence of ultrasonic extraction followed by hydro-distillation of *Elettariacardamomum* L. seeds was investigated [26]. To attain the maximum extraction performance, the optimum operating conditions included water-to-fine dried seeds ratio of 12, sonication at 10% of the maximal power and a sonication time of 30 min prior to hydro distillation for another 30 min. Effects of solvent, powder granularity, ultrasonic power, solid–liquid ratio, extraction time and extraction temperature on the yield of essential oil from cinnamon bark was explored [27]. Results indicated that ultrasonic power had major effect on the oil yield, followed by powder granularity, time, and solid–liquid ratio. The optimized extraction conditions are petroleum ether (b.p. 60–90) as extraction solvent, 165 W, 80–100 mesh cinnamon power, 40 min, and 6:60 (W/V) solid–liquid ratio. Under these optimal conditions, cinnamon oil yield reached 14.8%, in which the content of trans-cinnamaldehyde, analyzed by HPLC, was 82.62%.

Ultrasound assisted extraction (solvent concentration, pH, time of extraction) of anthocyanins and phenolics from Jabuticaba peels were evaluated by response surface methodology [28]. The extracted components were analyzed by LC–MS and HPLC and identified as cyanidin-3-O-glucoside (anthocyanin) and ellagic acid (phenol). For the maximized extraction of targeted compounds, the peels required sonication of 10 min in a 46% (v/v) ethanol – water solution acidified at pH of 1. It was estimated that 4.8 mg/g dry peel of monomeric anthocyanin; 92.8 mg/g dry peel of gallic acid equivalent; 4.9 mg/g dry peel of cyanidin-3-O-glucoside and 7.8 mg/g dry peel of ellagic acid was yielded from the extraction. The impact of ultrasonic intensity (764–1528, W/m^2), temperature (25–45°C) and extraction time (10–30 min) on the total carotenoids from peach palm fruit by-products using response surface methodology was carried out [29]. The maximum extraction of total carotenoids as 163.47 mg/100 g dried peel was attained at ultrasonic intensity of 1528 W/m^2, extraction temperature of 35°C and extraction time of 30 min. Ultrasound assisted extraction was compared with conventional solvent extraction for maximization of the yield of steviol glycosides [30]. The highest content of steviol glycosides, total phenolic compounds, and flavonoids in stevia extracts were obtained when ultrasound assisted extraction was used: extraction time 10 min, probe diameter 22 mm, and temperature 81.2°C. The application of ultrasound for improved extraction of total phenolics, fucose, and uronic acid from brown seaweed (*Ascophyllumnodosum*) was studied [31]. Process parameters such as ultrasonic intensity, extraction time, and solvent type were investigated to optimize extraction yields. The maximum yields on dry basis of total phenolics, fucose, and uronic acid were 82.70 mg GAE/g_{db}, 135.76 mg/g_{db}, and 197.19 mg/g_{db}, respectively. Maximum bioactive yields were obtained using 0.03 M HCl as solvent at an ultrasonic intensity of 75.78 W cm^{-2}.

5.3.3 SONIC FILTRATION

Separation of solids from liquids is an essential process to produce solid-free liquid (a solid isolated from the liquid). However, fouling is one of the main problems encountered in membrane filtration process. Ultrasonic energy can be applied to enhance the flux by moving through the concentration polarization and cake layer at the surface of the membrane; thereby maintaining the intrinsic permeability of membrane [32]. Acoustic filtration has been effectively used to increase the filtration rate of industrial wastewater, which is otherwise considered as difficult to process [33]. Furthermore, optimized ultrasound intensity is vital to avoid the damage of filters [34].

Ultrasonic waves can enhance the efficiency of the continuous belt drying process of fruit pulps and slurries. While conventional filtration for dewatering of coal slurry reduced the moisture content by 10%, ultrasound reduced the moisture content up to 50% [35]. The use of ultrasound in combination with membrane filtration was investigated and the research yielded positive results. The researchers used ultrasonic irradiation at low power level to assist the filtration of whey. The results showed a considerable improvement of flow rate, as ultrasound waves helped in preventing filter blockage and flow through it, by reducing the compressibility of the protein deposit and filter cake [36]. Additionally, the combination of filter with sonic waves augments the span of the filter, as continuous cavitation at the surface of the filter prevents clogging and caking. All the above-described factors are of significance to commercialize the filtration processes. However, there is still need in its development at laboratory and at commercial level.

Clarification of pomegranate juice by microfiltration at 17 mL s^{-1} flow rate, and 0.5 bar transmembrane pressure was carried out both in the presence and the absence of ultrasonic treatment [37]. The application of the ultrasonic treatment decreased the total resistance by more than four times, due to the reduction of reversible fouling and cake resistance. The irreversible fouling resistance created using reaction of juice components with membrane material was about 2.5×10^{12} m^{-1} in both processes with and without ultrasonic treatment, because ultrasound does not have an effect on bonding between the membrane and juice components. The ultrasonic filtration of whey solution was studied in order to determine the main parameters affecting the flux, retention and membrane fouling [38]. It was found that the water fluxes through the membrane were not affected by ultrasound in case of 3 kDa membrane, but in case of 100 kDa, the water fluxes were slightly increased by ultrasound. This phenomenon can probably be explained by the fact that acoustic streaming and/or cavitation causes turbulence, which may cause decreased friction of water in membrane pores, resulting in increased water permeability.

The effect of ultrasonic parameters on permeates flux in the ultrasound-assisted ultrafiltration (UAUF) process of *Radix astragalus* extracts was inspected, at ultrasound frequency of 45 kHz and 100 kHz [39]. Highly significant effects of 45 kHz ultrasound irradiation on both flux reduction and flux enhancement were observed than that of 100 kHz. The flux enhancements of 45 kHz ultrasound irradiation at power of 30 W and 90 W were 11% and 28%, respectively. Similar to that, flux enhancements of 100 kHz at power of 360 W and 600 W were 10% and 27%, respectively. It can be concluded that ultrasound at low frequency is highly effective on improving the flux when a constant output power was applied. A comparative study of ultrasound assisted and stirred dead-end microfiltration of grape pomace extracts was performed [40]. The permeate fluxes were 48%, 45% and 50% for ultrasound irradiation of 80, 120 and 160 W, which was higher than treatment with stirring speed of 1583, 1812, and 1995 rpm, respectively. Ultrasound with higher power possesses stronger effects of cavitation and acoustic streaming, which have been confirmed to be the dominant mechanisms affecting microfiltration performances in ultrasonic field.

5.3.4   ULTRASONIC FREEZING AND THAWING

Freezing is one of the most familiar methods of food preservation. The formation of ice crystals during the freezing of water present in the food material is an important parameter that determines the quality of food. The problems associated with conventional freezing are non-uniform crystal development, destruction of food material structure and loss in sensory food quality. However, with the aim of putting an end to these problems, innovative technologies such as air blast, plate contact, immersion freezing, cryogenic freezing, fluidized-bed freezing, high-pressure freezing and their combinations have emerged quickly.

The important step in freezing is the initial nucleation followed by crystallization [41]. Ultrasonic waves are capable of enhancing the nucleation rate and rate of crystal growth by creating numerous sites of nucleation in a saturated or super-cooled medium [42]. This may be due to cavitation bubbles acting as nuclei for crystal growth within the medium and increasing the number of nucleation sites. Conventional cooling assisted with ultrasound leads to rapid and even seeding, resulting in a much shorter dwell time [43]. Moreover, greater number of seeds lead to smaller ice crystals thereby reducing cell damage. Another reason for increased nuclei formation is acoustic cavitation. This phenomenon leads to fine ice crystals and shortening of the time between the onset of crystallization and the complete formation of ice. An extensive range of food materials has been frozen under the assistance of ultrasound [44]. Power ultrasound is a promising tool to support food freezing, particularly for premium food and pharmaceutical products [45].

The success of freezing depends on the optimized thawing conditions [46]. Thawing of frozen foodstuffs is intrinsically slow however it can be made more rapid by using microwave, dielectric or resistive thawing methods. The utilization of ultrasound to thaw frozen food was investigated for over years but the negative aspects such as poor penetration; restricted or localized heating and high power requirements hindered the development. Researchers have worked on the relaxation mechanism and proved that that large amount of acoustic energy is absorbed by frozen foods when the relaxation frequency is in range of ice crystals in the food [47]. Therefore, the thawing process in the relaxation frequency was quick and used only conductive heating. Experiments have shown that thawing cod blocks needed almost 71% less time by using acoustically assisted water immersion. Acoustic thawing is a promising technology if appropriate frequencies and ultrasonic power are chosen. Ultrasonic thawing can reduce thawing time and drip loss while improving product quality.

The effects of slow freezing (SF), immersion freezing (IF) and ultrasound-assisted freezing (UAF) on physico-chemical characteristics and volatile compounds of red radish were investigated [48]. When compared to IF and SF, the freezing time of ultrasound application at 0.26 W/cm^2 was shortened by 14% and 90%, respectively. These results indicated that ultrasound usage significantly improved the freezing rate. Reduced cell separation and disruption was observed under ultrasonic power intensities of 0.17 and 0.26 W/cm^2, which confirmed better cellular structures of radish tissues.

Freezing attributes were investigated and these were: Time of freezing, latent heat of fusion of ice, and freezable water content and quality parameters such as drip loss, color, firmness, and L-ascorbic acid content of broccoli which were osmodehydrofrozen and ultrasound-assisted osmodehydrofrozen during storage [49]. The ultrasound-assisted osmotic dehydration shortened the dehydration time and retained the firmness and L-ascorbic acid content after osmotic dehydration pretreatment. Also, the ultrasound-assisted osmotic dehydration reduced the drip loss and loss of L-ascorbic acid content with better color and firmness at −25°C for 6 months. The impact of ultrasound-assisted freezing on few physico-chemical characteristics of three mushroom varieties were studied [50]. The application of ultrasound reduced nucleation time by 24%, 53% and 34% in *Lentinula edodes, Agaricusbisporus* and *Pleurotuseryngii* respectively, at 0.39 Wcm^{-2}. Similarly, drip loss during the thawing process was reduced 10%, when compared to control samples.

The inclusion of ultrasound during immersion freezing of potato cubes was examined [51]. A 35 kHz ultrasound was applied to the potato sample at its geometrical center when the temperature was in the range from −0.1 to −3.0°C. The impact of different freezing and thawing methods on the physicochemical and nutritive properties of edamame (*Glycine max* L. Merrill) were investigated [52]. Ultrasound-assisted thawing significantly minimized the thawing time, compared to conventional water immersion thawing. Ultrasound-assisted thawing showed the best retention of ascorbic acid and chlorophyll and also reduced the drip loss of thawed samples at a power level of 900 W, whereas 1,200 W power level caused the most distinct loss of ascorbic acid. The effect of ultrasound-assisted freezing on the quality and microstructure of frozen dough were investigated [53]. Dough cylinders were immersion frozen in an ultrasonic bath at five different power levels at 25 kHz frequency. A significant reduction of more than 11% in total dough freezing time was noted at power levels of 288 or 360 W. The assistance of ultrasound was found to induce nucleation in model food and a linear relationship between ultrasound irradiation and nucleation temperatures. The time taken was found to be 3–6 s from the onset of ultrasound irradiation to the beginning of the nucleation.

5.3.5   ACOUSTIC DRYING

Traditional methods for dehydrating food products force a stream of hot air and are reasonably economical. However, the elimination of interior moisture relatively consumes long time and high temperatures can damage the food leading to change in color, taste and nutritional value of the rehydrated product. Alternative methods like freeze and spray drying may eliminate these disadvantages, but they are expensive and in certain cases are applicable only to liquids. Supplying vibrational energy stimulates drying thereby preventing these disadvantages. Ultrasonic drying continues to be the focus of considerable attention. Ultrasonic osmotic dehydration is one of the new technologies that employ low solution temperatures to attain high water loss and solute gain rates [54]. As the temperature is low during the drying process and the treatment time is short, organoleptic properties like flavor, color and nutritional value remain unaltered. A hydrodynamic mechanism of mass transfer is examined in this process and a significant increase in the water loss and solute gain is observed. Ultrasound has also been utilized as a pretreatment method prior to the drying of foodstuffs such as sea cucumber [55], pineapple [56], melon [57], and strawberry [58]. The pretreatment is related to a reduction in time in the drying process and also in rehydration properties.

Ultrasound as a pretreatment for convective drying of Andean blackberry (*Rubusglaucus Benth*) was performed and the parameters such as amplitude (0–90), time of sonication (10–30 min) and air temperature (40–60°C) were selected for the study [59]. Five times higher drying rate was found in ultrasound treated sample (90 min). In addition to continuous ultrasound, intermittent ultrasound applied for apple drying reduced the total sonication time to 50 and 10% [60]. Drying time savings of 23% (red bell pepper) and 27% (apple) were calculated for continuous ultrasound. Intermittent ultrasound at 50% net sonication time resulted in drying time savings of 18% in apple. Air-borne ultrasonic application was explored as a means of improving the convective drying of strawberry [61]. Experiments were conducted by setting the acoustic power applied (0, 30 and 60 W) and the air temperature (40, 50, 60 and 70°C). The increase in both the applied acoustic power and temperature gave rise to a significant reduction of drying time (13–44%). The application of power ultrasound involved a significant ($p < 0.05$) improvement in the effective moisture diffusivity and the mass transfer coefficient, the effect being less intense at high temperatures.

Application of ultrasound to osmotic dehydration of guava slices via indirect sonication using an ultrasonic bath system and direct sonication using an ultrasonic probe system was studied [62]. Pre-treatments were designed in three osmotic solution concentrations of 0, 35, and 70° Brix at indirect ultrasonic bath power from 0 to 2.5 kW for immersion times ranging for 20–60 min and direct ultrasonic probe amplitudes from 0 to 35% for immersion times of 6–20 min. Applying ultrasound pre-osmotic treatment in 70 °Brix prior to hot-air drying reduced the drying time by 33%, increased the effective diffusivity by 35%, and decreased the total color change by 38%. A remarkable decrease of hardness to 4.2 N obtained was also comparable to the fresh guava at 4.8 N.

The effect of ultrasound-assisted osmotic dehydration pretreatment on drying kinetics and effective moisture diffusivity of Mirabelle plum during drying was investigated [63]. The variable such as ultrasonication time (10 and 30 min), osmotic solution concentration (50 and 70% sucrose) and immersion time (60, 120, 180 and 240 min) were selected. It was inferred that the pretreatment (ultrasound-assisted osmotic dehydration) significantly increased the effective moisture diffusivity from 5.84×10^{-9} to $7.36 \times 10^{-9} \, m^2/s$ and also resulted in a 20% decrease in drying time. The impact of high-intensity ultrasound as pretreatment on the duration of drying and texture characteristics of infrared-dried pear slices using different amplitudes was analyzed [64]. Ultrasound device with a power capacity of 400 W at a frequency of 24 kHz and amplitudes of 25, 50, 75 and 100%, was used for pretreatment. Drying was performed in an infrared dryer at 70°C. The results showed that hardness of samples gradually decreased (from untreated sample [104.72 N], 50% of amplitude [93.461 N] to 100% amplitude [62.206 N]) with an increase in ultrasound intensity.

5.3.6 CUTTING OF FOODS

Introduction of sonic waves in cutting of foods has enhanced the performance of processing by streamlining production, reducing product waste and minimizing maintenance costs. Ultrasonic cutting uses a knife-type blade attached through a shaft to an ultrasonic source. The ultrasonic cutting characteristics depend on the food type and condition like if the food is frozen or thawed. Widespread application of ultrasound is in the cutting of fragile and heterogeneous products like cakes, pastry, cheese and other sticky products [65]. The unique characteristic of sonic cutting lies in improvements in hygiene as the vibration avoids the adherence of the product on the blade preventing the growth of microbes on the surface. The accuracy and repetitiveness of the cut produce a reduction in losses relative to the cutting and a better optimization of the weight and dimensions of portions.

Eight cheese varieties with varying moisture content (357–488 g kg^{-1}), fat (183–335 g kg^{-1}) and protein content (202–292 g kg^{-1}) were exposed to ultrasonic cutting using a 40 kHz guillotine sonotrode [66]. Cutting force curves were evaluated based on ultrasound-induced cutting work reduction and energy demand of the sonotrode. During conventional cutting without ultrasonic excitation, actual cutting force F increased continuously with cutting depths. Actual force was 53.1 ± 1.6 N (Gruyère), 30.6 ± 2.1 N (Edam), 22.5 ± 1.2 N (Tilsit), and 18.7 ± 1.7 N (Feta) at s = 15 mm for the control experiments. In the experiments of cutting with ultrasonic excitation, there was a significant leveling of the slope when a cutting depth of around 3 mm was exceeded, mainly because of reduced friction. The force values at s = 15 mm were again taken as a degree of experimental repeatability: F = 10.9 ± 0.6 N for Gruyère, F = 7.56 ± 0.51 N for Tilsit cheese, F = 8.52 ± 0.44 N for Edam cheese and F = 3.18 ± 0.42 N for Feta.

Cheese samples (Gouda and Gruyère), ham sausage, short pastry, malted bread, pumpernickel, pound cake, and milk rolls were subjected to cutting-friction sequences using a longitudinally oscillating 40 kHz titanium cutting device. The study was carried out without and with ultrasonic excitation of the blade (vibration amplitude a: 12 µm) at a vertical velocity of 300 mm min^{-1} and at 22 ± 1°C. The coefficient of variation was below 0.1. For Gouda cheese, at s = 15 mm, the mean force in the cutting sequence (F$_C$) was 73.5 ± 4.9 N and 8.88 ± 0.81 N for samples without and with ultrasound excitation, respectively. Similarly, the corresponding values of subsequent friction test (F$_F$) were 22.7 ± 1.8 N and 4.23 ± 0.38 N. For all samples, the F$_C$ value decreased when working with the vibrating cutting device. The most prominent reduction in cutting and friction work, indicating a high impact of ultrasound, was observed for Gouda cheese (K_C = 0.15), followed by pound cake (K_C = 0.27). Except ham sausage, for all other products, the influence of ultrasonic vibration on the cutting sequence were significantly higher than for friction, indicating that friction work is of minor importance on these products. However, friction work was more affected than cutting work ($P < 0.05$) in case of ham sausage, presumably because of a thin layer of water, which was visually observed at the cut surface of these samples.

5.4   SUMMARY

State of the art in sonication techniques can achieve worthwhile efficiency and economic gains. This technology is expected to reduce cost, save energy and ensure safety of food and bioproducts. For the extraction and oxidation processes by ultrasound, non-parametric simulative models can be used whereas for other processes, it is appropriate to employ the models which can express the physical meaning of the process. Several empirical models can be considered to model the drying process in the presence of ultrasound. More researches are expected to elucidate the ultrasonic effects on oxidation, filtration, brining, freezing, and thawing process. For designing the equipments at the industrial scale, it is required to model the ultrasonic effects through numerical stimulation. While developing new and custom-made equipment, these issues need to be addressed by engineers, physicists and food technologists.

KEYWORDS

- **cutting**
- **drying**
- **extraction**
- **filtration**
- **food technologists**

- **freezing**
- **microbial inactivation**
- **sonication**
- **thawing**
- **ultrasound**

REFERENCES

1. Cullen, P. J., Tiwari, B. K., & Valdramidis, V. P. *Novel Thermal and Non-Thermal Technologies for Fluid Foods*. 2011, Academic Press.
2. Vilkhu, K., Applications and opportunities for ultrasound assisted extraction in the food industry—A review. *Innovative Food Science & Emerging Technologies*, 2008, 9(2), 161–169.
3. Awad, T., Applications of ultrasound in analysis, processing and quality control of food: A review. *Food Research International*, 2012, 48(2), 410–427.
4. Legay, M., Enhancement of heat transfer by ultrasound: review and recent advances. *International Journal of Chemical Engineering*, 2011.
5. Mahapatra, A. K., Muthukumarappan, K., & Julson, J. L. Applications of ozone, bacteriocins and irradiation in food processing: a review. *Critical Reviews in Food Science and Nutrition*, 2005, 45(6), 447–461.
6. Soria, A. C., Villamiel, M. Effect of ultrasound on the technological properties and bioactivity of food: a review. *Trends in Food Science & Technology*, 2010, 21(7), 323–331.
7. Mizrach, A., Ultrasonic technology for quality evaluation of fresh fruit and vegetables in pre-and postharvest processes. *Postharvest Biology and Technology*, 2008, 48(3), 315–330.
8. McClements, D. J., Advances in the application of ultrasound in food analysis and processing. *Trends in Food Science & Technology*, 1995, 6(9), 293–299.
9. Mudhafar Kareem, A., & Koc, A. B. Ultrasound Assisted Oil Extraction from Date Palm Kernels for Biodiesel Production. In: *2010 Int. Conference Pittsburgh, Pennsylvania, June 20–June 23,* 2010.
10. Mason, T. J., & Lorimer, J. P. *Applied Sonochemistry.* The uses of power ultrasound in chemistry and processing, 2002, 1–48.
11. Suslick, K. S., & Flannigan, D. J. Inside a collapsing bubble: sonoluminescence and the conditions during cavitation. *Annu. Rev. Phys. Chem.*, 2008, 59, 659–683.
12. Piyasena, P., Mohareb, E., & McKellar, R. Inactivation of microbes using ultrasound: a review. *International Journal of Food Microbiology*, 2003, 87(3), 207–216.
13. Jones, S., Evans, G., & Galvin, K. Bubble nucleation from gas cavities—a review. *Advances in Colloid and Interface Science*, 1999, 80(1), 27–50.

14. Mason, T. J., & Peters, D. Practical sonochemistry: Power ultrasound uses and applications. 2002, Woodhead Publishing.

15. Baumann, A. R., Martin, S. E., & Feng, H. Removal of Listeria monocytogenes biofilms from stainless steel by use of ultrasound and ozone. *Journal of Food Protection*, 2009, 72(6), 1306–1309.

16. Pala, , Zorba, N. N. D., & Özcan, G. Microbial inactivation and physicochemical properties of ultrasound processed pomegranate juice. *Journal of Food Protection*, 2015, 78(3), 531–539.

17. Gabriel, A. A., Inactivation behaviors of foodborne microorganisms in multi-frequency power ultrasound-treated orange juice. *Food Control*, 2014, 46, 189–196.

18. Bermúdez-Aguirre, D., & Barbosa-Cánovas, G. V. Inactivation of Saccharomyces cerevisiae in pineapple, grape and cranberry juices under pulsed and continuous thermo-sonication treatments. *Journal of Food Engineering*, 2012, 108(3), 383–392.

19. Haughton, P., An evaluation of the potential of high-intensity ultrasound for improving the microbial safety of poultry. *Food and Bioprocess Technology*, 2012, 5(3), 992–998.

20. Silva, F. V., Use of power ultrasound to enhance the thermal inactivation of Clostridium perfringens spores in beef slurry. *International Journal of Food Microbiology*, 2015, 206, 17–23.

21. Herceg, Z., The effect of high intensity ultrasound treatment on the amount of Staphylococcus aureus and Escherichia coli in milk. *Food Technology and Biotechnology*, 2012, 50(1), 46.

22. Gao, S., Inactivation of Enterobacter aerogenes in reconstituted skim milk by high- and low-frequency ultrasound. *Ultrasonics Sonochemistry*, 2014, 21(6), 2099–2106.

23. Cregenzán-Alberti, O., Suitability of ccRSM as a tool to predict inactivation and its kinetics for Escherichia coli, Staphylococcus aureus and Pseudomonas fluorescens in homogenized milk treated by manothermosonication (MTS). *Food Control*, 2014, 39, 41–48.

24. Samaram, S., Optimisation of ultrasound-assisted extraction of oil from papaya seed by response surface methodology: oil recovery, radical scavenging antioxidant activity, and oxidation stability. *Food Chemistry*, 2015, 172, 7–17.

25. Hashemi, S. M. B., Kolkhoung (Pistacia khinjuk) kernel oil quality is affected by different parameters in pulsed ultrasound-assisted solvent extraction. *Industrial Crops and Products*, 2015, 70, 28–33.

26. Morsy, N. F., A short extraction time of high quality hydrodistiled cardamom (Elettaria cardamomum L. Maton) essential oil using ultrasound as a pretreatment. *Industrial Crops and Products*, 2015, 65, 287–292.

27. Li, P., L. Tian, & T. Li, Study on Ultrasonic-Assisted Extraction of Essential Oil from Cinnamon Bark and Preliminary Investigation of Its Antibacterial Activity. In: *Advances in Applied Biotechnology*. 2015, Springer. p. 349–360.

28. Rodrigues, S., Ultrasound extraction of phenolics and anthocyanins from jabuticaba peel. *Industrial Crops and Products*, 2015, 69, 400–407.

29. Ordóñez-Santos, L. E., Pinzón-Zarate, L. X., & González-Salcedo, L. O. Optimization of ultrasonic-assisted extraction of total carotenoids from peach palm fruit (Bactris gasipaes) by-products with sunflower oil using response surface methodology. *Ultrasonics Sonochemistry*, 2015.

30. Šic Žlabur, J., Optimization of ultrasound assisted extraction of functional ingredients from Stevia rebaudiana Bertoni leaves. *International Agrophysics*, 2015, 29(2), 231–237.

31. Kadam, S. U., Effect of Ultrasound Pretreatment on the Extraction Kinetics of Bioactives from Brown Seaweed (Ascophyllum nodosum). *Separation Science and Technology*, 2015, 50(5), 670–675.

32. Girard, B., Fukumoto, L., & Sefa Koseoglu, S. Membrane processing of fruit juices and beverages: a review. *Critical Reviews in Biotechnology*, 2000, 20(2), 109–175.

33. Mahmoud, A., Electrical field: A historical review of its application and contributions in wastewater sludge dewatering. *Water Research*, 2010, 44(8), 2381–2407.

34. Chemat, F., & Khan, M. K. Applications of ultrasound in food technology: processing, preservation and extraction. *Ultrasonics Sonochemistry*, 2011, 18(4), 813–835.

35. Rastogi, N. K., Opportunities and challenges in application of ultrasound in food processing. *Critical Reviews in Food Science and Nutrition*, 2011, 51(8), 705–722.

36. Muthukumaran, S., Mechanisms for the ultrasonic enhancement of dairy whey ultrafiltration. *Journal of Membrane Science*, 2005, 258(1), 106–114.

37. Aliasghari Aghdam, M., The effect of ultrasound waves on the efficiency of membrane clarification of pomegranate juice. *International Journal of Food Science & Technology*, 2015, 50(4), 892–898.

38. Ábel, M., Ultrasonically assisted ultrafiltration of whey solution. *Journal of Food Process Engineering*, 2015.

39. Cai, M., Li, W., & Liang, H. Effects of ultrasound parameters on ultrasound-assisted ultrafiltration using cross-flow hollow fiber membrane for Radixastragalus extracts. *Chemical Engineering and Processing: Process Intensification*, 2014, 86, 30–35.

40. Liu, D., Comparative study of ultrasound-assisted and conventional stirred dead-end microfiltration of grape pomace extracts. *Ultrasonics Sonochemistry*, 2013, 20(2), 708–714.

41. Matsumoto, M., Saito, S., & Ohmine, I. Molecular dynamics simulation of the ice nucleation and growth process leading to water freezing. *Nature*, 2002, 416(6879), 409–413.

42. Mason, T., Power ultrasound in food processing—the way. *Ultrasound in Food Processing*, 1998, 105.

43. Da-Wen, S., *Acceleration of Immersion Freezing of Potato By Using Power Ultrasound*, 2002.

44. Sun, D.-W., & Li, B. Microstructural change of potato tissues frozen by ultrasound-assisted immersion freezing. *Journal of Food Engineering*, 2003, 57(4), 337–345.

45. Rawson, A., Effect of thermal and non-thermal processing technologies on the bioactive content of exotic fruits and their products: Review of recent advances. *Food Research International*, 2011, 44(7), 1875–1887.

46. Ragoonanan, V., Hubel, A., & Aksan, A. Response of the cell membrane–cytoskeleton complex to osmotic and freeze/thaw stresses. *Cryobiology*, 2010, 61(3), 335–344.

47. Li, B., & Sun, D. W. Novel methods for rapid freezing and thawing of foods—a review. *Journal of Food Engineering*, 2002, 54(3), 175–182.

48. Xu, B.-G., Effect of ultrasound-assisted freezing on the physico-chemical properties and volatile compounds of red radish. *Ultrasonics Sonochemistry*, 2015.

49. Xin, Y., Zhang, M., & Adhikari, B. Freezing characteristics and storage stability of broccoli (Brassica oleracea L. var. botrytis L.) under osmodehydrofreezing and ultrasound-assisted osmodehydrofreezing treatments. *Food and Bioprocess Technology*, 2014, 7(6), 1736–1744.

50. Islam, M. N., The effect of ultrasound-assisted immersion freezing on selected physi-cochemical properties of mushrooms. *International Journal of Refrigeration*, 2014, 42, 121–133.
51. Comandini, P., Effects of power ultrasound on immersion freezing parameters of potatoes. Innovative Food Science & Emerging Technologies, 2013, 18, 120–125.
52. Cheng, X.-F., Zhang, M., & Adhikari, B. Effects of ultrasound-assisted thawing on the quality of edamames [Glycine max (L.) Merrill] frozen using different freezing methods. *Food Science and Biotechnology*, 2014, 23(4), 1095–1102.
53. Hu, S. Q., An improvement in the immersion freezing process for frozen dough via ultrasound irradiation. *Journal of Food Engineering*, 2013, 114(1), 22–28.
54. Rastogi, N., Recent developments in osmotic dehydration: methods to enhance mass transfer. *Trends in Food Science & Technology,* 2002, 13(2), 48–59.
55. Duan, X., Ultrasonically enhanced osmotic pretreatment of sea cucumber prior to microwave freeze drying. *Drying Technology*, 2008, 26(4), 420–426.
56. Fernandes, F. A., Linhares, F. E. Jr., & Rodrigues, S. Ultrasound as pre-treatment for drying of pineapple. *Ultrasonics Sonochemistry*, 2008, 15(6), 1049–1054.
57. Fernandes, F. A., Gallão, M. I., & Rodrigues, S. Effect of osmotic dehydration and ultrasound pre-treatment on cell structure: Melon dehydration. *LWT – Food Science and Technology*, 2008, 41(4), 604–610.
58. Garcia-Noguera, J., Ultrasound-assisted osmotic dehydration of strawberries: Effect of pretreatment time and ultrasonic frequency. *Drying Technology*, 2010, 28(2), 294–303.
59. Romero, J. C., & Yépez, V. B. Ultrasound as pretreatment to convective drying of Andean blackberry (Rubus glaucus Benth). *Ultrasonics Sonochemistry*, 2015, 22, 205–210.
60. Schössler, K., Jäger, H., & Knorr, D. Effect of continuous and intermittent ultrasound on drying time and effective diffusivity during convective drying of apple and red bell pepper. *Journal of Food Engineering*, 2012, 108(1), 103–110.
61. Gamboa-Santos, J., Air-borne ultrasound application in the convective drying of strawberry. *Journal of Food Engineering*, 2014, 128, 132–139.
62. Kek, S. P., Chin, N. L., & Yusof, Y. A. Direct and indirect power ultrasound assisted pre-osmotic treatments in convective drying of guava slices. *Food and Bioproducts Processing*, 2013, 91(4), 495–506.
63. Dehghannya, J., Gorbani, R., & Ghanbarzadeh, B. Effect of ultrasound-assisted os-motic dehydration pretreatment on drying kinetics and effective moisture diffusivity of mirabelle plum. *Journal of Food Processing and Preservation*, 2015.
64. Dujmić, F., Ultrasound-assisted infrared drying of pear slices: Textural issues. *Journal of Food Process Engineering*, 2013, 36(3), 397–406.
65. Zahn, S., Impact of excitation and material parameters on the efficiency of ultrasonic cutting of bakery products. *Journal of Food Science*, 2005, 70(9), E510–E513.
66. Arnold, G., Ultrasonic cutting of cheese: Composition affects cutting work reduction and energy demand. *International Dairy Journal*, 2009, 19(5), 314–320.

CHAPTER 6

MINIMAL PROCESSING OF FRUITS AND VEGETABLES

JAGBIR REHAL, DEEPIKA GOSWAMI, HRADESH RAJPUT, and HARSHAD M. MANGDE

CONTENTS

6.1   INTRODUCTION

Minimally processed fruits and vegetables emerged to fulfill new consumer's demands of healthy, palatable and easy to prepare plant foods. Although between processing and consumption a span of several days occurs, consumers still want to have fresh or fresh-like vegetables on their dishes or pans [31]. Their popularity is related to the convenience of these ready-to-eat products and an increasing awareness of the health benefits associated with eating fresh produce. The minimal processing is a method of preparing convenient fresh products that can be consumed, utilized straight away in less time with two objectives. Firstly, it keeps the produce fresh which is yet to supply in convenient form without losing its nutritional quality. Secondly, the minimally processed products have a shelf life sufficient to make distribution feasible and acceptable to its intended consumers [37].

Minimal processing is described as non-thermal technologies to process food in a manner to guarantee the food safety and preservation as well as to maintain as much as possible the fresh-like characteristics of fruits and vegetables [41]. Other terms used to refer to minimally processed products are "lightly processed," "partially processed," "fresh processed," "preprepared," and "Grade 4."

Minimally processed fruits and vegetables include peeled and diced potatoes; shredded lettuce, spinach and cabbage; de-stoned peach, mango; sliced melon and other fruit slices; vegetable snacks, such as carrot and celery sticks, and cauliflower and broccoli florets; packaged mixed salads; peeled and chopped onions; peeled and cored pineapple; fresh sauces; peeled citrus fruits; and microwaveable fresh vegetable trays (Figure 6.1). Minimally processed products may be prepared at the source of production or at regional and local processors. Whether a product may be processed at source or locally depends on the perishability of the processed form relative to the intact form, and on the quality required for the designated use of the product.

6.2   UNIT OPERATIONS FOR MINIMAL PROCESSING

Minimally processed foods do not require further preparation before consumption except probably for addition of dressing. The various operations involved are:

- **Sorting:** it ensures the separation of unacceptable material and is usually accomplished manually though sorting by machinery is also employed by larger establishments
- **Peeling:** the products which require peeling to ensure acceptance of the final product undergo this operation, e.g., potatoes, carrots,

FIGURE 6.1   Minimally processed fruits and vegetables.

onion, etc. This is usually accomplished manually or by abrasive methods, lye peeling, etc.

- **Chopping/shredding:** it ensures that all the inedible and undesirable parts like seeds and stems of the produce are removed from the final product. This should be accomplished by sharp blades to minimize damage to the tissues and retain the quality of the end product.

- **Washing and draining:** it ensures the removal of dirt and decrease in the microbial load of the material. The quality of water, its temperature, pH and quantity have effect on the final quality of the minimally processed foods. The gentle draining of water from the product is very important for getting good quality product. Centrifugation is the best way to achieve this. Delicate textured fruits require passage through semi-fluidized beds with forced air to remove moisture.

- **Packaging:** technologies like modified atmosphere packaging, shrink wrap, active packaging, etc. are employed for packaging of minimally processed fresh cut fruits and vegetables.

- **Storage:** While conventional food-processing methods extend the shelf-life of fruit and vegetables, the minimal processing to which fresh-cut fruit and vegetables are subjected renders products highly perishable, requiring chilled storage to ensure a reasonable shelf-life [29]. Temperature is a critical factor for the shelf life of minimal processed produce and so a cold chain needs to be maintained where the ideal temperature is closer to 1°C. Storage at 10°C or above allows most bacterial pathogens to grow rapidly on fresh cut produce.

6.3   PHYSIOLOGICAL CHANGES OF MINIMAL PROCESSED PRODUCE

The major obstacle to overcome in the production of minimally processed products is that of activated respiration of live tissues which results in reduced shelf life. Moreover, it is well-known that processing of vegetables promotes a faster physiological deterioration, biochemical changes and microbial degradation of the product even when only slight processing operations, through slicing or cutting tissues [43], may result in degradation of the color, texture and flavor [36]. Naturally, accelerated quality loss through senescence and enzymatic browning occur.

Every step the produce undergoes, from cultivation to the shelf, is important from the point of view of quality and safety. Wounding and other minimal processing procedures can cause physiological effects, including ethylene production, increase in respiration, membrane deterioration, water loss, susceptibility to microbiological spoilage, loss of chlorophyll, formation of pigments, loss of acidity, increase in sweetness, formation of flavor volatiles, tissue softening, enzymatic browning, lipolysis and lipid oxidation [56].

The amount of ethylene also increases during minimal processing and since ethylene contributes to the synthesis of enzyme responsible for maturation so it promotes physiological changes such as ripening, softening, and senescence. Green color of vegetative tissue is another quality attribute that is affected to a great extent by ethylene. Senescence causes oxidative breakdown of membrane lipids. As the senescing process progresses the cellular structure and membrane integrity are decreased and the tissue become more susceptible to the stress caused by the unit operations of minimal processing. Wounding, cutting, peeling also causes decompartmentalization of enzymes and their substrates, e.g., a group of enzymes called polyphenol oxidases are responsible for the discoloration known as enzymatic browning. Dehydration due to peel removal may be partially responsible for some of the softening that is observed in fresh-cut produce.

To retard the respiration rate and to prevent the growth of aerobic spoilage microorganisms in minimally processed products, packaging is of paramount importance. Some packaging processes used for fresh produce are also used on minimally processed products. Since respiration continues in such processed products, considerations are similar in both whole and processed products. Guidelines for packing fresh or minimally processed fruit and vegetables generally specify a washing or sanitizing step to remove dirt, pesticide residues, and microorganisms responsible for quality loss and decay [50].

6.4 QUALITY OF MINIMALLY PROCESSED FRUITS AND VEGETABLES

Post-harvest quality loss is primarily due to transpiration, respiration, ripening, bruising and ensuing enzymatic activity, microbial decay, senescence, handling and mechanical damage during transportation. Hence the quality of minimally processed products is determined by the raw materials used as well as the production process itself. The cultivar selected is important and it should be compatible with the processing methods and the desired end product, e.g., texture retention, number of seeds, sugar and acid ratio, susceptibility to browning. Their main quality characteristic is their visual appeal along with texture, flavor and nutrition. As fruits and vegetables contain over 90% water, a loss of 5% or more water is visually noticeable, lowering the grade of the produce and resulting in a decrease in its commercial value. Major effects of water loss are a reduction in weight and a wilted appearance. Defects during the pre-harvest stage also directly affect their quality, for, e.g., Field defects such as tip-burn on lettuce can reduce the quality of the processed product because the brown tissue is distributed throughout the packaged product. The produce should be harvested at their best stage of maturity for processing.

To reduce the risks associated with the production of minimally processed vegetables, it is important to use high quality raw materials. This can be ensured by the producer by following Good Agricultural Practices (GAPs) during the farm activities and Good Manufacturing Practices (GMPs) in producing minimally processed produce. During harvest, the superficial microbiota of vegetables comprises mainly Gram negative saprophytes, but pathogenic microorganisms can also be found. Vegetables may harbor pathogenic Escherichia coli and Salmonella spp. e enteric bacteria involved in large foodborne outbreaks worldwide, causing symptoms of gastroenteritis, and even chronic infections [22, 28]. *Listeria monocytogenes* can also contaminate RTE vegetables and it is a psychro-tolerant and ubiquitous microorganism that causes listeriosis, an atypical infection with low mortality but high fatality rates among the elderly, pregnant women and immuno-compromised individuals [1, 10].

In conjunction with GMPs, the use of Sanitation Standard Operating Practices (SSOPs), and the implementation of a HACCP food safety plan will minimize the food safety risks. In addition, food safety needs to be maintained through the distribution and marketing channels to ensure the consumer is receiving safest product possible. Storage temperature is the single most important factor affecting quality of minimally processed fruit and vegetables. The good sensory quality is meaningless if the safety of the product is overlooked.

6.5   PRESERVATION FACTORS FOR MINIMAL PROCESSING

There are many other preservation techniques that are currently being used by the fresh-cut industry such as antioxidants, chlorines and modified atmosphere packaging (MAP). Nonetheless, new techniques for maintaining quality and inhibiting undesired microbial growth are demanded in all the steps of the production and distribution chain as microorganisms adapt to survive in the presence of previously effective control methods [2]. The various steps which need to be taken into consideration for obtaining optimum quality are discussed below:

6.5.1   CUTTING AND SHREDDING

To minimize the number of broken cells during various operations like peeling, cutting, shredding, these must be performed with knives or blades as sharp as possible made from stainless steel. The use of blunt blades causes excessive physical injury to the produce especially next to the cut accelerating deterioration. The market value pertaining to visual appeal is more in sharp cut pieces whereas the blunt cut pieces develop a translucency, a visual defect. Barry et al. [8] found that slicing affected ($p < 0.05$) ascorbic acid retention in the order: manual tearing > manual slicing > machine slicing for Iceberg lettuce. Moreover, flushing with 100% nitrogen increased retention (~5%, p<0.05) over packages with product modified atmospheres. Storage at 3°C increased retention (~20%, p<0.05) compared with storage at 8°C. Work done by [58] Woo et al. (1998) showed that dry peeling of garlic reduced microbial load more effectively than wet peeling and preserved ascorbic acid content, while removal of the garlic root also led to a reduction in microbial populations. Microbial quality was reduced in garlic and green onion according to severity of cutting, while washing of vegetables reduced total microorganism counts; repeated washing and use of chilled water resulted in a further reduction of microbial load.

However, many different solutions have been tested to avoid the acceleration of decay due to peeling, cutting or slicing. The newest tendency is called the immersion therapy [5]. Cutting a fruit while it is submerged in water will control turgor pressure, due to the formation of a water barrier that prevents movement of fruit fluids while the product is being cut. Additionally, the watery environment also helps to flush potentially damaging enzymes away from plant tissues. On the other hand, UV-C light has been also used while cutting fruit to cause a hypersensitive defense response to take place within its tissues, reducing browning and injury of in fresh-cut products [38]. Another alternative could be the use of water-jet cutting, a non-contact cutting method which utilizes a concentrated stream of high-pressure water to cut through a wide range of foodstuffs. However, the main steps throughout the processing chain of minimal processing fruits and vegetables are washing and disinfection.

6.5.2   WASHING AND DISINFECTION

Guidelines for packing fresh or minimally processed fruits and vegetables generally specify a washing or sanitizing step to remove dirt, pesticide residues, and microorganisms responsible for quality loss and decay [50]. The contact period of the water and its temperature decide the effectiveness of the operation. It should be near to chilling temperature to cool the products as well as avoid spoilage before processing and packing.

Optimum concentration of chlorine needs to be used which is between 50 to 100 ppm. Additionally, different washing and disinfection treatments may negatively affect the nutritional and sensory quality of the product [39]. However, little is known about nutritive value content of minimally processed produce [48]. The current published data suggest that none of the available washing and sanitizing methods, including some of the newest sanitizing agents such as chlorine dioxide and ozone, can guarantee the microbiological quality of minimally processed vegetables without compromising their sensorial quality [11, 13, 46, 50].

The best mixed immersion solution ($P \leq 0.05$) was determined to be 2 g/L citric acid, 1 g/L calcium chloride and 250 g/L garlic extract [35], which showed a positive effect on shelf-life of minimally processed lettuce, controlling enzymatic browning, chlorophyllase activity and weight loss. The microorganisms' growth was not significantly controlled, but the fact that the organoleptic results of immersion solution-treated lettuce showed about 80% acceptability during 9 days of storage, suggests that some slight modifications on immersion solution could be the basis of a promising formula for minimally processed lettuce.

The effects of an antioxidant dipping treatment (in an aqueous solution of 1% ascorbic acid (AA) and 1% citric acid for 3 min) and of modified atmosphere (90% N_2O, 5% O_2 and 5% CO_2) packaging (MAP) on some functional properties of minimally processed apples have been investigated by Cocci et al. [20]. Color, texture and some chemical indices associated with the ripening stage of the product (titrable acidity and soluble solids content) were also evaluated. As a consequence of the anti-browning treatment, the AA content of dipped samples was about 20-fold higher than not treated samples at the beginning of storage and remained higher until the sixth day of refrigeration. Moreover, the dipping treatment resulted in an increase in the apple slice antioxidant activity, while MA had a negative effect on AA levels. Ultrasound has great potential to be used either alone or in association with other non-thermal technologies [27]. Areas such as sanitization and washing may highly benefit, since ultrasound enhances the removal of pesticides and also lowers the amount of microorganism on fruits and vegetables. Different treatments have been evaluated to reduce browning in fresh-cut products. The most common is to dip or immerse them in anti-browning agent solutions. The use of citric acid alone seemed to be enough for a significant inhibitory effect on PPO when very high oxygen partial pressures were used for extending the shelf life of minimally processed potatoes [40].

A newer tendency has been reported by Bari et al. [7], who combined the efficacy of chemical disinfectant with the antimicrobial effect of bacteriocins produced by lactic acid bacteria. They investigated the efficacy of nisin and pediocin treatments in combination with EDTA, citric acid, sodium lactate, potassium sorbate and phytic acid in reducing *Listeria monocytogenes* on fresh-cut produce. They concluded that pediocin and nisin applications in combination with organic acids caused a significant reduction of native microflora and inoculated populations on fresh produce.

6.5.3   COOLING

To remove the tissue fluids which are released during the cutting and shredding of fruits and vegetables, their washing and cooling is important to avoid any undesirable microbial or chemical reactions. This is achieved by use of water at the temperature preferably 0°C, which can be chlorinated at low levels. This helps to arrest the surface browning a well. Maintaining the cold chain throughout the marketing is imperative to slow the adverse effects of minimal process. This should be optimized to keep the cells alive as well as preserve quality of produce.

6.5.4   COATINGS

Edible coatings are made from biopolymeric substances and can be applied on the produce directly or used as films. Edible coatings are prepared as solutions and emulsions from proteins, lipids, and polysaccharides and are applied on produce surfaces by different mechanical procedures, such as dipping, spraying, and brushing, or by electrostatic deposition [4].

Edible coatings may contribute to prolong the shelf-life of minimally processed foods by working as a barrier against water vapor, gases, etc. and also safeguard against microbiological growth as these usually grow on food surfaces. The barrier ability of applied coating and film increases the CO_2 level and hence retard the internal oxidation process thereby changing the inner atmosphere and extending the shelf life of the commodities. The three main types of macromolecules used for the fabrication of edible films are polysaccharides, proteins, and lipids and resins as well. The natural waxes like beeswax, paraffin wax and carnauba wax are very often used for reducing the weight loss of different fruits. The incorporation of antimicrobial agents into packaging flexible films (coatings) is, therefore, an alternative to this problem. Benitez et al. [9] found that edible Aloe Vera gel coating seems to have a beneficial effect on the retention of quality in minimally processed kiwifruit slices by retarding the yellowing process, reducing microbial growth and improving total pectin and texture retention. Moreover, all Aloe Vera concentrations reduced O_2 consumption and CO_2 production, thereby preventing anaerobic conditions in the trays. The sensory evaluation of the samples coated with 5% Aloe also yielded the highest score for overall quality. Aloe Vera based edible coatings improve the quality of minimally processed 'Hayward' kiwifruit.

Rojas-Graü *et al.* [49] and Oms-Oliu *et al.* [45] observed that apple or melon wedges coated with alginate, gellan or pectin edible coatings cross-linked with calcium salts outstandingly maintained their initial firmness during refrigerated storage. Chien *et al.* [19] maintained the vitamin C content of sliced dragon fruit coated with low molecular weight chitosan. Tapia *et al.* [53] reported that the addition of ascorbic acid to the alginate edible coating helped to preserve the natural vitamin C content in fresh-cut papaya. Hernández-Muñoz *et al.* [33] indicated that chitosan-coated strawberries retained more calcium gluconate (3079 g/kg dry matter) than strawberries dipped into calcium solutions (2340 g/kg). Tapia *et al.* [54] maintained counts of *Bifidobacterium lactis* Bb-12 above 10^6 cfu/g on papaya and apple pieces coated with alginate or gellan solutions containing the probiotic microorganism during 10 days of refrigerated storage. Geraldine et al. [30] studied the physical properties of the agar-agar based (1%) coatings incorporated with 0.2% chitosan and 0.2% acetic acid, as well as their effects on coating of minimally processed garlic cloves. Average moisture loss of coated garlic cloves was three times lower when compared to the control samples (no coated garlic cloves).

One of the major advances in employing edible coatings is that they can be used to deliver active substances such as antioxidants, antimicrobials, vitamins, etc. through encapsulation technique. The trigger of their release can be changes in temperature, pH etc. The successful application of coating depends on different factors that can influence the effectiveness of the films is coating thickness, type of produce, area covered and temperature.

6.5.5 IONIZING RADIATION

Low-dose gamma irradiation is very effective in reducing bacterial, parasitic, and protozoan pathogens in raw foods. It has already been tested in minimally processed fruit and vegetables observing that dose of 2.0 kGy strongly inhibited the growth of aerobic mesophilic and lactic microflora in shredded carrots [18].

Most of the studied minimally processed vegetables can be irradiated with doses up to 2 kGy. These doses are effective in reducing the initial microflora in 4–5 logs and at the same time extending the shelf-life of the products without adverse effect on their sensory characteristics.

Irradiation of sprouts rather than seeds is recommended as a final treatment, as irradiation of the seeds is not sufficient to guarantee sufficient reduction of pathogens. Based on D10 values observed for the most resistant organism studied (*L. monocytogenes*), irradiation with 2.5 kGy is recommended to ensure the microbiological safety and inactivate vegetative pathogenic bacteria by 5 log-cycles. Chaudry et al. [16] concluded from their studies that a dose of 2 kGy is sufficient to maintain the textural and sensorial quality, and the reduction of bioload of minimally processed carrots for 14 days at 5°C. All the samples were free of coliforms and E.coli after Gamma radiation initially and throughout storage period.

The effect of irradiations on the furan formation in fresh-cut fruits and vegetables was studied by Fan and Sokorai [25]. Furan is regarded as a possible carcinogen and is commonly found in foods that have been treated with traditional heating techniques. They reported that almost all fruits and vegetables produced non-detectable levels of furan hence irradiation induced furan is unlikely concern for fresh-cut produce according to them.

Fungi and viruses are typically more resistant to radiation than are bacteria so irradiation is therefore clearly best suited to control bacterial pathogens. The advantage of irradiation as quarantine treatment is that unlike heat treatment, irradiations can treat fully mature produce without compromising its quality and appearance. This in turn enables to fetch better price to the producers.

6.5.6   HURDLE TECHNOLOGY

Hurdle technology can be applied several ways in the design of preservation systems for minimally processed foods at various stages of the food chain:, i.e., use of natural antimicrobials or other stress factors, in addition to refrigeration, use of heat co-adjuvants to reduce the severity of thermal treatments. A synergist effect could work if the hurdle in a food hits different targets (e.g., cell membrane, DNA, enzyme systems, pH, a_w, Eh) within the microbial cell, and thus disturbs the homeostasis of the microorganisms present in several aspects. Therefore, employing different hurdles in the preservation of a particular food should be an advantage, because microbial stability could be achieved with a combination of gentle hurdles. In practical terms, this could mean that it is more effective to use different preservatives in small amounts in a food than only one preservative in large amounts, because different preservatives might hit different targets within the bacterial cell, and thus act synergistically.

Another methodology employed was based on combinations of mild heat treatments, such as blanching for 1–3 minutes with saturated steam, slightly reducing the a_w (0.98–0.93) by addition of glucose or sucrose, lowering the pH (4.1–3.0) by addition of citric or phosphoric acid, and adding antimicrobials (1000 ppm of potassium sorbate or sodium benzoate, as well as 150 ppm of sodium sulphite or sodium bisulphite) to the product syrup. During storage of HMFP, the sorbate and sulphite levels decreased, as well as a_w levels, due to hydrolysis of glucose [3]. Goyeneche et al. [32] studied the effects of two hurdle technologies, citric acid application at 0.3%, 0.6% and 0.9% and thermal treatments for 1, 2 and 3 min at 50 °C, on the color of radish slices over 10 d of refrigerated storage and found that treating for 1 min thermal treatment with 0.3% Citric acid was the best treatment to improve the retention of typical natural color of the minimally processed sliced radishes.

6.5.7   STORAGE CONDITIONS

The effect of allyl isothiocyanate (AITC), in combination with low temperature (10°C) storage on postharvest quality of minimally processed shredded cabbage was investigated by Banerjee et al. [6] who reported that an optimum concentration of 0.05 LL/mL for AITC was found to be effective in maintaining the microbial and sensory quality of the product for a period of 12 days.

The shelf life of fresh cut cactus pears (peeled intact or in halves), packed in bi-directional polypropylene bags 25 μm thick, sealed and stored at 4°C, can be extended up to 20 days without affecting their quality [21]. Also in nopalitos, the lowest ethanol production, smallest change in color and firmness were obtained at 4°C with sealed 25 μm-thick polypropylene bags. Tsouvaltzis et al. [57] studied the effect of storage atmosphere on the main quality attributes of minimally processed leeks and stated that storage of minimally processed leek stalks in 1% O_2 at 6.5°C for 14 days minimized leaf and root growth as well as color changes at the center of the basal cut surface, but did not prevent peripheral discoloration of the basal cut surface; the other reduced O_2 and elevated CO_2 treatments were less effective than 1% O_2 in reducing leaf and root growth and cut surface discoloration. Storage in 1% O_2 + 14% CO_2, however, resulted in an additional beneficial effect compared with 1% O_2 alone by preventing the appearance of peripheral discoloration on the basal cut surface.

To maintain high quality and to extend the shelf life of intact and minimally processed (removal of roots and compressed stem) green onions (*Allium cepa A. fistulosum*), the potential benefits of controlled atmospheres (CA) and heat treatment were evaluated by Hong et al.[34]. They found that atmospheres of 0.1–0.2% O_2 or 0.1–0.2% O_2 containing 7.5–9% CO_2 were the CA conditions best maintained the visual appearance and prolonged shelf life to more than 2 weeks at 5°C in both intact and cut onions. Heat treatment (55°C water for 2 min) of the white leaf bases effectively controlled 'telescoping' of cut onions stored at 5°C. Total soluble sugars generally decreased in intact and minimally processed green onions, but were maintained in heat-treated cut onions.

6.6   PACKAGING OF MINIMALLY PROCESSED FOODS

The latest technologies for the packaging of fresh cut fruits and vegetables is active packaging, shrink wrap packaging, modified atmosphere packaging, etc. Modified atmosphere packaging (MAP) is a technique where the air surrounding the food in the package is changed to another composition to preserve the initial fresh state of the product. Among fresh-cut produce equilibrium modified atmosphere packaging (EMAP) is the most commonly used packaging technology. When packaging vegetables and fruits the gas atmosphere of package is not air whose general composition is; O_2 – 21%; CO_2 – 0.01%; N_2 – 78% but consists usually of a lowered level of O_2 and increased level of CO_2 which slows down the normal respiration of the product and prolongs its shelf life. To prevent anaerobic respiration some O_2 needs to be present otherwise acetaldehyde and alcohol are produced which cause physiological disorders and quality loss. There are two types of MAP-passive and active. Passive MAP involves placing produce in a gas permeable package, sealing the package and allowing produce respiration to reduce oxygen and increase carbon dioxide in the package to desired steady-state equilibrium [51]. In active MAP, the O_2 and CO_2 concentrations are modified initially and changes dynamically depending on the respiration rate of the commodities and the permeability of the film surrounding the produce [17, 24]. Passive MAP is applied only to fruits and vegetables. However, active MAP could be applied to all kind of food products. So the knowledge about rate of respiration of the produce and the permeability of the packaging membrane is important to maintain an equilibrium atmosphere within the package.

Active packaging involves the use of antimicrobial packaging. Some antimicrobials act as plasticizers and improve the flexibility of films as shown by Marcos et al. [42] in case of polyvinyl alcohol (PVOH) modified with enterocin. The deterioration due to O_2 can be minimized by the oxygen absorbing systems used in conjunction with MAP or vacuum packaging. Moisture absorbing materials are also used to absorb excess moisture released by the minimal processed fruits and vegetables. Similarly, ethylene scavengers like potassium permanganate embedded in silica, are used to reduce the ripening of fruits and vegetables. Active packaging also involves the incorporation of UV filters in transparent packaging materials to retain the color.

The various machines which are used for gas flushing and sealing of the product are horizontal/vertical form-fill-seal (HFFS/VFFS), thermoform-fill-seal, preformed tray and lidding film (PTLF), three-web thermoform-fill-seal (TWTFFS). The product is hermetically sealed, pack integrity is maximized, liquid leakage is minimized and optimum retail display is possible by the use of these machines. In vacuum packaging of products a retractable snorkels pull a vacuum and then back-flush a desired gas mixture before heat sealing. Individual shrink wrap packaging of fruits and vegetables helps in preventing the moisture loss and reducing the respiration rate which helps to maintain the texture of the produce and extend the shelf life. It reduces the transpiration rate thereby reducing condensation of droplets within the package and hence shelf life is extended without any refrigeration. Polyvinyl chloride (PVC), used primarily for over wrapping, and polypropylene (PP) and polyethylene (PE), used for bags, are the films most widely used for packaging minimally processed products. Multilayered films, often with ethylene vinyl acetate (EVA), can be manufactured with differing gas transmission rates.

6.7 OUTBREAKS DUE TO MINIMALLY PROCESSED FOODS

Outbreaks are defined as the occurrence of two or more cases of similar illness resulting from the ingestion of a common food. Consumption of fruit and vegetables is associated with a healthy lifestyle. A large portion of this produce is consumed raw, and the number of foodborne outbreaks associated with these products has increased correspondingly [44]. A number of important human pathogens can contaminate fresh-cut produce and there has been an augment in the number of food produce-linked foodborne outbreaks in recent years. Minimally processed food is easily contaminated by food borne pathogens either directly or via cross-contamination during food preparation.

Minimally processed vegetables have been implicated in a number of foodborne outbreaks, most commonly involving *Salmonella* and *E. coli*. In addition to *Salmonella* and *E. coli*, the foodborne pathogens *Listeria monocytogenes* and *Shigella* have also been associated with minimally processed vegetables. Other pathogens are also potential contaminants from production of raw products through to consumption.

According to Centre for Disease Control (CDC) and Prevention's Outbreak net Foodborne Outbreak Online Database [15] and EFSA Summary Reports [23], among all microorganisms, Norovirus was the main pathogen responsible (59% in the United States and 53% in the European Union) followed by Salmonella (18% in the United States and 20% in European Union). Norovirus is shown to be responsible for most of the produce-related outbreaks, followed by Salmonella. Norovirus is mainly linked with the consumption of salad in the United States and of berries in the European Union, as demonstrated by the Multiple Correspondence Analysis [14]. Salmonella was the most common bacterial pathogen responsible for produce outbreaks [52], accounting for nearly half of the outbreaks due to bacteria (53% in the United States and 50% in the European Union). Regarding food vehicle, E. coli was associated with the consumption of various fresh vegetables, fruits, and sprouts, but especially with lettuce and unpasteurized apple juices [47]. *Campylobacter jejuni* was involved in produce outbreaks linked to the consumption of salad, lettuce, tomato, and melon.

Numerous data confirmed that vegetables have been implicated in outbreaks of food-borne listeriosis [12, 26, 55] suggesting that cabbage, celery, lettuce, cucumber, onion, leeks, tomatoes and fennel can have a high incidence of L. monocytogenes. Among the greatest concerns with human pathogens on fresh and fresh-cut fruits and vegetables are enteric pathogens such as Escherichia coli O157:H7 and Salmonella that have fast growth rates and low infectious dose. For instance, salmonellosis has been linked to tomatoes, seed sprouts, cantaloupe, apple or orange juice; *E. coli* O157:H7 infection has been associated with lettuce, sprouts, and apple juice; and enterotoxigenic E. coli has been linked to lettuce and carrots. Furthermore, listeriosis remains a great public health concern, as it has one of the highest case fatality rates of all the foodborne infections in Europe (20–30%). Documented associations of shigellosis, parasitic diseases, and virus infections have also been made. Similar problems have arisen with meat and fish products. However, the public health importance of these infections is not always recognized. For example, little attention has been paid to listeriosis, particularly because it is a relatively rare disease in comparison with other common foodborne illnesses such as salmonellosis. On the contrary, consumers concern for outbreaks related to toxicity of chemical substances is further amplified, although epidemiological investigation of foodborne illnesses in Europe has shown that only 0.5% of outbreaks are due to chemical sources.

6.8 SUMMARY

Minimally processed fruits and vegetables are those which are prepared for the convenience of the consumers and are distributed in fresh like state. The unit operations used for the preparations of the produce accelerate the physiological changes in them like, ripening and ethylene production and so additional attention is needed to maintain their quality traits. Chemical treatments, edible coatings, cold chain maintenance, modified atmosphere packaging, active packaging, etc. are employed to achieve this. Various emerging technologies are being explored for the preparation of minimally processed fruits and vegetables like high hydrostatic pressure application, pulsed electric field treatment, pulsed light technology, cold plasma treatment, etc. More work is needed to understand the implications of these treatments on the quality of fruits and vegetables. Moreover, work needs to be done regarding the safety aspects of minimal processed products as reports of outbreaks due to their consumption are often reported.

KEYWORDS

- **cooling**
- **disinfection**
- **fresh processed**
- **fruits and vegetables**
- **Grade 4**
- **hurdle technology**
- **ionizing radiations**

- **lightly processed**
- **minimal processing**
- **outbreaks**
- **packaging**
- **physiological changes**
- **quality aspects**
- **storage conditions**
- **unit operations**
- **washing**

REFERENCES

1. Abadias, M., Usall, J., Anguera, M., Solsona, C., & Viñas, I. (2008). Microbiological quality of fresh, minimally-processed fruit and vegetables, and sprouts from retail establishments. *International Journal of Food Microbiology*, 123 (1–2), 121–129.
2. Allende, A., & Artes, F. (2003). Combined ultraviolet-C and modified atmosphere packaging treatments for reducing microbial growth of fresh processed lettuce. *Lebensmittel. Wiss. U. Technology*, 36, 779–786.
3. Alzamora, S. N., Cerrutti, P., Guerrero, S., Lopez-Malo, A. (1995). Minimally processed fruits by combined methods. In: *Food preservation by Moisture Control, Fundamentals and Applications*. Barbosa-Canovas, G. V., Welti-Chanes, J. (Eds.); Technomic: Lancaster, PA, pp. 463–492.
4. Amefia, A. E., Abu-Ali, J. M., & Barringer, S. A. (2006). Improved functionality of food additives with electrostatic coating. *Innovat Food Sci. Emerg. Tech.*, 7(3), 176–181.
5. AR-USDA. (2005). Fresh-cut fruit moves into the fast lane. *Agricultural Research Magazine*, 53(8).
6. Banerjee, A., Penna, S., & Variyar, P. S. (2015). Allyl isothiocyanate enhances shelf life of minimally processed shredded cabbage. *Food Chemistry*, 183, 265–272.
7. Bari, M. L., Ukuku, D. O., Kawasaki, T., Inatsu, Y., Isshiki, K., & Kawamoto, S. (2005). Combined efficacy of nisin and pediocin with sodium lactate, citric acid, phytic acid, and potassium sorbate and EDTA in reducing the Listeria monocytogenes population of inoculated fresh-cut produce. *Journal of Food Protection*, 68, 1381–1387.
8. Barry-Ryan, C., & Beirne, D. (1998). Ascorbic Acid Retention in Shredded Iceberg Lettuce as Affected by Minimal Processing. *Journal of Food Science*, 64, 498–500.

9. Benitez, S., Achaerandio, I., Sepulcre, F., & Pujola M. (2013). Aloe vera based edible coatings improve the quality of minimally processed 'Hayward' kiwifruit. *Postharvest Biology and Technology*, 81, 29–36.

10. Beuchat, L. R. (1996). Listeria monocytogenes incidence on vegetables. *Food Control*, 7 (4/5), 223–228.

11. Beuchat, L. R., Adler, B. B., Clavero, M. R. S., & Nail, B. V. (1998). Efficacy of spray application of chlorinated water in killing pathogenic bacteria on raw apples, tomatoes and lettuce. *Journal of Food Protection*, 61, 1305–1311.

12. Beuchat, L. (2002). Ecological factors influencing survival and growth of human pathogens on raw fruits and vegetables. *Microbes Infection*, 4, 413–423.

13. Brackett, R. E. (1999). Incidence, contributing factors, and control of bacterial pathogens in produce. *Postharvest Biology and Technology*, 15, 305–311.

14. Callejo'n, R. M., Rodrı'guez-Naranjo, M. I., Ubeda, C., Hornedo-Ortega, R., Garcia-Parrilla, M. C., & Troncoso, A. M. (2015). Reported Foodborne Outbreaks Due to Fresh Produce in the United States and European Union: Trends and Causes. *Foodborne Pathogens and Disease*, 12, 32–38.

15. Centers for Disease Control and Prevention. [CDC] CDC's Outbreak Net Foodborne Outbreak Online Database. 2013. http://wwwn.cdc.gov/foodborneoutbreaks/, accessed September 2, 2014.

16. Chaudry, M. A., Bibi, N., Khan, M., Khan, M., Badshah, A., & Qureshi, M. J. (2004). Irradiation treatment of minimally processed carrots for ensuring microbiological safety. *Radiation Physics and Chemistry*, 71, 169–173.

17. Chauhan, O. P., Raju, P. S., Dasgupta, D. K., & Bawa, A. S. (2006). Instrumental textural changes in banana (var. Pachbale) during ripening under active and passive modified atmosphere. *Int. J. Food Prop*, 9 (2), 237–253.

18. Chervin, C., & Boisseau, P. (1994). Quality maintenance of ready-to-eat shredded carrots by gamma irradiation. *Journal of Food Science*, 59, 359–362.

19. Chien, P. J., Sheu, F., & Yang, F. H. (2007). Effects of edible chitosan coating on quality and shelf life of sliced mango fruit. *J. Food Eng,* 78, 225–229.

20. Cocci, E., Rocculi, P., Romani, S., & Rosa, M. D. (2006). Changes in nutritional properties of minimally processed apples during storage. *Postharvest Biology and Technology*, 39, 265–271.

21. Corrales, G, J., Ayala, V, G., Franco, E, A. M., & Garcia, O. P. (2006). Minimal processing of cactus pear and tender cactus cladodes. *Acta Hort. (ISHS)*, 728, 223–230.

22. D'Aoust, J. Y. (2007). Current foodborne pathogens: salmonella. In: *Food safety handbook: Microbiological Challenges*, M. Storrs, M. C. Devoluy, P. Cruveiller (Eds.), BioMérieux Education, France (2007), pp. 128–141

23. EFSA, (2013). *European Union Summary Reports*. accessed September 12, 2015. www.efsa.europa.eu/en/zoonosesscdocs/zoonosescomsumrep.htm

24. Erkan, M., Wang, C. Y. (2006). Modified and controlled storage of subtropical crops. *Postharv. Biol. Technol*, 5(4), 1–8.

25. Fan, X., & Sokorai, K. J. B. (2008). Effect of ionizing radiations on furan on fresh-cut fruits and vegetable. *J Food Sci*, 73(2), C79–C83.

26. Farber, J. M., Wang, S. L., Cai, Y., & Zhang, S. (1998). Changes in populations of Listeria monocytogenes inoculated on packaged fresh-cut vegetables. *Journal of Food Protection*, 61, 192–195.

27. Fernandes, F. A. N., & Rodrigues, S. (2009). Ultrasound applications in minimal processing. *Stewart Postharvest Review*, 5, 1–7.
28. Francis, G. A., Thomas, C., & O'Beirne, D. (1999). The microbiological safety of minimally processed vegetables. *International Journal of Food Science and Technology*, 34 (1), 1–22.
29. Garcia, E., & Barrett, D. M. (2002). Preservative treatments for fresh cut fruits and vegetables. In: *Fresh-Cut Fruits and Vegetables. Science, Technology and Market*. O. Lamikanra (Ed.), Boca Raton, FL: CRC Press
30. Geraldine, R. M., Soares, N., de F. F., Botrel, D. A., & Gonc,alves, L.de A. (2008). Characterization and effect of edible coatings on minimally processed garlic quality. *Carbohydrate Polymers,* 72, 403–409.
31. Gomez-Lopez, V. M., Devlieghere, F., Bonduelle, V., & Debevere, J. (2005). Intense light pulses decontamination of minimally processed vegetables and their shelf-life. *International Journal of Food Microbiology*, 103, 79–89.
32. Goyeneche, R., Agüero, M. V., Roura, S., & Scala, K. D. (2014). Application of citric acid and mild heat shock to minimally processed sliced radish: Color evaluation. *Postharvest Biology and Technology*, 93, 106–113.
33. Hernández-Muñoz, P., Almenar, E., Ocio, M. J., & Gavara, R. (2006). Effect of calcium dips and chitosan coatings on postharvest life of strawberries (*Fragaria×ananassa*). *Postharv. Biol. Technol,* 39, 247–253.
34. Hong, G., Peiser G., & Cantwell, M. I. (2000). Use of controlled atmospheres and heat treatment to maintain quality of intact and minimally processed green onions. *Postharvest Biology and Technology*, 20, 53–61.
35. Ihl, M., Aravena, L., Scheuermann, E., Uquiche, E., & Bifani, V. (2003). Effect of immersion solutions on shelf-life of minimally processed lettuce. *Lebensm.-Wiss. U.-Technol,* 36, 591–599.
36. Kabir, H. (1994). Fresh-cut vegetables. In: *Modified Atmosphere Food Packaging*. A. L. Brods, & V. A. Herndon (Eds.), (pp. 155–160). Institute of Packaging Professionals.
37. Kaur, C., & Kapoor, H. C. (2000). Minimal processing of fruits and vegetables. *Indian Food Pac*, 156.
38. Lamikanra, O., & Bett-Garber, K. (2005). Fresh-cut fruit moves into the fast lane. Agricultural Research 2005 (http://www.ars.usda.gov/is/AR/archive/aug05/fruit0805.pdf).
39. Laurila, E., & Ahvenainen, R. (2002). Minimal processing of fresh fruits and vegetables. In: *Fruit and Vegetable Processing*. W. Jongen (Ed.), Cambridge, UK/Boca Raton, FL: Woodhead Publishing Limited/CRC Press LLC.
40. Limbo, S., & Piergiovanni, L. (2006). Shelf life of minimally processed potatoes: Part 1. Effects of high oxygen partial pressures in combination with ascorbic and citric acids on enzymatic browning. *Postharvest Biology and Technology*, 39(3), 254–264.
41. Manvell, C. (1997). Minimal processing of food. *Food Science and Technology Today*, 11, 107–111.
42. Marcos, B., Aymerich, T., Monfort, J. M., Garriga, M. (2010). Physical performance of biodegradable film intended for antimicrobial food packaging. *J Food Sci.* 75, 502–507.

43. O'Beirne, D., & Francis, G. A. (2003). Reducing pathogen risk in MAP-prepared produce. In: *Novel Food Packaging Techniques.* R. Ahvenainen (Ed.), (pp. 231–232). Cambridge, UK/Boca Raton, FL: Woodhead Publishing Limited/CRC Press LLC.

44. Olaimat, A. O., & Holley, R. A. (2012). Factors influencing the microbial safety of fresh produce: A review. *Food Microbiol,* 32,1–19.

45. Oms-Oliu, G., Soliva-Fortuny, R., & Martín-Belloso, O. (2008). Using polysaccharide-based edible coatings to enhance quality and antioxidant properties of fresh-cut melon. *LWT-Food Sci, Technol,* 41, 1862–1870.

46. Ongeng, D., Devlieghere, F., Debevere, J., Coosemans, J., & Ryckeboer, J. (2006). The efficacy of electrolyzed oxidizing water for inactivating spoilage microorganisms in process water and on minimally processed vegetables. *International Journal of Food Microbiology,* 109(3), 289–291.

47. Orue, N., Garcı'a, S., Feng, P., & Heredia, N. (2013). Decontamination of Salmonella, Shigella, and *Escherichia coli* O157:H7 from leafy green vegetables using edible plant extracts. *J Food Sci,* 78, M290–M296.

48. Qiang, Z., Demirkol, O., Ercal, N., & Adams, C. (2005). Impact of food disinfection on beneficial biothiol contents in vegetables. *Journal of Agricultural and Food Chemistry,* 53, 9830–9840.

49. Rojas-Graü, M. A., Tapia, M. S., & Martin-Belloso, O. (2008). Using polysaccharide-based edible coatings to maintain quality of fresh-cut Fuji apples. *LWT-Food Sci. Technol,* 41, 139–147.

50. Sapers, G. M. (2003). Washing and sanitizing raw materials for minimally processed fruit and vegetable products. In: *Microbial Safety of Minimally Processed Foods* J. S. Novak, G. M. Sapers, & V. K. Juneja (Eds.), (p. 222). Boca Raton, FL: CRC Press LLC.

51. Schlimme, D. V., & Rooney, M. L. (1994). Packaging of minimally processed fruits and vegetables. In: *Minimally Processed Refrigerated Fruits and Vegetables,* edited by R. C. Wiley. Chapman & Hall, New York. pp. 135–182.

52. Sivapalasingam, S., Friedman, C. R., Cohen, L., & Tauxe, R. (2004). Fresh produce: A growing cause of outbreaks of foodborne illness in the United States, through 1997. *J Food Prot,* 67, 2342–2353.

53. Tapia, M. S., Rojas-Grau, MA, Carmona, A., Rodriguez, F. J., Soliva-Fortuny, R., & Martin-Belloso, O. (2008) Use of alginate and gellan-based coatings for improving barrier, texture and nutritional properties of fresh-cut papaya. *Food Hydrocoll,* 22, 1493–1503.

54. Tapia, M. S., Rojas-Graü, M. A., Rodríguez, F. J., Ramírez, J.., Carmona, A., & Martin-Belloso, O. (2007). Alginate- and gellan-based edible films for probiotic coatings on fresh-cut fruits. *J. Food Sci,* 72, E190–E196.

55. Thunberg, R. L., Tran, T. T., Bennett, R. W., Matthews, R. N., & Belay, N. (2002). Microbial evaluation of selected fresh produce obtained at retail markets. *Journal of Food Protection,* 65, 677–682.

56. Toivonen, P. M. A., & De-Ell, J. R. (2002). Physiology of fresh-cut fruits and vegetables. Fresh-cut fruits and vegetables. In: *Science, Technology and Market.* O. Lamikanra (Ed.), Boca Raton, FL: CRC Press.

57. Tsouvaltzis, P., Brecht, J. K., Siomos, A. S., & Gerasopoulos, D. (2008). Responses of minimally processed leeks to reduced O_2 and elevated CO_2 applied before processing and during storage. *Postharvest Biology and Technology*, 49, 287–293.
58. Woo, P. P., Sung, H. C., & Dong, S. L. (1998). Effect of minimal processing operations on the quality of garlic, green onion, soybean sprouts and watercress. *Journal of the Science of Food and Agriculture*, 77(2), 282–286.

PART III

BIOPROCESS ENGINEERING AND BIOTECHNOLOGY: APPLICATIONS AND PRODUCTS

CHAPTER 7

APPLICATION OF BIOCOMPOSITE POLYMERS IN FOOD PACKAGING: A REVIEW

ARIJIT NATH, ARPIA DAS, and CHIRANJIB BHATTACHARJEE

CONTENTS

7.1   INTRODUCTION

The outstanding development of biotechnology is the result of judicious combination of basic and applied science. As food is the main interest around the world, researches related with food technology are considered as a great challenge from a long prior. Among different branches of food technology, development of food packaging material is one of the considerable great interests.

The purpose of food packaging is to maintain the quality and safety of food products during transportation and storage by preventing unfavorable conditions [1], such as: spoilage by microorganisms, chemical contaminants, permeation of water vapor, oxygen, carbon dioxide, flavors, etc. (Figure 7.1).

In order to provide these functions, packaging materials plays a crucial role [2–4]. In the logical sense, food-packaging material is quite different from other durable items, such as home appliances, electronics and other equipments, because of its safety aspects are very much important issue. Basic packaging materials are made by paperboard, paper and plastic etc. From a long prior, there is a great concern to develop different kinds of packaging materials in order to enhance their functionalities for maintaining and improve the food qualities. From the middle of twenties century, petroleum based plastics are popularly used because of their low cost and excellent physico-chemical properties. Generally synthetic polymers, such as polypropylene, polyethylene, polyurethane are used for food packaging. It has been reported that more than 40% of plastics are used for packaging and almost half of them are used for food packaging. Although the synthetic plastic have been received popularly used for different types of food packaging but they create sever disposal problem because they are not easily degraded in environment after using. Moreover it has been

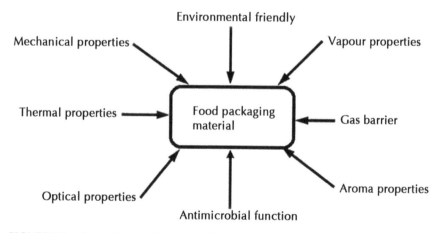

FIGURE 7.1   General properties required for food packaging materials.

reported that petroleum based plastics is not suitable for food packaging because it creates several health hazards [5, 6].

Recently due to concern on environmental impact of conventional petroleum based food packaging materials, more attentions have been placed to invent the biodegradable polymers, such as poly(-glycolic acid), poly(-lactic acid), poly(-hydroxybutyrate), poly(e-caprolactone) which are considered for food packaging material. The biodegradable polymers are derived from renewable biological resources, which have excellent gas and solute barriers and mechanical properties. Biopolymers can be broadly classified into different categories based on the origin of the feedstock and their manufacturing processes (Figure 7.2). The present chapter describes the applications of different types of biocomposite polymer in food packaging in a comprehensive way.

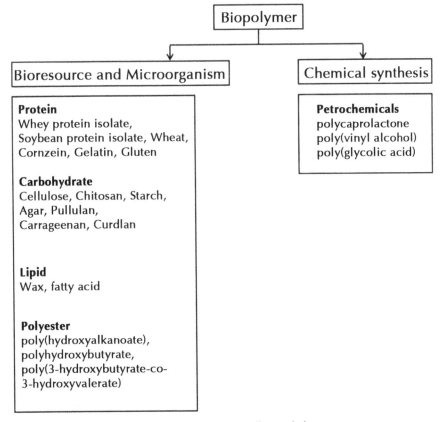

FIGURE 7.2   Classification of biopolymers according to their source.

Biopolymers are polymeric materials in which at least one step in the degradation process is done by naturally occurring organisms. Under appropriate oxygen, moisture and temperature, biodegradation of biopolymers cannot produce any toxic or environmentally harmful residue [7–16]. Moreover, biopolymer-packaging materials are excellent vehicles for incorporating a wide variety of additives (antioxidants, antifungal agents, antimicrobials, colors and other nutrients), extending quality and shelf-life of foods [7–21].

There are several concerns have been addressed for improvement of bio-based polymer prior to their commercial use in food packaging technology [22–24]. These concerns include increase of degradation rates at the end, improves mechanical properties and so on. Generally carbohydrate and protein derived polymers are good barriers against oxygen at low to intermediate relative humidity and have good mechanical properties. Contradictorily, they act as a poor barrier against water vapor due to their hydrophilic nature. Recently, a new category of materials, known as, biocomposite polymers with enhanced barrier, mechanical and thermal properties, antimicrobial activity have been considered as a promising option for food packaging materials. They are exhibiting much improved properties as compared to the original biopolymers [25–33].

7.2   BIOCOMPOSITE POLYMERS

Composite biopolymer consists of two or more chemically different materials or phases (matrix phase and dispersed phase), separated by proper interface. Matrix is the primary phase, usually uniform, more ductile and soft. Contradictorily, dispersed phase, usually stronger than matrix phase, is embedded in the matrix phase in a discontinuous form [34, 35]. Depending upon the feedstock ingredients, many different types of composite polymers are synthesized [36]. Different types of polymers, such as polyester, epoxy, polyamide, phenolic, polyimide, vinyl ester, polypropylene, polyether ether ketone, and others are prevalently used for matrix preparation [37]. Biocomposite polymers are low weight and good functional properties, such as corrosion resistance, high fatigue strength and mechanical strength than its original [38, 40]. Biocomposite polymers which are used in food packaging can withstand at the stress at high temperature, storage and transportation

[38–41]. Mostly, nanomaterials are introduced in biobased polymer phase because of high aspect ratio and high surface area of nanoparticles, which lead to increase the antimicrobial activity and mechanical barrier property of food packaging materials [29]. Bio-nanocomposite polymer is able to create a bridge between high rigid large packaging materials to flexible pouches. It also solves the global recycling issues, as well as packaging waste disposal. Since 1950s, clay based green-composite polymers have been received a great interest [42]. Toyota and his research team prepared polyamide 6/montmorillonite nanocomposites by in situ polymerization of ε-caprolactam [43]. After that with time progress extensive research works have been performed by both academic and industrial groups. To synthesis green polymer composites, various kinds of natural fibers were employed with biodegradable resins, such as poly(-hydroxybutyrate), poly(-lactic acid), polysaccharides, hyaluronate and modified starch [44–46]. A classical example is the use of nano sized montmorillonite clay to improve mechanical and thermal properties of nylon [47]. The main disadvantages of polymer matrix composites are high coefficient of thermal expansion and low thermal resistance [48].

7.3 FOOD PACKAGING APPLICATIONS

In food packaging industry, use of proper packaging materials and methods is always been the main interests because it minimizes food losses, provide safety and maintain the food quality. Examples of improved performance of nanocomposite packaging materials are: (i) gas (carbon dioxide, oxygen) and water vapor barrier properties, (ii) thermal stability, (iii) high mechanical strength, (iv) biodegradability and recyclability, (v) chemical stability, (vi) dimensional stability, (viii) heat resistance, (ix) good optical clarity, as well as (x) developing active antimicrobial and antifungal surfaces; and these have made their popularity in food packaging purposes. This material also leads to lower weight of packages because less material is required to obtain the same or even better barrier properties [49].

Presence of oxygen in the packaged foods causes many undesirable reactions, such as nutrient losses, color changes, off-flavor development and microbial growth. The nanocomposite material considerably enhances the shelf-life of many types of food, specifically, meats, cheese, confectionery, cereals and boiling package foods, fruit juice and so on [50–52].

Nanocomposites have been considered as antimicrobial activity as growth inhibitors [53], antimicrobial agents [54], antimicrobial carriers [55] or antimicrobial packaging films [56, 57]. It acts as antimicrobial compounds in or coated onto the packaging materials. They are able to control the growth of undesirable microorganisms, such as food pathogen [5, 58–60]. Nanocomoposite are particularly effective to control the growth of undesirable microorganisms because of the high surface-to-volume ratio and enhanced surface reactivity of the nano-sized antimicrobial agents. The antimicrobial activities of silver nanoparticles have been exploited by synthesizing non-cytotoxic coating for methacrylic thermosets (nanocomposite material based on a lactose modified chitosan and silver nanoparticles) [61, 62].

Polymer/clay nanocomposite with improved barrier properties is commonly used as a barrier layer in a multilayer packaging material with other structural layer. Different types of multilayer nanocomposite are used in packaging technology, including carbonated beverages, beer bottles and thermoformed containers. First type of multilayer is composed with a barrier layer in the middle and outside structural layers. The second type of nanocomposite has a passive barrier in which the middle layer is reinforced with nanocomposite film. The third type of nanocomposite is an active barrier composed of gas scavenger incorporated film layer. Nylon 6 is commonly used in co-extrusion with other plastic materials, which provides both toughness and strength to the structure. It is used as a high barrier packaging material. Polyamide 6 is another most widely used plastic materials used to produce laminated films, sheets and bottles because it is inexpensive, clear, thermo-formable tough and strong over a broad range of temperatures with good chemical resistance.

High barrier nanocoating consists of hybrid organic–inorganic nanocomposite materials by sol–gel process [63]. The coatings are developed through atmospheric plasma technology using dielectric barrier discharges. It was reported that nanoscale silicate and alumina particles increase the scratch and abrasion resistance of coatings without hindering the transparencies [64]. In the year 2009, Applerot with his co-worker reported about the preparation method of Zinc oxide coated glass using ultrasonic irradiation. They also demonstrated its (the glass slide coated with low level of Zinc oxide coating) significant antibacterial effect against both Gram-positive and Gram negative bacteria [65]. Also it was reported that Titanium oxide coated oriented polypropylene films has strong antibacterial activity against *Escherichia coli* and reduced the microbial contamination of lettuce [66]. In 2010, Vartiainen with his co-worker reported that bio-hybrid nanocomposite (bentonite and chitosan) coated on argon-plasma activated low-density polyethylene paper has improve barrier properties against oxygen, water vapor, grease and ultraviolet light transmission [67].

Easy biodegradation is one of the unique features of biocomposite polymer in the context of sustainable environment. The biodegradation of polymers is a complex process, which is catalyzed by enzymes and oxidation [68]. It was reported that polycaprolactone in clay-based nanocomposite polymer is attributed to catalytic role of the organo-clay during the biodegradation process. Also it was reported that there was no significant difference between pure poly-lactic acid and poly-lactic acid nanocomposite until a month of composting but later there is a high degradation rate of poly-lactic acid nanocomposite [69–71]. It was reported that degradation of poly-lactic acid follows the series of biochemical steps and finally they are converted to carbon dioxide and water [72]. Nieddu with his co-worker reported that degradation rate of nanocomposite is more than the pure pure-lactic acid [73]. Some researcher reported that the enhanced biodegradation of poly-lactic acid based nanocomposite is caused by high relative hydrophilicity of the clays (dispersed phase), which allow an easier permeability of water into polymer matrix phase [74].

Other advantages of bionanocomposite polymer are that they acted as an antioxidant releasing films, color containing films, anti-fogging and anti-sticking films, light absorbing/regulation systems, susceptors for microwave heating, gas permeable/breathable films, bioactive agents for controlled release and insect repellant materials [75]. Smart or intelligent packaging is projected to monitor and provide information about the quality of the packaged food or its surrounding to predict or decide the shelf life. They alert consumer about contamination by pathogens, detect harmful chemicals, food deterioration, and food quality [76, 77].

Different aspects of biocomposite polymers for food packaging are discussed in subsequent sections.

7.4   MECHANICAL AND BARRIER PROPERTIES

Development of nanocomposite with organoclays has been proven to improve the mechanical properties of various bio-polymers, even with a low filler contain (<5% wt). It was observed that mechanical properties of polymer/clay nanocomposites are strongly dependent on filler content. In Figure 7.3, typical structures of multilayer nanocomposite packaging materials for gas barrier property has been depicted.

Lee with his co-worker reported on the biodegradable polymer/montmorillonite nanocomposites. In that case, poly-(butylene succinate) and Cloisite 30B were used as a biodegradable polymer matrix and montmorillonite respectively [78]. Huang and Yu reported about tensile properties of starch/montmorillonite nanocomposites, prepared with various filler concentrations of 0–11 wt% to the starch. They proposed that strength and Young's modulus are increased monotonically with increase the filler content up to 8%, then leveled off, while the tensile strain is decreased [79]. Also Huang with his co-worker reported that increase of tensile strength and strain of corn starch/montmorillonite nanocomposites by 450% and

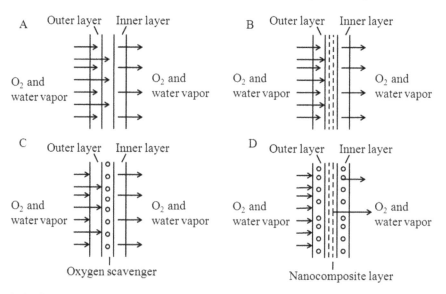

FIGURE 7.3   Schematic diagram of multilayer nanocomposite oxygen and water vapor barrier packaging materials: (A) No barrier, (B) Passive barrier, (C) Active barrier, (D) Passive-Active barrier.

20%, respectively with addition of 5% clay [80]. Chen and Zhang reported that percent elongation at break of the soy protein/montmorillonite bio-nanocomposite sheets is decreased with increase of montmorillonite content, whereas tensile strength of nanocomposite sheets is increased from 8.77 MPa to 15.43 MPa with increase of montmorillonite content up to 16% [81]. Similar results of the tensile testing of bio-nanocomposites, based on other biopolymer have been reported by many investigators [82–90]. The substantial improvement of mechanical properties of polymer nanocomposites can be attributed to high rigidity and aspect ratio of nanoclay. It is well reported that polymer nanocomposites have excellent barrier properties against oxygen, carbon dioxide and water vapor. This depends on the type of clay, i.e., compatibility between clay and polymer matrix, structure of nanocomposites and aspect ratio of clay platelets. Generally good gas barrier properties of polymer nanocomposite are archived by exfoliated clay minerals with large aspect ratio [91].

Also Rhim with his co-worker reported that water vapor transmission rate of agar/unmodified montmorillonite (Cloisite Na^+) nanocomposite films is significantly lower than agar/organically modified montmorillonite (Cloisite 30B, 20A) nanocomposite [88, 89]. They also reported that water vapor permeability of agar/clay (Cloisite Na^+) nanocomposite films decreases exponentially with increase in clay content from 0 to 20 wt% [88]. Yano with his co-worker developed polyimide/clay nanocomposite films with four different sizes of clay minerals, such as hectorite, montmorillonite, saponite and synthetic mica to understand the effect of aspect ratio on the barrier properties of nanocomposite films. They found that at constant clay content (2% wt), the relative permeability coefficient decreases with increase the length of the clay [92]. Contradictorily, Cloisite 30B is more effective in reducing the water vapor permeability of poly-lactic acid based nanocomposite films and chitosan based nanocomposite film [56, 90]. Similar results about water vapor permeability of bio-nanocomposite films based on other biopolymers, such as whey protein isolate [93, 94], starch [95], soy protein isolate [96, 97], poly(caprolactone) [29] and wheat gluten [98] have been reported. Park with his co-worker investigated water vapor transmission rate of thermoplastic starch/clay nanocomposite films with respect to Cloisite Na^+ or Cloisite 30B in the thermoplastic starch matrix [99]. In case of thermoplastic starch hybrid nanocomposite films, water vapor transmission rate decreases considerably even there are small amounts of clays of both Cloisite 30B and Cloisite Na$^+$, which signifies that layered structure of clay defends the transmission of moisture vapor through the film matrix [100–102]. Park with his co-worker used melt intercalation method to develop thermoplastic starch/Poly(butylene succinate)/clay ternary nanocomposites [103]. It was reported that water vapor transmission rate of thermoplastic starch/Poly(butylene succinate)/clay ternary nanocomposite film decreases with increase in Poly(butylene succinate) contents, although the water vapor transmission rate of thermoplastic starch films is much higher than Poly(butylene succinate) films [78, 103]. Park with his co-worker investigated the relative water vapor permeability of cellulose acetate/clay nanocomposites with different triethyl citrate plasticizer and organoclay content in a controlled temperature and relative humidity chamber [104]. The composite, starch filled with cellulose whiskers, water sensitivity is decreased and its thermomechanical properties were increased [105].

It is proven that addition of cellulose nanofibers is effective to improve water vapor barrier of the film (water vapor permeability decreases from 2.66 to 1.67 g mm/kPa.h.m^2). Poly-lactic acid/layered silicate nanocomposite membrane was developed by Koh with his co-worker. They reported that in case of poly-lactic acid/layered silicate nanocomposite membrane, with increase of organoclay content, gas permeabilities is decreased. Furthermore, they reported that gas permeability barrier feature of poly-lactic acid/layered silicate nanocomposite membrane with Cloisite 30B is very high compared with other types of nanoclay [19]. It is also reported that organoclay also behave as the barrier against gas molecules to pass through polymer matrix [101].

Different types of hybrid poly-lactic acid had been developed, filled with many organic clays, such as dodecyltrimethyl ammonium bromide-montmorillonite, hexadecyl amine-montmorillonite, Cloisite 25A and so on [106]. Cabedo with his co-worker developed nanocomposites by amorphous poly-lactic acid and chemically modified kaolinite and they reported that good interaction between polymer and clay leads to an increase in oxygen barrier properties [107]. In 2002, Gorrasi with his co-worker developed different compositions of Polycaprolactone/organically modified clay nanocomposites by melt blending or catalyzed ring-opening polymerization of caprolactone [108]. Meera with his co-worker developed exfoliated nanocomposites by situ ring opening polymerization of caprolactone with organically modified clay by using dibutylin dimethoxide as an initiator/catalyst. The barrier properties were tested for water vapor and dichloromethane, and zero concentration diffusion coefficients (D_o) were found for both vapors [109]. The water sorption potentiality of nanocomposite increases with increase of montmorillonite content. The thermodynamic diffusion parameters, D_o, of parent Polycaprolactone, both micro-composites and intercalated nanocomposites were found to be same. Contradictorily, exfoliated nanocomposites have very low values, even for low montmorillonite content [109, 110].

In this context, an innovative procedure was published by Sorrentino with his co-worker [111]. Particularly, they developed an antimicrobial benzoic acid, on an magnesium/aluminium layered double hydroxide by ionic bonds, followed by incorporation to a polycaprolactone matrix [109]. Systematic study on synthesis of polycaprolactone nanocomposites is performed with different percentages of montmorillonite, degree of montmorillonite intercalation and different organic modifiers of montmorillonite on the diffusion coefficient of dichloromethane [112]. For the organic solvent (dichloromethane) diffusion, the intercalated samples showed lower values of the diffusion parameters. This result signifies that beside the content of clay, size and type of dispersion of inorganic component in polymer phase is important for improving the barrier properties of polymer nanocomposite. For the diffusion of dichloromethane in the poly-caprolactone composite samples with 3 wt% of montmorillonite, the diffusion parameter decreased from micro-composites to exfoliated nanocomposites [113].

7.5 ANTIMICROBIAL PROPERTIES OF BIO-NANOCOMPOSITES

Antimicrobial properties of nanoparticle or nanocomposite polymer has long been recognized and exploited in food packaging technology [54, 55, 57, 114–120]. Nanocomoposite have antimicrobial property because they have high surface-to-volume ratio which enhanced the antimicrobial surface activity [119]. Various items, including metal ions (silver, copper, gold, platinum), metal oxide (titanium oxide, zinc oxide, magnesium oxide), organically modified nanoclay (quaternary ammonium modified montmorillonite, silver-zeolite), natural biopolymers (chitosan), natural antimicrobial agents (nisin, thymol, carvacrol, isothio-cyanate, antibiotics), enzymes (peroxidase, lysozyme) and synthetic antimicrobial agents (quaternary ammonium salts, ethylene-di amine-tetra acetic acid, propionic, benzoic, sorbic acids) have been used in nano range with combined form to synthesis nanocomposite.

7.5.1   METAL AND METAL OXIDE

Among numerous nanoparticles, silver nanoparticles have been used popularly for development of packaging materials. This is not only due to their unique properties, such as electric, optical, catalytic, thermal stability, but also for its antimicrobial activity against broad spectrum of bacteria, viruses and fungus [120, 122]. It was reported that silver ions interact with negatively charged biomacromolecular components (sulfhydryl or disulfide groups of enzyme) and nucleic acids, change the structure of bacterial cell walls which lead to disruption of metabolic processes followed by cell death [123, 126].

In many cases silver nano-particles is conjugated with zeolite [127], silicate [128] and nanoclay [129] for development of composite nanomaterial. These have strong antimicrobial property with a broad spectrum but it was reported that Ag-zeolite have not any antimicrobial activity against spores of heat-resistant bacteria [127]. Also silver-zeolite incorporated chitosan film showed strong antimicrobial activity against both Gram-negative and Gram-positive bacteria [56]. Silver-silicate nanocomposite was synthesized by flame spray pyrolysis process and incorporated into polystyrene, which showed significant antibacterial activity against both *Staphylococcus aureus* and *Escherichia coli* [130]. Silver-montmorillonite nanoparticles synthesized by exchanging the Na^+ of natural montmorillonite with silver ions (from silver nitrate solution) showed antimicrobial activity against *Pseudomonas sp.* [131].

A green synthetic approach for preparing antimicrobial silver nanoparticles was reported. Varieties of carbohydrates, such as sucrose, chitosan, waxy and soluble corn starch were used for this purpose. The carbohydrates act as both reducing and stabilizing agents and also as a template for carrying silver nanoparticles. Various nanoparticles, based on Cu^{2+}, Zn^{2+}, Ag^+ and Mn^{2+} showed a significantly increased antimicrobial activity against *Escherichia coli*, *Salmonella choleraesuis* and *Staphylococcus aureus* [132]. It was reported that silver nanoparticles based on metallic silver or silver salts may be readily incorporated into thermoplastic packaging polymeric materials, such as polystyrene, polyethylene, propylene and nylon [58]. Silver nanoparticles have been incorporated into biopolymer films, such as chitosan and starch, and they had strong antimicrobial activity against both Gram-positive and Gram-negative bacteria [55, 133–135]. The efficiency of antimicrobial function of these polymeric nanocomposites is greatly influenced by various factors, such as silver content, particle size, size distribution, degree of particle agglomeration, interaction of silver surface with the base polymer [125]. The great advantage of silver nanoparticles in silver nanocomposite is that they are able to disperse on the surface of polymer matrix without the formation of aggregation. Due to antifungal and bacteriostatic properties, copper nanoparticle has been proposed to develop nanocomposite polymer for food packaging application [12]. However, copper is not considered as a safe in food material because it is toxic and it promotes biochemical deterioration of food staffs [127].

Metal oxides, such as zinc oxide, titanium oxide and magnesium oxide have been exploited for synthesis of antimicrobial packaging films due to their strong antimicrobial activity compared with organic antimicrobial compounds [115]. These metallic oxides are used as photocatalysts. When the photocalyst is irradiated with ultraviolet radiation, highly reactive oxygen species are generated, which seems to be one of the cause of their antimicrobial activity [65, 136, 137]. Among all metallic oxides, titanium oxide has been widely used in paints, foods and cosmetics as well as food packaging materials. Titanium oxide is inert, non-toxic, inexpensive and environmentally-friendly with antimicrobial activity against a wide variety of microorganisms [138]. When titanium oxide is incorporated into a polymer matrix, it will provide a protection against food pathogen as well as odor, staining, deterioration and allergens on the presence of radiation of relatively low wavelength (close to ultra-violet wavelength). TiO_2 thin films is generally prepared using several substrates by various techniques, such as chemical vapor deposition, magnetron sputtering, evaporation, ion beam technique, electro-deposition, chemical spray pyrolysis and sol–gel method [139]. But due to its low photon utilization efficiency and necessity of the ultraviolet as an excitation source, its practical applications are limited. To overcome these problems, the modification of TiO_2 has also been attempted by doping them with metallic ions or oxides, such as ferric ion, silver or tin oxide [140, 141]. Thin TiO_2 film has excellent mechanical and chemical durability in the visible and near-infrared region. Moreover incorporation of TiO_2 into synthetic plastic matrix increases the biodegradability [142].

7.5.2   BIOPOLYMER/CLAY NANOCOMPOSITE

It was reported that bio-nanocomposite prepared with some organically modified nanoclay has antimicrobial activity [56, 57]. Chitosan/clay nanocomposite films with two different types of nanoclay (i.e., a natural montmorillonite and an organically modified montmorillonite, Cloisite 30B) were tested for the antimicrobial activity against pathogenic Gram-positive bacteria (*Listeria monocytogenes* and *Staphylococcus aureus*). They found positive response (antimicrobial activity) of organically modified montmorillonite, Cloisite 30B, while the natural montmorillonite did not show any antimicrobial activity. They suggested that antimicrobial activity mainly attributed to the quaternary ammonium salt of organically modified montmorillonite (Cloisite 30B) [56, 135, 143, 144]. In 2008, Hong and Rhim reported that some organically modified clays, such as Cloisite 20A and Cloisite 30B have strong antimicrobial activity against both Gram-negative and Gram-positive bacteria. They also demonstrated that inhibition of Gram-positive pathogenic bacteria, *Listeria monocytogenes* in presence of organically modified montmorillonite attributed to the destruction of cell membrane by quarternary ammonium salt in organically modified nanoclay [54].

Wang with his co-worker performed a couple of antimicrobial studies with chitosan/organic rectorite modified by cetyltrimethyl ammonium bromide nanocomposite and chitosan/unmodified Ca^{2+}-rectorite films prepared by a solution intercalation method [57, 144–145]. They reported that both chitosan-based nanocomposites showed stronger antimicrobial activity than pure chitosan, particularly against Gram-positive bacteria, where as the pristine unmodified Ca^{+2}-rectorite could not inhibit the growth of bacteria. They also reported that antibacterial activity of nanocomposites increases with the inter-layer distance of layered silicates in nanocomposite and the increase of amount of clay in nanocomposite [144]. In this context, it was reported that degree of antimicrobial activity depends on the type of organoclay and polymer matrix used to prepare the nanocomposite [59, 89, 90, 93].

7.5.3   NANOCLAYS

It was reported that mineral clay has biocide, as well as antimicrobial carrier property [146–149]. Biocidal metals, such as silver, zinc, copper and magnesium (as charge compensating ions) can be incorporated into the clay matrix by ion exchange. Also nanoparticles of neutral metals can be formed inside the clay by reduction of metal ions [56]. Aminopropyl functionalized magnesium phillosilicates is synthesized by sol–gel technique, which strongly inhibits the growth of *Staphylococcus aureus, Candida albicans, Escherichia coli* [150]. The inhibition of microbial growth by Aminopropyl functionalized magnesium phillosilicates clay attributes to the presence of amino propyl groups and their charge interactions. Quaternary phosphonium salt functionalized few-layered graphite has been suggested for food packaging material because it was proven that they have excellent thermal stability and long-acting antimicrobial activity against *Staphylococcus aureus* and *Escherichia coli* [151].

7.5.4   NANOCOMPOSITES WITH COMBINED ANTIMICROBIALS

Development of antimicrobial biopolymer based nanocomposite films, such as chitosan with silver nanoparticles has been received a great attention because of strong antimicrobial function of silver ions. Silver ions adhere to the negatively charged bacterial cell wall and change the cell wall permeability, additionally modify the DNA replication mechanisms, make the abnormal cell size, cytoplasm contents, cell membrane and outer cell layers of sensitive cells, and protein denaturation, which induces cell lysis and death [122]. Two different types of method for the preparation of chitosan/silver nanoparticle composite films have been used, i.e., use of reduction of silver ion from silver nitrate [152–155] and direct use of silver nanoparticles [56].

Rhim with his co-worker prepared chitosan/silver nanoparticle composite films by direct addition of Ag-nanoparticles and chitosan/silver-adsorbed zeolite (silver-Ion) into chitosan film forming solution and tested their antimicrobial activity. They reported that both of these biocomposite films exhibited strong antibacterial activity against both Gram-positive and Gram-negative bacteria [56]. Most of the chitosan/silver nanoparticle composites have been prepared using reduction of silver ion from silver nitrate. Silver nanoparticles are conventionally synthesized by the reduction of silver salt precursors using chemical reducing agents, such as triethanolamine, formamide, $NaBH_4$, hydrazine, etc. Contradictorily, physical methods for reduction of silver salts by ray irradiation, microwave irradiation, ultraviolet irradiation, thermal treatment, photochemical process and sonochemical process have been recommended as green technology for nanocomposite synthesis. Yoksan and Chirachanchai developed nanocomposite using silver nanoparticles by ray irradiation reduction of silver nitrate in a chitosan solution, which has been introduced into chitosan–starch based films synthesis. They found this composite film enhances the antimicrobial activity against *Staphylococcus aureus, Escherichia coli* and *Bacillus cereus* [133].

Li with his co-worker developed three component films, i.e., chitosan/ silver/ zinc oxide (CS/Ag/ZnO) blend films by one-step sol–cast transformation method. In this method silver nanoparticles were synthesized using chitosan as reducing agent under hot alkaline condition and at the same time zinc oxide nanoparticles NPs has been formed in the composite. It was reported that chitosan/ silver/ zinc oxide blend film has excellent antimicrobial activities against variety of microorganisms, such as *Bacillus subtilis, Staphylococcus aureus, Escherichia coli, Aspergillus* sp, *Rhizopus* sp, *Penicillium* sp and yeast [156].

7.6 BIODEGRADATION

Biodegradability of bio-nanocomposite is one of the most important issues. Biodegradation of polymer signifies the fragmentation of its skeleton, loss of mechanical properties and so on through the action of microorganisms, including fungi, bacteria and actinomycetes. Tetto with his co-worker first investigated on biodegradability of polycaprolactone derived nano-composites and reported that polycaprolactone/clay nanocomposites has improved biodegradability compared to pure polycaprolactone [157]. They explained that such improved biodegradability of polycaprolactone in clay-based nanocomposites may be faster biodegradation organoclay. Sinha Ray with his co-worker performed a series of biodegradation tests of nanocomposite made by poly-lactic acid and organoclay [69, 71]. They compared biodegradability of both poly-lactic acid and poly-lactic acid nanocomposite films prepared by melt blending using soil compost test at 58°C. They found no significant difference between pure poly-lactic acid and poly-lactic acid nanocomposite until a month of composting but significantly higher degradation (complete degradation) of poly-lactic acid nanocomposite was observed within two months. They assumed that enhancement of biodegradation of poly-lactic acid nanocomposite films is attributed to the presence of hydroxylated terminal group in the clay layers [70].

Jong with his co-worker reported that the degradation of poly-lactic acid follows hydrolysis of ester bond, fragmentation into oligomer, solubilization of oligomer fragments, diffusion of soluble oligomers, and finally converted to carbon dioxide and water [72]. Nieddu with his co-worker also reported the similar types of results. They prepared five different types of nanoclays and different levels of clay contented nonocomposite using melt intercalation method. They reported the degradation rate of nanocomposite is more than 10 times (with respect of lactic acid release) or 22 times (with respect of weight change) higher than pure lactic acid. The degree of degradation is higher for intercalated nanocomposite which is dependent on the type of clays. They also reported that fluorohectorite with a dihydroxy organic modifier is more compatible with ploy-lactic acid than montmorillonite clays [73].

Paul with his co-worker tested the effect of clay type on hydrolytic degradation of poly-lactic acid. Three different types of organoclays, such as Cloisite Na⁺, Cloisite 30B and Cloisite 25A were used for experimental purposes. They found that biodegradability of poly-lactic acid nanocomposites is high compared to pure poly-lactic acid [68]. In 2009, Fukushima with his co-worker also observed the enhanced biodegradability of poly-lactic acid based nanocomposite and they concluded that the higher rates of biodegradation of poly-lactic acid nanocomposite is attributed to the high relative hydrophilicity of the clays, which allow an easier permeability of water into polymer matrix [74].

Similar results were reported for other bio-nanocomposite polymers, including soy protein-based nanocompsite [158], polyhydroxybutyrate [159], nano-silica/starch/polyvinyl alcohol films [160], etc. Contradictory results were also published in this direction. Lee with his co-worker prepared melt intercalated aliphatic polyester, such as polybutylene succinate/organoclay (Cloisite 30B) nanocomposites polymers with different content of nanoclay and performed soil compost test. They found that rate of biodegradability of nanocomposite decreases compared to the pure polymer butylene succinate. It was reported that nanocomposites developed by the intercalated clays with high aspect ratio hinders the microorganism to enter into the bulk of film, as a consequence the degradation is hampered [78]. Similar observations for biodegradation tendency of poly-hydroxybutyrate nanocomposite and poly-lactic acid/chitosan-organically modified montmorillonite nanocomposite were reported by few researchers [161, 162]. It was also reported that antimicrobial action (against food pathogenic bacteria, including Gram-positive bacteria) of chitosan-based nanocomposite polymer attributes the quaternary ammonium group in the modified organoclay [56].

Hong and Rhim proved that due to the presence of quaternary ammonium group in the organoclay (Cloisite 30B); it has also a strong bactericidal activity against Gram-positive bacteria and bacteriostatic activity against Gram-negative bacteria [54]. Also Someya with her co-worker investigated on biodegradability of polybutylenes adipate-co-butylene terephthalate based nanocomposites, prepared by melt blending with two different types of layered silicates, i.e., non-modified montmorillonite and octadecylamine-modified montmorillonite [163]. The biodegradability of these in the aqueous medium (by determining biochemical oxygen demand) showed that the addition of montmorillonite to poly(butylenes adipate-co-butylene terephthalate) endorsed biodegradation, whereas the presence of octadecylamine-modified montmorillonite did not promote the biodegradation.

7.7   GELATIN-BASED ZINC OXIDE NANOPARTICLE: CASE STUDY

In 2015, Shankar with his co-workers have reported about the preparation of gelatin-based zinc oxide nanoparticles composite films and its application in food packaging technology [164]. Gelatin-based zinc oxide nanoparticle composite films have been synthesized using various types of ZnO NPs, developed by zinc nitrate and zinc acetate with or without capping agent (carboxymethyl cellulose). The zinc oxide nanoparticles have been various shape and sizes ranging from 200–400 nm depending on the preparation method. Subsequently the gelatin/ZnO NPs nanocomposite films have been characterized by ultraviolet-visible spectroscopy, field emission scanning electron microscope, Fourier transform infrared spectroscopy, thermal gravimetric analysis and X-ray powder diffraction analysis. The effect of different types of ZnO NPs on the morphology, thermal stability, crystallization behavior, mechanical, water vapor barrier, water contact angle and antibacterial properties of the gelatin-based nanocomposite films have been investigated. It has been reported that gelatin/ZnO NPs nanocomposite films exhibited three absorption peaks around 220–230 nm, 270–280 nm and 360 nm. Also it has been reported that apparent surface color and transmittance of gelatin film is greatly influenced by the types of ZnO NPs. Fourier transform infrared spectroscopy spectra indicated interaction between ZnO NPs and N-H groups of gelatin. X-ray powder diffraction result depicts the crystalline structure of ZnO NPs with dominant [100] facet in the gelatin/ZnO NPs nanocomposite films. It has been reported that synthesized nanocomposite (ZnO NPs incorporated) films exhibited enhanced thermal stability, high barrier property against ultraviolet, moisture content, water contact angle, water vapor permeability and elongation at break of ZnO NPs. Contradictorily tensile strength and modulus of elasticity of composite film have been decreased ($p < 0.05$) compared with control gelatin film. Moreover it has been reported that the nanocomposite films exhibited strong antibacterial activity against both Gram-negative and Gram-positive food pathogenic bacteria. Developed nanocomposite films were more active against Gram-positive bacteria than Gram-negative bacteria. Furthermore, it has been claimed that developed nanocomposite films have a high potentiality to become a food packaging material.

7.8   CONCLUSIONS

The present chapter describes the applications of different types of bio-composite polymer in food packaging in a comprehensive way. It has been observed that biocomposite polymer has high specific strength, specific stiffness, abrasion, fractures and corrosion resistance potentiality. Its weight is low and fatigue strength is more than its parent. As a food packaging materials, biocomposite materials are received great attention because of their unique properties, such as high barrier properties against the diffusion of carbon dioxide, oxygen, water vapor, flavor compounds. They are easily biodegradable and have superior biocompatible properties. Some pioneer researcher developed nanostructure materials and used for development of biocomposite polymer. They reported that nanostructure in biocomposite have more smart properties. They provide antimicrobial activities, oxygen scavenging ability, indication of degree of exposure of some factors, such as oxygen levels or inadequate temperatures. Wrapping up all information, it is concluded that application of biocomposite polymer opens a new arena in food packaging technology.

7.9   SUMMARY

Recently development of biocomposite polymer and its applications have been received a great attention due to depletion and negative impact of fossil fuel. Biocomposite polymer consists of two or more chemically defined materials or phases, i.e., dispersed phase and matrix phase, among them, any of has biological origin. Biocomposite polymer has high specific strength, low weight and cost. It has specific stiffness, abrasion, fractures and corrosion resistance potentiality. Its fatigue strength is more than its parent.

The unique properties of biocomposite polymer, such as high barrier properties against the diffusion of carbon dioxide, oxygen, water vapor, flavor compounds, superior biodegradability has exploited for development of food packaging materials. In many cases, nanostructure materials have been used for development of biocomposite polymer. Without any contradiction, nanostructure in biocomposite provides active and/or smart properties to food packaging systems, including antimicrobial activities, oxygen scavenging ability, enzyme immobilization or indication of degree of exposure to some detrimental factors (oxygen levels or inadequate temperatures). The present chapter represents application of different types of biocomposite polymer in food packaging technology with special interest on synthesis of gelatin/ZnO green composite polymer, its characterization and application in food packaging.

KEYWORDS

- antimicrobial properties
- biocomposite polymer
- biodegradation
- biopolymer/clay nanocomposite
- food packaging
- green material
- mechanical and barrier properties
- metal and metal oxide
- nanoclays
- nanocomposite material
- nanocomposites combined with antimicrobials

REFERENCES

1. Rhim, J. W., Park, H. M., & Ha, C. S. (2013). Bio-nanocomposites for food packaging applications. *Progress in Polymer Science, 38*, 1629–1652.
2. Singh, R. P., & Anderson, B. A. (2004). The major types of food spoilage: an overview. In: S. R. Woodhead (Ed.), Understanding and measuring the shelf-life of food, Woodhead Publishing Ltd., Cambridge, UK.
3. Singh, R. K., & Singh, N. (2005). Quality of packaged food. In: J. H. Han (Ed). *Innovations in Food Packaging*, Elsevier Academic Press, San Diego, USA.

4. Brown, H., & Willims, J. (2003). Packaged product quality and shelf life. In: R. Coles, D. Mcdowell, M. J. Kirwan (Ed.), *Food Packaging Technology*, Blackwell Publishing Ltd., UK.

5. Suppakul, P., Miltz, J., Sonnenveld, K., & Bigger, S. W. (2003). Active packaging technologies with an emphasis on antimicrobial packaging and its applications, *Journal of Food Science*. 68, 408–420.

6. Marsh, K., & Bugusu, B. (2007). Food packaging – roles, materials, and environmental issues. *Journal of Food Science*, 72, R38–55.

7. Weber, C. J., Haugaard, V., Festersen, R., & Bertelsen, G. (2002). Production and application of biobased packaging materials for the food industry. *Food Additive and Contamination*, 19, 172–177.

8. Sorrentino, A., Gorrasi, G., & Vittoria, V. (2007). Potentilal perspectives of bionanocomposites for food packaging applications. *Trends in Food Science and Technology*, 18, 84–95.

9. Siracusa, V., Rocculi, P., Romani, S., & Rossa, M. D. (2008). Biodegradable polymers for food packaging: a review. *Trends in Food Science and Technology,* 19, 634–643.

10. Zhao, R., Torley, P., and Halley, P. J. (2008). Emerging biodegradable materials: starch- and protein-based bio-nanocomposites. *Journal of Material Science*, 43, 3058–3071.

11. Bordes, P., Pollet, E., Avérous, L. (2009). Nano-biocomposites: biodegradable polyester/nanoclay systems. *Progress in Polymer Science*, 20, 125–155.

12. Chandra, I., & Rustgi, R. (1998). Biodegradable polymers. *Progress in Polymer Science*, 23, 1273–1335.

13. Scott, G. (2000). 'Green' Polymer, Polymer Degradation and Stability, 68, 1–7.

14. Trznadel, M. (1995). Biodegradable polymer materials. *International Polymer Science and Technology*, 22, 58–65.

15. Krochta, J. M., Mulder-Johnston, C. De (1997). Edible and Biodegradable Polymer Films: Challenges and Opportunities. *Food Technology,* 51, 61–74.

16. Siracusa, V., Rocculi, P., Romani, S., & Rossa, M. D. (2008). Biodegradable polymers for food packaging: a review. *Trends in Food Science and Technology*, 19, 634–643.

17. Bharadwaj, R. K. (2001). Modeling the barrier properties of polymer-layered silicate nanocomposites, *Macromolecules*, 34, 9189–9192.

18. Neilsen, L. E. (1967). Models for the permeability of filled polymer systems. *Journal of Macromolecular Science: Part A – Chemistry,* 1, 929–942.

19. Koh, H. C., Parka, J. S., Jeong, M. A., Hwang, H. Y., Hong, Y. T., Ha, S. Y., & Nam, S. Y. (2008). Preparation and gas permeation properties of biodegradable polymer/layered silicate nanocomposite membranes. *Desalination*, 233, 201–209.

20. Sorrentino, A., Tortora, M., and Vittoria, V. (2006). Diffusion behavior in polymer–clay nanocomposites. *Journal of Polymer Science Part B: Polymer Physics*, 44, 265–274.

21. Han, J. H. (2000) Antimicrobial food packaging. *Food Technology*, 54, 56–65.

22. Clarinval, A. M., & Halleux, J. (2005). Classification of biodegradable polymers. In: R. Smith (Eds.), *Biodegradable polymers for industrial applications*, Woodhead Publishing Ltd., Cambridge, UK.

23. Akbari, Z., Ghomashchi, T., & Moghadam S. (2007). Improvement in food packaging industry with biobased nanocomposites. *Journal of Food Engineering*, 3, 1–24.
24. Peterse, K., Nielsen P. V., Bertelsen, G., Lawther, M., Olsen, M. B., Nilsson, N. H., & Mortensen G. (1999). Potential of biobased materials for food packaging. *Trends in Food Science and Technology*, 10, 52–68.
25. Arora, A., & Padua, G. W. (2010). Review: nanocomposite in food packaging, *Journal of Food Science*. 75, 43–49.
26. Zhao, R., Torley, P., & Halley P. J. (2008). Emerging biodegradable materials: starch- and protein-based bio-nanocomposites, *Journal of Material Science*, 43, 3058–3071.
27. Bordes, P., Pollet, E., & Avérous, L. (2009). Nano-biocomposites: biodegradable polyester/nanoclay systems. *Progress in Polymer Science*, 20, 125–155.
28. Pandey, J. K., Kumar, A. P., Misra, M., Mohanty, A. K., Drzal, L. T., & Singh, R. P. (2005). Recent advances in biodegradable nanocomposites. *Journal of Nanoscience and Nanotechnology*, 5, 497–526.
29. Sinha Ray, S., & Bousmina, M. (2005). Biodegradable polymers and their layered silicate nanocomposites: in greening the 21st century materials world. *Progress in Materials Science*, 50, 962–1079.
30. Rhim, J. W., & Ng, P. K. (2007). Natural biopolymer-based nanocomposite films for packaging applications. *Critical Review in Food Science*, 47, 411–433.
31. Yang, K. K., Wang, X. L., & Wang, Y. Z. (2007). Progress in nanocomposite of biodegradable polymer, *Journal of Industrial and Engineering Chemistry*. 13, 485–500.
32. Pavlidou, S., & Papaspyrides, C. D. (2008). A review on polymer-layered silicate nanocomposites. *Progess in Polymer Science*, 33, 1119–1198.
33. Imran, H., Revol-Junelles, A. M., Martyn, A., Tehrany, E. A., Jacquot, M., Linder, M., & Desobry, S. (2010). Active food packaging evolution: transformation from micro- to nanotechnology. *Critical Review in Food Science and Nutrition*, 50, 799–821.
34. Huber, T., Müssig, J., Curnow, O., Pang, S., Bickerton, S., & Staiger, M. P. (2012). A critical review of all-cellulose composites. *Journal of Material Science*, 47, 1171–1186.
35. Faruk, O., Bledzki, A. K., Fink, H. P., & Sain, M. (2000–2010). Biocomposites reinforced with natural fibers. *Progress in Polymer Science*, 37, 1552–1596.
36. Wang, M. (2003). Developing bioactive composite materials for tissue replacement. *Biomaterials*, 24, 2133–2151.
37. Dicker, M. P. M., Duckworth, P. F., Baker, A. B., Francois, G., Hazzard, M. K., & Weaver, P. M. (2014). Green composites: A review of material attributes and complementary applications. *Composites: Part A*, 56, 280–289.
38. Sinha Ray, S., & Okamoto, M. (2003). Polymer/layered silicate nanocomposites: a review from preparation to processing. *Progress in Polymer Science*, 28, 1539–641.
39. Sinha Ray, S., Easteal, A., Quek, S. Y., & Chen, X. D. (2006). The potential use of polymer–clay nanocomposites in food packaging. *Journal of Food Engineering*, 2, 1–11.
40. Giannelis, E. P. (1996) Polymer layered silicate nanocomposites. *Advanced Material*, 8, 29–35.
41. Thostenson, E. T., Li, C. Y., & Chou, T. W. (2005). Nanocomposites in context. *Composite Science and Technology*, 65, 491–516.

42. Carter, L. W., Hendricks, J. G., & Bolley, D. S. (1950). Elastomer reinforced with a modified clay. US Patent 2,531,396, National Lead Co.

43. Deguchi, R., Nishio, T., & Okada, A. Polyamide composite material and method for preparing the same. US Patent 5,102,948,07.04.1992, Appl. 02.05.1990; EP 398 551 B1, 15.11.1995, Appl. 02.05.1990, Ube Industries, Ltd., and Toyota Jidosha Kabushiki Kaisha, Aichi, Japan.

44. Abdul Khalil, H. P. S., Hanida, S., Kang, S. C. W., & Nik Fuaad N. A. (2007). Agro-hybrid composite: The effects on mechanical and physical properties of oil palm fiber (EFB)/glass hybrid reinforced polyester composites. *Journal of Reinforced Plastic Composite*, 26, 203–218.

45. Abdul Khalil, H. P. S., Kumar, R. N., Asri, S. M., Nik Fuaad, N. A., & Ahmad, M. N. (2007). Hybrid thermoplastic pre-preg oil palm frond fibers (OPF) reinforced in polyester composites. *Polymer-Plastics Technology and Engineering*, 46, 43–50.

46. Abdul Khalil, H. P. S., Ismail, H., Ahmad, M. N., Ariffin, A., & Hassan K. (2001). Conventional agro-composites from chemically modified fibers. *Polymer International*, 50, 1.

47. Cho, J. W., & Paul, D. R. (2001). Nylon 6 nanocomposites by melt compounding. *Polymer*, 42, 1083–1094.

48. Abdul Khalil, H. P. S., Bhat, A. H., & Ireana Yusra, A. F. (2012). Green composites from sustainable cellulose nanofibrils: A review. *Carbohydrate Polymer*, 87, 963–979.

49. Smolander, M., & Chaudhry, Q. (2010). Nanotechnologies in food packaging. In: Chaudhry, Q., Castle, L., & Watkins R. (Ed.) Nanotechnologies in food. Cambridge, RSC Publishing, UK.

50. Moreira, M. R., Pereda, M., Marcovich, N. E., & Roura, S. I. (2011). Antimicrobial effectiveness of bioactive packaging materials from edible chitosan and casein polymers: assessment on carrot, cheese, and salami. *Journal of Food Science*, 76, 54–63.

51. de Oliveira, T. M., Soares, N. F. F., Pereira, R. M., & Fraga, K. F. (2007). Development and evaluation of antimicrobial natamycin-incorporated film in Gorgonzola cheese conservation. *Packaging Technology and Science*, 20, 147–53.

52. Kerry, J. P., O'Grady, M. N., & Hogan, S. A. (2006). Past, current and potential utilization of active and intelligent packaging systems for meat and muscle-based products: a review. *Meat Science*, 74, 113–130.

53. Cioffi, N., Torsi, L., Ditaranto, N., Tantillo, G., Ghibelli, L., Sabbatini, L., Bleve-Zacheo, T., D'alessio, M., Zambonin, P. G., & Traversa, E. (2005). Copper nanoparticle/polymer composites with antifungal and bateriostatic properties. *Chemistry of Materials*, 17, 5255–5262.

54. Hong, S. I., & Rhim, J. W. (2008). Antimicrobal activity of organically modified nanoclays. *Journal of Nanoscience and Nanotechnology*, 8, 5818–24.

55. Bi, L., Yang, L., Narsimhan, G., Bhunia, A. K., & Yao, Y. (2011). Designing carbohydrate nanoparticles for prolonged efficacy of antimicrobial peptide. *Journal of Controlled Release*, 150, 150–156.

56. Rhim, J. W., Hong, S. I., Park, H. M., Ng, P. K. W. (2006). Preparation and characterization of chitosan-based nanocomposite films with antimicrobial activity. *Journal of Agricultural and Food Chemistry*, 54, 5814–5822.

57. Wang, X., Du, Y., Yang, J., Wang, X., Shi, X., & Hu Y. (2006). Preparation, characterization and antimicrobial activity of chitosan/layered silicate nanocomposites. *Polymer, 47*, 6738–44.

58. Appendini, P., & Hotchkiss, J. H. (2002). Review of antimicrobial food packaging. *Innovative Food Science and Emerging Technology, 3,*113–126.

59. Nigmatullin, R., Gao, F., & Konovalova, V. (2008). Polymer-layered silicate nanocomposites in the design of antimicrobial materials. *Journal of Material Science, 43,* 5728–5733.

60. Persico, P., Ambrogi, V., Carfagna, C., Cerruti, P., Ferrocino, I., & Mauriello, G. (2009). Nanocomposite polymer films containing carvacrol for antimicrobial active packaging. *Polymer Engineering and Science.* 49, 1447–1455.

61. Travan, A., Marsich, E., Donati, I., Benicasa, M., Giazzon, M., Felisari, L., & Paoletti S. (2011). Silver-polysaccharide nanocomposite antimicrobial coatings for methacrylic thermosets. *Acta Biomaterialia, 7,* 337–346.

62. Carneiro, J. O., Texeira, V., Carvalho, P., & Azevedo, S. (2011). Self-cleaning smart nanocoatings. In: A. S. H. Makhlouf, I. Tiginyanu (Ed.) Nanocoatings and ultra-thin films, Woodhead Publishing Ltd., Cambridge, UK.

63. Garland, A. (2004). Nanotechnology in plastics packaging: commercial applications in nanotechnology. Pira International Limited, UK.

64. Selke, S. (2009). Nanotechnology and packaging. In: Yam, K. L. (Ed.) The Wiley encyclopedia of packaging technology. 3rd ed., John Wiley & Sons. Inc., Hoboken, New Jersey, USA.

65. Applerot, G., Lipovsky, A., Dror, R., Perkas, N., Nitzan, Y., Luhart, R., & Gedanken, A. (2009). Enhanced antibacterial activity of nanocrystalline ZnO due to increased ROS-mediated cell injury, *Advanced Functional Materials, 19*, 842–952.

66. Chawengkijwanich, C., & Hayata, Y. (2008). Development of TiO_2 powder coated food packaging film and its ability to inactivate *Esherrichia coli* in vitro and in actual test. *Internatonal Journal of Food Microbiology, 123,* 288–292.

67. Vartiainen, J., Tuominen, M., & Nättinen, K. (2010). Bio-hybrid nanocomposite coatings from sonicated chitosan and nanoclay. *Journal of Applied Polymer Science,* 116, 3638–47.

68. Paul, M. A., Delcourt, C., Alexandre, M., Degée, Ph., Monteverde, F., & Dubois, Ph. (2005). Polylactide/montmorillonite nanocomposites: study of the hydrolytic degradation. Polymer Degradation and Stability, 87, 535–542.

69. Sinha Ray, S., Yamada, K., Okamoto, M., & Ueda, K. (2003). Biodegradable polylactide/montmorillonite nanocomposites, J. Nanosci. Nanotech. 3, 503–510.

70. Sinha Ray, S., Yamada, K., Okamoto, M., & Ueda, K. (2003). New polylactidelayered silicate nanocomposites. 2. Concurrent improvements of materials properties, biodegradability and melt rheology. *Polymer,* 44, 857–866.

71. Sinha Ray, S., & Okamoto, M. (2003). New polylactide/layered silicate nanocomposites: open a new dimension for plastics and composites. *Macromolecular Rapid Communications*, 24, 815–840.

72. Jong, S. J. de, Arias, E. R., Rijkers, D. T. S., Van Nostrum, C. F., Kettenes-van, J. J., Kettenes-van den Bosch, J. J., & Hennik, W. E. (2001). New insights into the hydrolytic degradation of poly(lactic acid): participation of the alcohol terminus. *Polymer,* 42, 2795–2802.

73. Nieddu, E., Mazzucco, L., Gentile, P., Benko, T., Balbo, V., Mandrile, R., & Ciardelli G. (2009). Preparation and biodegradation of clay composite of PLA, *Reactive and Functional Polymers.* 69, 371–379.

74. Fukushima, K., Abbate, C., Tabuani, D., Gennari, M., & Camino, G. (2009). Biodegradation of poly(lactic acid) and its nanocomposites. *Polymer Degradation and Stability*, 94, 1646–1655.

75. Ozdemir, M., & Floros, J. D. (2004). Active food packaging technology. *Critical Review in Food Science*, 44, 185–193.

76. Han, J. H., Ho, C. H. L., & Rodrigues, E. T. (2005). Intelligent packaging. In: Han J. H. (Ed.) Innovations in food packaging, Elsevier Academic Press, San Diego, USA.

77. Yam, K. L., Takhistov, P. T., & Miltz, J. (2005). Intelligent packaging: concepts and applications. *Journal of Food Science*, 70, 1–10.

78. Lee, S. R., Park, H. M., Lim, H. T., Kang, K. Y., Li, L., Cho, W. J., & Ha, C. S. (2002). Microstructure, tensile properties, and biodegradability of aliphatic polyester/clay nanocomposites. *Polymer*, 43, 2495–2500.

79. Huang, M., & Yu, J. (2006). Structure and properties of thermoplastic corn starch/montmorillonite biodegradable composites. *Journal of Applied Polymer Science*, 99, 170–176.

80. Huang, M., Yu, J., & Ma, X. (2005). High mechanical performance MMT-urea and formamide-plasticized thermoplastic cornstarch biodegradable nanocomposites. *Carbohydrate Polymers,* 62, 1–7.

81. Chen, P., & Zhang, L. (2006). Interaction and properties of highly exfoliated soy protein/montmorillonite nanocomposites. *Macromolecules*, 7, 1700–1706.

82. Wang, S. F., Shen, L., Tong, Y. J., Chen, L., Phang, I. Y., Lim, P. Q., & Liu, T. X. (2005). Biopolymer chitosan/montmorillonite nanocomposites: preparation and characterization. *Polymer Degradation and Stability,* 90, 123–131.

83. Rao, Y. Q. (2007). Gelatin–clay nanocomposites of improved properties. Polymer. 48, 5369–5375.

84. Chivrac, F., Pollet, E., Schmutz, M., & Avérous, L. (2008) New approach to elaborate exfoliated starch-based nanocomposites. *Macromolecules*, 9, 896–900.

85. Tang, X. G., Kumar, P., Alavi, S., & Sandeep, K. P. (2012). Recent advances in biopolymers and biopolymer-based nanocomposites for food packaging materials. *Critical Review in Food Science*, 52, 426–442.

86. Rimdusit, S., Jingjid, S., Damrongsakkul, S., Tiptipakorn, S., & Takeichi, T. (2008). Biodegradability and property characterization of methyl cellulose: effect of nanocompositing and chemical cross-linking, *Carbohydrate Polymers.* 72, 444–455.

87. Roohani, M., Habibi, Y., Belgacem, N. M., Ebrahim, G., Karimi, A. M., & Dufresne, A. (2008). Cellulose whiskers reinforced polyvinyl alcohol copolymers nanocomposites. *European Polymer Journal,* 44, 2489–2498.

88. Rhim, J. W. (2011). Effect of clay contents on mechanical and water vapor barrier properties of agar-based nanocomposite films. *Carbohydrate Polymers*, 86, 691–699.

89. Rhim, J. W., Lee, S. B., & Hong, S. I. (2011). Preparation and characterization of agar/clay nanocomposite films: the effect of clay type. *Journal of Food Science*, 76, 40–48.

90. Rhim, J. W., Hong, S. I., & Ha, C. S. (2009). Tensile, water barrier and antimicrobial properties of PLA/nanoclay composite films. *Food Science and Technology-LWT*, 42, 612–617.

91. Choudalakis, G., & Gotsis, A. D. (2009). Permeability of polymer/clay nanocomposites: a review. *European Polymer Journal*, 45, 967–984.

92. Yano, K., Usuki, A., & Okad, A. (1997). Synthesis and properties of polyimide–clay hybrid films. *Journal of Polymer Science and Polymer Chemistry*, 35, 2289–2294.

93. Sothornvit, R., Rhim, J. W., & Hong, S. I. (2009). Effect of nano-clay type on the physical and antimicrobial properties of whey protein isolate/clay composite film. *Journal of Food Engineering*, 91, 468–473.

94. Sothornvit, R., Hong, S. I., An, D. J., & Rhim, J. W. (2010). Effect of clay content on the physical and antimicrobial properties of whey protein isolate/organo-clay composite films. *Food Science and Technol-LWT*, 43, 279–284.

95. Tang, X., Alavi, S., Herald, T. J. (2008). Barrier and mechanical properties of starch–clay nanocomposite films. *Cereal Chemistry*, 85, 433–439.

96. Kumar, P., Sandeep, K. P., Alavi, S., Truong, V. D., & Gorga, R. E. (2010). Preparation and characterization of bio-nanocomposite films based on soy protein isolate and montmorillonite using melt extrusion. *Journal of Food Engineering*, 100, 480–489.

97. Kumar, P., Sandeep, K. P., Alavi, S., Truong, V. D., & Gorga, R. E. (2010). Effect of type and content of modified montmorillonite on the structure and properties of bio-nanocomposite films based on soy protein isolate and montmorillonite. *Journal of Food Science*, 75, 46–56.

98. Tunc, S., Angellier, H., Cahyana, Y., Chalier, P., Gontard, N., & Gastaldi, E. (2007). Functional properties of wheat gluten/montmorillonite nanocomposite films processed by casting. *Journal of Membrane Science*. 289, 159–168.

99. Park, H. M., Lee, W. K., Park, C. Y., & Ha, C. S. (2003). Environmentally friendly polymer hybrids. Part I mechanical, thermal, and barrier properties of thermoplastic starch/clay nanocomposites. *Journal of Materials Science*. 38, 909–915.

100. Kim, M., & Pometto III, O. R. (1994). Food packaging potential of some novel degradable starch–polyethylene plastics. *Journal of Food Protection*, 57, 1007–1012.

101. Gacitua, W. E., Ballerini, A. A., & Zhang J. (2005). Polymer nanocomposites: synthetic and natural filers a review. *Maderas. Ciencia y tecnología*, 7, 159–178.

102. Cussler, E. L., Highes, S. E., Ward, W. J., & Aris, R. (1998). Barrier membranes. Journal of Membrane Science, 38, 161–174.

103. Park, H. M., Kim, G. H., & Ha, C. S. (2007). Preparation and characterization of biodegradable aliphatic polyester/thermoplastic starch/organoclay ternary hybrid nanocomposites. Composite Interfaces, 14, 427–438.

104. Park, H. M., Mohanty, A., Misra, M., & Drzal, L. T. (2004). Green nanocomposites from cellulose acetate bioplastic and clay: effect of ecofriendly triethyl citrate plasticizer. Biomacromolecules, 5, 2281–2288.

105. Dufresne, A., Dupeyre, D., & Vignon, M. R. (2000). Cellulose microfibrils from potato tuber cells: processing and characterization of starch–cellulose microfibril composites. Journal of Applied Polymer Science, 76, 2080–2092.

106. Biswas, M., & Ray, S. S. (2001). Recent progress in synthesis and evaluation of polymer-montmorillonite nanocomposites. Advanced Polymer Science, 155, 167–221.

107. Cabedo, L., Feijoo, J. L., Villanueva, M. P., Lagarón, J. M., & Giménez, E. (2006). Optimization of biodegradable nanocomposites based on PLA/PCL blends for food packaging applications. Macromolecular Symposia. 233, 191–197.

108. Gorrasi, G., Tortora, M., Vittoria, V., Galli, G., & Chiellini, E. (2002). Transport and mechanical properties of blends of poly(epsilon-caprolactone) and a modified montmorillonite-poly(epsilon-caprolactone) nanocomposite. Journal of Polymer Science, Part B: Polymer Physics. 40, 1118–1124.

109. Meera, A. P., Thomas, P. S., & Thomas, S. (2012). Effect of organoclay on the gas barrier properties of natural rubber nanocomposites. Polymer Composite, 33, 524–531.

110. Gorrasi, G., Tortora, M., Vittoria, V., Pollet, E., Alexandre, M., & Dubois, P. (2004). Physical properties of poly(-caprolectone) layered silicate nanocomposites prepared by controlled grafting polymerization, *Journal of Polymer Science; Part B.* 42, 1466–1475.

111. Sorrentino, A., Gorrasi, G., Tortora, M., Vittoria, V., Costantino, U., Marmottini, F., & Pasella, F. (2005). Incorporation of Mg–Al hydrotalcite into a biodegradable poly (epsilon, caprolectone) by high energy ball milling, *Polymer*. 46, 1601–1608.

112. Tajima, T., Suzuki, N., Watanabe, Y., & Kanzaki, Y. (2005). Intercalation compound of diclofenac sodium with layered inorganic compounds as a new drug material. *Chemical and Pharmaceutical Bulletin.* 53, 1396–401.

113. Gorrasi, G., Tortora, M., Vittoria, V., Pollet, E., Lepoittenvin, B., & Alexandre, M. (2003). Vapor barrierproperties of polycaprolactone montmorillonite nanocomposites: effect of clay dispersion. *Polymer*, 44, 2271–2279.

114. Quintavalla, S., & Vicini, L. (2002). Antimicrobial food packaging in meat industry, *Meat Science.* 62, 373–380.

115. Zhang, L., Jiang, Y., Ding, Y., Daskalakis, N., Jeuken, L., Povey, M., O'Neill, A. J., & York, D. W. (2010). Mechanistic investigation into antibacterial behavior of suspensions of ZnO nanoparticles against E. coli. *Journal of Nanoparticle Research*, 12, 1625–1636.

116. Emamifar, A., Kadivar, M., Shahedi, N., & Soleimanian-Zad, S. (2011). Effect of nanocomposite packaging containing Ag and ZnO on inactivation of Lactobacillus plantarum in orange juice. *Food Control,* 22, 408–413.

117. Friedman, M., & Junesa, V. K. (2010). Review of antimicrobial and antioxidative activities of chitosans in food, *Journal of Food Protection.* 73, 1737–1761.

118. Cioffi, N., Torsi, L., Ditaranto, N., Tantillo, G., Ghibelli, L., Sabbatini, L., Bleve-Zacheo, T., D'alessio, M., Zambonin, P. G., & Traversa, E. (2005). Copper nanoparticle/polymer composites with antifungal and bateriostatic properties. *Chemistry Materials,* 17, 5255–5262.

119. Damm, C., Münsted, H., & Rösch, A. (2008). The antimicrobial efficacy of polyamide 6/silver nano- and microcomposites. *Materials Chemistry and Physics,*108, 61–66.

120. Carlson, C., Hussain, S. M., Schrand, A. M., Braydich-Stolle, L. K., Hess, K. L., Jones, R. L., & Schlager, J. J. (2008). Unique cellular interaction of silver nanoparticles: size-dependent generation of reactive oxygen species. *The Journal of Physical Chemistry,* 112, 13608–13619.

121. Dallas, P., Sharma, V. K., & Zboril, R. (2011). Silver polymeric nanocomposites as advanced antimicrobial agents: classification, synthetic paths, applications, and perspectives. *Advances in Colloid and Interface Science,* 166, 119–135.

122. Russell, A. D., & Hugo, W. B. (1994). Antimicrobial activity and action of silver. *Progress in Medical Chemistry,* 31, 351–370.

123. Butkus, M. A., Edling, L., & Labare, M. P., (2003). The efficacy of silver as a bactericidal agent: advantages, limitations and considerations for future use. *Journal of Water Supply: Research and Technology*. 52, 407–416.

124. Feng, Q. L., Wu, J., Chen, G. Q., Cui, F. G., Kim, T. N., & Kim, J. O. (2000). A mechanistic study of the antibacterial effect of silver ions on *Escherichia coli* and *Staphylococcus aureus*. *Journal of Biomedical Material Research*, 52, 662–668.

125. Kim, J. S., Kuk, K. E., Yu, K. N., Kim, J. H., Park, S. J., Lee, H. J., Kim, S. H., Park, Y. K., Park, Y. H., Hwang, C. Y., Kim, Y. K., Lee, Y. S., Jeong, D. H., & Cho, M. H. (2007). Antimicrobial effects of silver nanoparticles. *Nanomedicine*, 3, 95–101.

126. Sondi, I., & Salopek-Sondi, B. (2004). Silver nanoparticles as antimicrobial agent: a case study of E. coli as a model for Gram-negative bacteria. *Journal of Colloid and Interface Science*, 275, 177–182.

127. Fernández, A., Soriano, E., Hernández-Munoz, P., & Gavara, R. (2010). Migration of antimicrobial silver from composite of polylactide with silver zeolite. *Journal of Food Science*, 75, 186–193.

128. Egger, S., Lehmann, R. P., Height, M. J., Loessner, M. J., & Schuppler, M. (2009). Antimicrobial properties of novel silver-silica nanocomposite material. *Applied Environmental Microbiology*, 75, 2973–2976.

129. Incoronato, A. L., Buonocore, G. G., Conte, A., Lavorgna, M., & Del Nobile, M. A. (2010). Active systems based on silver-montmorillonite nanoparticles embedded into bio-based polymer matrices for packaging applications. *Journal of Food Protection*, 73, 2256–2262.

130. Egger, S., Lehmann, R. P., Height, M. J., Loessner, M. J., & Schuppler, M. (2009). Antimicrobial properties of novel silver-silica nanocomposite material. *Applied Environmental Microbiology*, 75, 2973–2976.

131. Valodkar, M., Bhadoria, A., Pohnerkar, J., Mohan, M., Thakore, S. (2010). Morphology and antibacterial activity of carbohydrate-stabilized silver nanoparticles. *Carbohydrate Research*, 345, 1767–1773.

132. Du, W. L., Niu, S. S., Xu, Y. L., Xu, Z. R., Fan, C. L. (2009). Antibacterial activity of chitosan tripolyphosphate nanoparticles loaded with various metal ions. *Carbohydrate Polymers*, 75, 385–389.

133. Yoksan, R., & Chirachanchai, S. (2010). Silver nanoparticle-loaded chitosanstarch based films: fabrication and evaluation of tensile, barrier and antimicrobial properties. *Material Science and Engineering*, 30, 891–897.

134. Vimala, K., Mohan, Y. M., Sivudu, K. S., Varaprasad, K., Ravindra, S., Reddy, N. N., Padma, Y., Sreedhar, B., & Mohana Raju., K. (2010). Fabrication of porous chitosan films impregnated with silver nanoparticles: a facile approach for superior antibacterial application. *Colloids and Surfaces B: Biointerfaces*, 76, 248–258.

135. Tripathi, S., Mehrotra, G. K., & Dutta, P. K. (2011). Chitosan-silver oxide nanocomposite film: preparation and antimicrobial activity, B. Material Science, 34, 29–35.

136. Dong, C., Song, D., Cairney, J., Maddan, O. L., He, G., & Deng, Y. (2011). Antibacterial study of Mg(OH)$_2$ nanoplatelets. *Bulletin of Materials Science*, 46, 576–582.

137. Applerot, G., Perkas, N., Amirian, G., Girshevitz, O., Dedanken, A. (2009). Coating of glass with ZnO via ultrasonic irradiation and a study of its antibacterial properties. *Applied Surface Science*, 256, 3–8.

138. Fujishima, A., Rao, T. N., & Truk, D. A. (2000). Titanium dioxide photocatalysis. *Journal of Photochemistry and Photobiology C: Photochemistry Reviews*. 1, 1–21.

139. Celik, E., Gokcen, Z., Ak Azem, N. F., Tanoglu, M., & Emrullahoglu, O. F. (2006). Processing, characterization and photocatalytic properties of Cu doped TiO_2 thin films on glass substrate be sol–gel technique. *Materials Science and Engineering: B,* 132, 258–265.

140. Zhang, W., Chen, Y., Yu, S., Chen, S., & Yin, Y. (2008). Preparation and antibacterial behabior of Fe^{3+}-doped nanostructured TiO_2 thin films. Thin. Solid Films, 516, 4690–4694.

141. Sikong, L., Konggreong, B., Kantachote, D., & Sutthisripok, W. (2010). Photocatalytic activity and antibacterial behavior of Fe^{3+}-doped TiO_2/SnO_2 nanoparticles. *Journal of Energy Research*, 1, 120–125.

142. Kubacka, A., Serrano, C., Ferrer, M., Lunsdorf, H., Bielecki, P., Cerrada, M. L., Fernández-Garcia, M., & Fernández-Garcia, M. (2007). High-performance dualaction polymer-TiO_2 nanocomposite films via melting processing, *Nano Letters,* 7, 2529–2534.

143. Helander, I. M., Nurmiaho-Lassila, E. L., Ahvenainen, R., Rhoades, J., & Roller, S. (2001). Chitosan disrupts the barrier properties of the outer membrane of Gram-negative bacteria. *International Journal of Food Microbiology*, 71, 235–244.

144. Wang, X., Du, Y., Luo, J., Lin, B., & Kennedy, J. F. (2007). Chitosan/organic rectorite nanocomposite films: structure, characteristic and drug delivery behavior. *Carbohydrate Polymers*, 69, 41–49.

145. Wang, X., Du, Y., Luo, J., Yang, J., Wang, W., & Kennedy, J. F. (2009). A novel biopolymer/rectorite nanocomposite with antimicrobial activity. *Carbohydrate Polymers,* 77, 449–456.

146. Ohashi, F., Oya, A., Duclaux, L., & Beguin, F. (1998). Structural model calculation of antimicrobial and antifungal agents derived from clay minerals. *Applied Clay Science*, 12, 435–445.

147. Patakfalvi, R., & Dékány, I. (2004) Synthesis and intercalation of silver nanoparticles in kaolinite/DMSO complexes. *Applied Clay Science*, 25, 149–159.

148. Whilton, N. T., Burkett, S. L., & Mann, S. (1998). Hybrid lamellar nanocomposites based on organically functionalized magnesium phillosilicate clay with interlayer reactivity. *Journal of Materials Chemistry,* 8, 1927–1932.

149. Patil, A. J., Muthusamy, E., & Mann, S. (2005). Favrication of functional proteinorganoclay lamellar nanocomposites by biomolecule-induced assembly of exfoliated aminopropyl functionalized magnesium phillosilicates. *Journal of Materials Chemistry*, 15, 3838–3843.

150. Chandrasekaran, G., Han, H. K., Kim, G. J., & Shin, H. J. (2011). Antimicrobial activity of delaminated aminopropyl functionalized magnesium phyllosilicates. *Applied Clay Science*, 53, 729–736.

151. Xie, A. G., Cai, X., Lin, M. S., Wu, T., Zhang, X. J., Lin, Z. D., & Tan, S. (2011). Long-acting antibacterial activity of quaternary phosphonium salts fuctionalized few-layered graphite. *Material Science and Engineering*, 176, 1222–1226.

152. Sanpui, P., Murugadoss, A., Prasad, P. V. D., Ghosh, S. S., & Chattopadhyay, A. (2008). The antibacterial properties of a novel chitosan-Agnanoparticle composite. *International Journal of Food Microbiology,* 124, 142–146.

153. Wei, D., Sun, W., Qian, W., Ye, Y., & Ma, X. (2009). The synthesis of chitosan-based silver nanoparticles and their antibacterial activity. Carbohydrate Research, 344, 2375–2382.

154. Ahmad, M. B., Shameli, K., Darroudi, M., Yunus, W. M. Z. W., & Ibrahim, N. A. (2009). Synthesis and characterization of silver/clay/chitosan bionanocomposites by UV-irradiation method. *American Journal of Applied Sciences, 6*, 2030–2035.

155. Diaz-Visurraga, J., García, A., & Cárdenas, G. (2010). Lethal effect of chitosan-Ag (I) films on Staphylococcus aureus as evaluated by electron microscopy. *Journal of Applied Microbiology, 108*, 633–646.

156. Li, L. H., Deng, J. C., Deng, H. R., Liu, Z. L., & Li, X. L. (2010). Preparation, characterization and antimicrobial activities of chitosan/Ag/ZnO blend films. *Journal of Chemical Engineering, 160*, 378–382.

157. Tetto, J. A., Steeves, D. M., Welsh, E. A., & Powell, B. E. (1999). Biodegradable poly(ε-caprolactone)/clay nanocomposites, *ANTEC'99*. 1628–1632.

158. Sasmal, A., Nayak, P. L., & Sasmal, S. (2009). Degradability studies of green nanocomposites derived from soy protein isolate (SPI)-furfural modified with organoclay. *Polymer Plastic Technology, 48*, 905–909.

159. Maiti, P., Batt, C. A., & Giannelis, E. P. (2007). New biodegradable polyhydroxybutyrate/layered silicate nanocomposites. *Biomacromolecules, 8*, 3393–3400.

160. Tang, S., Zou, P., Xiong, H., & Tang, H. (2008). Effect of nano-SiO$_2$ on the performance of starch/polyvinyl alcohol blend films. *Carbohydrate Polymers, 72*, 521–526.

161. Maiti, P., Batt, C. A., & Giannelis, E. P. (2003). Renewable plastics: synthesis and properties of PHB nanocomposites. *Journal of Macromolecular Science – Reviews, 88*, 58–59.

162. Wu, T. M., & Wu, C. Y. (2006). Biodegradable poly(lactic acid)/chitosan-modified montmorillonite nanocomposites: preparation and characterization. *Polymer Degradation and Stability, 91*, 2198–2204.

163. Someya, Y., Kondo, N., & Shibata, M. (2007). Biodegradation of poly(butylenes adipate-co-butylene terephthalate)/layered-silicate nanocomposites. *Journal of Applied Polymer Science, 106*, 730–736.

164. Shankar, S., Teng, X., Li, G., & Rhim, J. W. (2015). Preparation, characterization, and antimicrobial activity of gelatin/ZnO nanocomposite films. *Food Hydrocolloid, 45*, 264–271.

CHAPTER 8

VALUE ADDED PRODUCT RECOVERY FROM FRUIT AND VEGETABLE WASTE THROUGH BIOTECHNOLOGICAL INTERVENTION

ANSHU SINGH

CONTENTS

8.1 INTRODUCTION

In 2011, the Food and Agriculture Organization of the United Nations reported that roughly 1.3 billion tons (= one-third of the edible parts of food produced globally) is lost or wasted every year. In 2013, the USDA and EPA launched the US Food Waste Challenge, which calls on entities along the food chain to join efforts to reduce, recover, and recycle food waste (Figure 8.1).

Food waste generation is becoming a serious problem for large-scale industries because of limited landfill sites and strict regulation to be followed for its proper disposal. There are two ways by which waste management at food industry level can be managed: one is by reducing at the processing level and second is through using the residual matter either as a raw material for secondary processes or as ingredients of new products. Reduction at source requires a redesigning of the whole process, which is costlier. Therefore in such scenario, process techniques to obtain value-added products from these waste opens up a sustainable new avenue for the industry.

This chapter focuses on the value added product recovery from fruit and vegetable waste through biotechnological intervention.

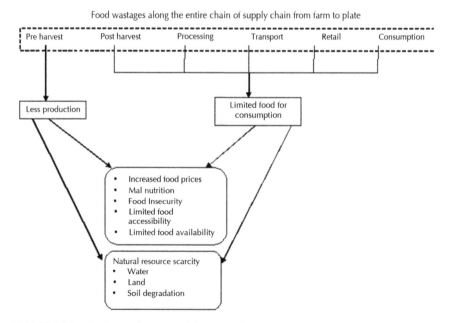

FIGURE 8.1   Socio–environmental impact of wastages generated during food supply chain.

8.2 SOURCES OF FOOD WASTES

The various sources of food wastes (Figure 8.2) are domestic/household, industrial and agricultural waste, which are generated during food supply chain [28, 29].

8.2.1 AGRICULTURAL WASTE

The lack of proper harvesting technique and improper postharvest operation is responsible for the release of agro-food waste.

8.2.2 INDUSTRIAL WASTE

During food manufacturing processes, both solid and liquid wastes are generated. Industrial waste generation covers a large section of waste generation as it includes major levels of food supply chain. Improper storage and transportation is responsible for rotten portion, peels, shells, scraped portions of fruits and vegetables forms the solid waste generated due to processing while slurries and washed water are liquid processing wastes. Improper packaging also leads to waste generation.

8.2.3 DOMESTIC/HOUSEHOLD WASTES

Domestic food wastes are generated through scraping and peeling. Apart from these factors over-purchasing, bulk-buying, cooking too much, over

FIGURE 8.2 Various type of waste generated during food supply chain from field to fork.

date based on labeling are also few reasons, which add to domestic food wastes. Often it has been observed that these household are discarded with other non-organic waste therefore their separation for efficient bioconversion is limited.

8.3   CHARACTERIZATION OF FOOD WASTE

The determination of the amount of waste, components and nutritional value is an important step of an overall waste management strategy. Various research studies have suggested that identification and quantification are critical step for selecting the processing steps [1, 9]. Pumpkin and spinach wastes anaerobic digestion produced 0.373 liter and 0.269 liter of methane, respectively [26]. This indicates that the difference in the production pattern to a large extent depends on the nature of the substrate. Smaller quantity of waste production will hamper the recycling as surplus supply of material will be a problem. Similarly identification provides the knowledge of components present in waste. Biochemical characterization is an advantage during the selection of substrate for microbial fermentation.

8.3.1   PHYSICAL CHARACTERIZATION OF FOOD WASTE

Physical characterization of waste includes estimation of weight, volume, moisture, ash and total solid (volatile solid + fixed solids). The moisture content and the bulk density of the sample is important because large fluctuations in either of these moisture content reflect a significant difference in waste composition. For waste rich in solid residue, the total solid content gives idea about the origin, age or previous treatment while for the liquid waste, estimation of dissolved and suspended solid is done. Carrot waste has maximum suspended solids than apple, green pea and tomato waste.

8.3.2   CHEMICAL CHARACTERIZATION OF FOOD WASTE

The study on chemical characterization includes: measurement of carbon to nitrogen ratio, calorific value, cellulose, hemicellulose, starch, reducing sugars, protein, total organic carbon, phosphorus, nitrogen, biological oxygen demand (BOD), chemical oxygen demand (COD), pH, halogens, toxic metals, etc. [28]. All these chemicals analyzes provide insight on the applicability of waste for value added product development. Most of the wastes are solid in nature therefore, depending upon biochemical composition they are used as substrates for fermentation. The rate and extent of microbial decomposition depends upon characteristics of initial feedstock. A common feature of various forms of food wastes includes high COD and BOD. Apple pomace have high BOD (9.6g O_2/L) and COD (18.7g O_2/L) than grapefruit pomace. Among vegetable waste, carrot waste has BOD of 1.82g O_2/L and COD of 2.3g O_2/L, which is greater than that of green peas and tomatoes.

8.4 BIOTECHNOLOGY INTERVENTION

The major biotechnological intervention in food waste processing includes fermentation and use of enzyme. Fermentation is whole cell mediated process where live microbes are used for the bioconversion. The Figure 8.3 provides an overview, how we can add value to the food waste by applying the biotechnological techniques.

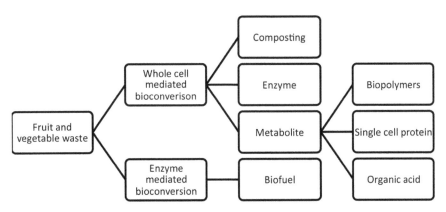

FIGURE 8.3   Value added product recovered through biotechnological intervention from food waste.

8.4.1 WHOLE CELL MEDIATED BIOCONVERSION, FERMENTATION TECHNOLOGY

Fermentation is the technique where microbial organisms are used for enhancing the properties. During the process, apart from increasing the nutritional quality of the food taste, aroma and even texture also gets enhanced. This process has also been used to enhance the shelf life of the products. Microbes are added as inoculants referred to as starter cultures.

It has been reported that the food waste is composed of 30–60% starch, 5–10% proteins and 10–40% (w/w) which make these perfect choice for fermentation. Value added products can be recovered from the waste through solid state fermentation or submerged fermentation as described in Figure 8.4.

8.4.1.1 Enzyme Productions Through Fermentation Using Food Waste as Raw Material

Most of the valuable enzyme production is done by the solid state fermentation process. The fermentation is performed on non-soluble food waste

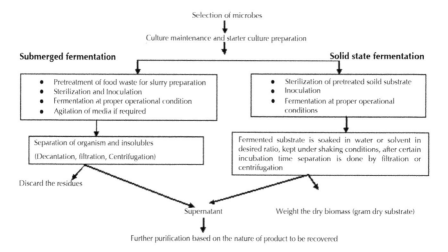

FIGURE 8.4 Process scheme from production of value added product from food waste through fermentation.

material which acts as a source of nutrient for the microbes (bacteria, fungus) in absence of free flowing media.

These porous substrates maintain a proper condition for microbial growth and while growing microbes secrete enzymes. Before using the waste as raw material as substrate pretreatment is necessary. The pretreatment can be mechanical or physical such as cooked, dried, and ground prior to use in solid state fermentation. The parameters which need to be considered during the process are particle size, moisture and fermentation parameters: pH, temperatures, nutrients, inoculation size, and incubation period. Various food wastes have been studied and used for the enzyme production (Table 8.1).

TABLE 8.1   Enzymes Produced From Food Waste

Enzymes	Agro-food waste
Amylase	Wheat bran, rice straw, rice bran, red gram husk, spent brewing grains, jowar straw, coconut oil cake, jowar spathe
Cellulase	Rice bran, wheat bran, orange peel, pineapple peel and pulp, rice straw, corn straw, wheat straw, spent hulls of pulses, rice husk
Chitinase	Wheat bran, rice husk, sugarcane bagasse, grape fruit peel
Laccase	Wheat straw, barley straw, apple peel, orange peel, potato peelings, barley bran, kiwi fruit waste, banana skin, rice straw, wheat bran
Lipase	Ground nut oil cake, Linseed and mustard oil cake, coconut oil cake, wheat bran, soybean oil cake, olive oil cake, sesame oil cake, rice bran
Pectinase	Orange peel, banana peel, lemon peel pomace, citrus peel, orange bagasse, mango peel, apple pomace, grape pomace, wheat bran, sugarbeet pulp,
Protease	Soybean cake, lentil husk, wheat bran, rice husk, rice bran, spent brewing grain, coconut oil cake, palm kernel cake, sesame oil cake, jackfruit seed powder, olive oil cake, green gram husk, chick pea husk, jatropha seed oil cake, red gram husk, black gram husks
Tannase	Pomegranate peel, spent tea leaves powder, tamarind seed powder, palm kernel, ber leaves, amla leaves, wheat bran, rice bran, rice straw dust, sugarcane bagasse, keekar leaves powder, jamun leaves powder, coconut coir, banana peel, orange bagasse, black grapes peel
Xylanase	Sugarcane bagasse, wheat bran, rice straw, rice bran, soybean hull, grapes pomace, sorghum straw, apple pomace, cranberry pomace, strawberry pomace

8.4.1.2   Metabolite Production

Currently a number of value added product has been recovered from the fruit and vegetable waste via fermentation process such as organic acids, amino acids, vitamins, single cell protein and bio polymers (Table 8.2).

8.4.1.3   Production of Different Biochemicals Utilizing Wastes Through Fermentation Techniques

Activities in this sector are still in the inceptive stages. Under mentioned are few of the citations indicating the wastes that have been tried and the array of biochemicals that have been retrieved from vegetable waste.

8.4.1.3.1   Asparagus waste

Asparagus officinalis L. is a well-known healthy vegetable for its anti-oxidant properties [33]. The dried old stalk of the asparagus is discarded as waste. Wang et al. [36] evaluated the potential of asparagus straw as compost for mushroom (*Agaricus blazei* Murrill) production. The results

TABLE 8.2   Various Types of Metabolite Produced From Fruit and Vegetable Wastes [29]

Type of metabolite		Waste used
Organic acid	Citric acid	Beet pulp, pineapple pulp, pineapple wastes, apple pomace, grape pomace, guava kiwi orange or prickly pear peels
	Lactic acid	Liquid potato waste, potato starch waste
	Oxalic acid	Sugar beet molasses
	Acetic acid	Carrot and radish waste
Single cell protein		Cucumber peels, mango and apple residues, banana peel, pomegranate peel, orange peels,
Biopolymer	Polyhydroxybutrate	Potato processing waste
	Pullulan	Grape pomace
	Xanthan	Grape pomace, orange peel, sugarbeet pulp, olive mill wastewater

showed that the combination of cottonseed hull or cow dung with aspar-agus waste is good manure for better polysaccharides yield in fruiting bodies.

8.4.1.3.2 Cabbage waste

The cabbage waste consists of raw rotten heads and the outer part of Chi-nese cabbage, which is usually removed before sale. Trimming process for product development in factories also generates waste. Chanda and Chakrabarti [3] reported that the juice from the cabbage waste is rich source of sugar than the deproteinized leaf juice of turnip, mustard, or radish wastes.

Major portion of Chinese cabbage waste contains water up to 90%, which was used by Choi et al. [6] for single cell protein production. The pretreated juice was evaluated for the yeast growth. The results suggested that pretreated juices are better substrate for production of microbial bio-mass rich in protein without any nutrient supplementation. Choi et al. [6] reported that the increase in cell biomass can be achieved by supplemen-tation of waste juice by ammonium sulfate, corn steep powder, and zinc sulfate.

Another form of cabbage waste is brine, produced from Kimchi, a fer-mented vegetable product. During the kimchi production, cabbage heads are soaked in brine for overnight. The vegetables are collected and the remaining part comes out as waste. Choi et al. [5] studied the growth of osmotolerant yeast for single cell protein (SCP) production. *Pichia guil-liermondii* A9 cell mass reached 0.69 g of dry cells/L containing 40% of protein.

8.4.1.3.3 Carrot waste

The carrot is a root vegetable often consumed as juice. During the juice extraction, the carrot pomace/pulp are produced by industries. With the rise in demand for fresh juices, the large quantities of carrot pomace/pulp make up the total waste. Vrije et al. [35] utilized carrot pulp for biohydrogen pro-duction by using two extreme thermophilic bacteria. *Caldicellulosiruptor saccharolyticus* actively utilized the carrot pulp while *Thermotoga neapolitana* was unable to grow under agitation. Pretreatment of carrot pulp before the organism inoculation resulted in 10% better biohydrogen yield. Now a days entomopathogenic fungi are in use as biopesticide for insect species. Tincilley et al. [31] reported that agricultural byproduct can be a substrate for *Nomuraea rileyi*, an entomopathogenic fungus. Sahayaraj and Namasivayam [27] also assessed the potential of several agricultural waste residues including vegetable wastes for production of entomopathogenic fungi. Carrot waste showed maximum mass production followed by jack seeds and lady's finger wastes, respectively.

8.4.1.3.4   Onion waste

Onion (*Allium cepa* L.) waste includes the outer dry layers and the apical parts of onion bulbs with some industrial wastewater. These wastes are not suitable for animal feed or landfill disposal because of its characteristic strong flavor. Romano et al. [23] investigated onion waste for biomethane production and utilized a two-phase anaerobic digestion system for the first time. Similarly, Romano and Zhang [24] explored anaerobic mixed biofilm reactor for biogas energy production, by co-digestion of onion waste juice and aerobic wastewater produced from onion industries. Bulk mass of rotten or deformed onions, rejected from the industries could be a source of vinegar production as reported by Horiuchi et al. [10, 12] *Acetobacter aceti* TUA549B converted the alcohol produced by action of *Saccharomyces cerevisiae* to vinegar found to be rich in potassium with 1.6–6.9 times amino acid content. The onion residues left after the onion juice extraction were used as compost by Horiuchi et al. [11]. Onion wastes were further characterized for their ability to produce biogas by Lubberding et al. [19]. The volatile fatty acid production was much higher during the anaerobic fermentation process.

8.4.1.3.5   Peas waste

Peas wastes mainly composed of shells rich in organic molecules have proved to be a promising substrate for biogas production. Kalia et al. [13]

used pea shell slurry at 1–5% total solid (TS) concentration for production of methane and hydrogen. Hydrogen was produced 33- 46%, while methane was 45 % of total biogas. Ensiled green pea shell slurry, after the removal of fibrous material, was used for methane production [20]. The output was 0.62 m³/kg VS biogas yield with 8% (TS) loading rate.

8.4.1.3.6   Tomato waste

Processing of tomato generates liquid and solid wastes. Solid wastes include skin and seeds of tomato which are found to be rich in nutrients including carotenoids. Tomato waste has been found to be a potential source of energy through anaerobic digestion. Trujillo et al. [32] performed anaerobic digestion of tomato plant waste mixed with rabbit wastes. It was found that addition of tomato plant waste to rabbit wastes in proportions higher than 40% improved the methane production. In a separate study by Viswanath [34], different vegetable wastes including tomato wastes were used for biogas production and a major yield of 74.5% was found within 12 h of feeding.

The solid tomato waste can be potential candidate for biotransformation into organic nutrient rich product for agricultural purposes. Fernandez-Gomez et al. [7] reported that tomato waste could be successfully converted into chemically stable and nutrient enriched biomass by vermicomposting. Freixo et al. [8] carried out production of laccase and xylanase from tomato pomace using a strain of *Coriolus versicolor*. Liu et al. [18] reported production of polyhydroxyalkanoate (PHA), biodegradable thermoplastics, during treatment of tomato cannery waste. A maximum of 20% PHA content was obtained using mixed microbial culture employing a two stage production process.

8.4.1.3.7   Sugar beet waste

The sugar industries using sugar beet as raw material generate organic wastes at various stages of sugar production. The major by-products are sugar beet tailings, beet pulp and beet molasses. The presence of high amount of organic substances- cellulose, sugars, etc. in the generated waste, makes it a suitable substrate for microbial conversion into useful product.

Fermentation technology has been used to improve the nutritional value (increase the protein content) of sugar beet residues to be used as feed. Nigam and Vogel [21] performed bioconversion of beet molasses and sugar beet pulp using 4 different *Candida* species (*utilis, tropicalis, parapsilosis* and *solani*) and found that *C. utilis* and *C. tropicalis* performed better. In 48 hours, maximum protein content of 37.5% and 43.4% was observed in molasses and sugar beet pulp, respectively. Sugar beet vinasse was used for production of high value protein rich mass of *Spirulina maxima* by Barrocal [2]. Both in batch and continuous methods, culture positive growth of *Spirulina* was observed and the biomass concentration was 3.5 g/L and 4.8 g/L, respectively. Fermentation of sugar beet lignocellulosics also yields enzymes with economic importance. Roche et al. [22] produced α-L- arabinofuranosidase using *Trichoderma reesei* by bioconversion of sugar beet pulp. At the end of fermentation (165 hours), the production of extracellular enzyme was found to be high and stabilized with activity of 433 IU/g of dry fermented medium.

Production of extracellular arabinanase using growth medium consisting of sugar beet pulp and corn steep liquor by *Fusarium oxysporium* was done by Cheilas et al. [4]. Maximum activity of 0.25 U/mL was observed. Production of pullulan was done by Roukas et al. [25] using sugar beet molasses as substrate in stirred tank bioreactor. This biopolymer is used for coating, packaging, food and cosmetic industries and for medicinal applications. Maximum polysaccharide concentration of 23 g/L was achieved by this process. Yoo and Harcum [37] reported production of xanthum gum from waste sugar beet pulp (WSBP). The amount of xanthum gum generated was 67 to 89% from degraded WSBP.

8.4.2   ENZYME MEDIATED BIOCONVERSION

In this process, enzymes are either used for extraction of biochemical present in waste or through biocatalysis they convert any substrate present in waste to high value biochemicals. Enzyme mediated extraction of carotenoids, from tomato waste resulted in better yields by the use of pectinase and cellulose enzymes in comparison to solvent extraction process [30].

Fruits and vegetables waste are becoming promising raw materials to recover phenolic compounds with good antioxidant activity because the discarded upper portions of vegetables/fruits such as skins, peels, hulls and seeds have better content of phenolic compounds. As these compounds mostly remain bounded to the other biomolecules in cell wall therefore, addition of cell wall degrading enzymes plays prime role in their release. Carbohydrate cleaving enzymes such as pectinases, cellulases and hemicellulases have been well studied for the purpose of maximum recovery of phenolic antioxidant (Table 8.3).

Enzymes are also used for the pretreatment process for the biofuel generation. Laccase, cellulase, hemicellulase enzymes have been well used during the bioethanol generation from the fruit or vegetable waste.

8.5 CONCLUSIONS

Proper waste handling and disposal system is important for the safety of human being and the environment. Recycling and use of these food wastes for production of useful products can be a better approach towards achieving it. This can be done by development of cheap, eco-friendly and simple techniques for better utilization of generated waste. Use of biotechnology fulfills the above requirement and its potential needs to be further explored.

TABLE 8.3 Waste Used For Phenolic Antioxidant Recovery

Type of Waste	Enzyme used	Ref.
Apple peel	Cellulase	[16]
Blackcurrant pomace	Cellulase	[15]
Citrus peels	Celluzyme MX	[18]
Grape pomace	Pectinolytic and cellulolytic enzymes	[14]
Raspberry pomace	Cellulase, hemicellulase and pectinase	[17]

8.6 SUMMARY

Foods being perishable in nature get easily spoiled from its transformation as raw material at farm to proper food till consumption at plate. 30% of incoming raw material is generated as a waste during product development. Harvesting, post-harvest, transportation, storage, marketing and product development a large amount and various types of residual matter are generated. Disposal of these sugar rich residues have direct impact on environment because of deterioration and spoilage. It has been estimated that 4.5 ton of carbon dioxide is released from per ton of food waste. Indirectly, it also affects overall processing cost of the industries due to its disposal cost. Biotechnological intervention provides a solution to recycle the waste into value added product by adding profitability to the process.

There are two strategies to reuse the waste, one is through whole cell mediated, i.e., use of microbial species to convert waste and the other is enzyme mediated. Integrated model where waste of one operation is raw material for another is a one of the well-accepted approach to derive value added product from food waste. Such approach has given a new dimension to extract high value compounds/ biochemicals from the waste in an eco-friendly and economically viable manner.

KEYWORDS

- **biotechnology**
- **enzymes**
- **fermentation**
- **food waste**
- **recovery**
- **recycling**
- **reuse**
- **solid state fermentation**
- **value added products**

REFERENCES

1. Abu-Qdais, H. A., Hamoda, M. F., & Newham, J. (1997). Analysis of residential solid waste at generation sites. *Waste Manage. Res.*, 15, 395–406.
2. Barrocal, V. M., Garcia-Cubero, M. T. Gonzalez-Benito, G., & Coca, M. (2010). Production of biomass by *Spirulina maxima* using sugar beet vinasse in growth media. *New Biotechnol.*, 27, 851–856.
3. Chanda, S., & Chakrabarti, S. (1996). Plant origin liquid waste, a resource for single-cell protein production by yeast. Bioresource Technol., 57, 51–54.
4. Cheilas, T., Stoupis, T., Christakopoulos, P., Katapodis, P., Mamma, D., Hatziniko-laou, D. G., Kekos, D., & Macris, B. J. (2000). Hemicellulolytic activity of *Fusarium oxysporum* grown on sugar beet pulp, production of extracellular arabinanase. *Proc. Biochem.,* 35, 557–561.
5. Choi, M. H., & Park, Y. H. (1999). Growth of *Pichia guilliermondii* A9, an osmotolerant yeast, in waste brine generated from kimchi production. Bioresource Technol., 70, 231–236.
6. Choi, M. H., Ji, G. E., Koh, K. H., Ryu, Y. W., Jo, D. H., & Park, Y. H. (2002). Use of waste Chinesecabbageas a substrate foryeast biomass production. Bioresource Technol., 83, 251–253.
7. Fernandez-Gomez, M. J., Nogales, R., Insam, H., Romero, E., & Goberna, M. (2010) Continuous-feeding vermicomposting as a recycling management method to revalue tomato-fruit wastes from greenhouse crops. *Waste Manage.,* 30, 2461–2468.
8. Freixo, M. R., Karmali, A., Frazao, C., & Arteiro, J. M. (2008). Production of laccase and xylanase from *Coriolus versicolor* grown on tomato pomace and their chromatographic behavior on immobilized metal chelates. *Proc. Biochem.,* 43, 1265–1274.
9. Green, J. H., & Kramer, A. (1979). Food processing waste management. The AVI publishing company, INC. Westport, Connecticut, USA, pp. 1–14.
10. Horiuchi, J. I., Kanno, T., & Kobayashi, M. (1999). New vinegar production from onions. *J. Biosci. Bioeng.,* 88, 107–109.
11. Horiuchi, J. I., Tada, K., Kobayashi, M., Kanno, T., & Ebie, K. (2004). Biological approach for effective utilization of worthless onions: vinegar production and composting. *Resource Conservat. Recyc.,* 40, 97–109.
12. Horiuchi, J. I., Yamauchi, N., Osugi, M., Kanno, T., Kobayashi, M., & Kuriyama, H. (2000). Onion alcohol production by repeated batch process using a flocculating yeast. *Bioresource Technol.,* 75, 153–156.
13. Kalia, M. K., Srilatha, H. R., Srinath, K., Bharathi, K., & Nand, K. (1997). Production of methane from green pea shells in floating dome digesters. *Proc. Biochem.,* 32, 509–513.
14. Kammerer, D., Claus, A., Schieber, A., & Carle, R. (2005). A novel process for the recovery of polyphenols from grape (*Vitis vinifera, L.*) pomace. *J. Food Sci.,* 70, C157–C163
15. Kapasakalidis, P. G., Rastall, R. A., & Gordon, M. H. (2009). Effect of a cellulase treatment on extraction of antioxidant phenols from black currant (*Ribes nigrum, L.*) pomace. *J. Agric. Food Chem.,* 57, 4342–4351.

16. Kim, Y. J., Kim, D. O., Chun, O. K., Shin, D. H., Jung, H., Lee, C. Y., & Wilson, D. B. (2005). Phenolic extraction from apple peel by cellulases from *Thermobifida fusca*. *J. Agric. Food Chem.*, 53, 9560–9565.

17. Laroze, L., Soto, C., & Zúñiga, M. (2010). Phenolic antioxidants extraction from raspberry wastes assisted by enzymes.*Electr. J. Biotechnol.*, 13, 1–11.

18. Li, B. B., Smith, B., & Hossain, M. D. M. (2006). Extraction of phenolics from citrus peels, II. Enzyme-assisted extraction method. *Separation and Purification Technol.*, 48,189–196.

19. Lubberding, H. J., Gijzen, H. J., Heck, M., & Vogels, G. D. (1988). Anaerobic digestion of onion waste by means of rumen microorganisms. *Biol. Wastes.*, 25, 61–67.

20. Marouani, L., Bouallagui, H., Ben, C. R., & Hamdi, M. (2002). Biomethanation of green wastes of wholesale market of Tunis. In: Proceedings of the international symposium on environmental pollution control and waste management, Tunis, pp. 318–323.

21. Nigam, P., & Vogel, M. (1991). Bioconversion of sugar Industry by-products-molasses and sugar beet pulp for single cell protein production by yeasts. *Biomass Bioenerg.*, 1, 339–345.

22. Roche, N., Berna, P., Desgranges, C., & Durand, A. (1995). Substrate use and production of α-L-arabinofuranosidase during solid-state culture of *Trichoderma reesei* on sugar beet pulp. *Enzyme Microb. Technol.*, 17, 935–941.

23. Romano, R., Zhang, R., & Hartman, K. (2004). Anaerobic digestion of onion wastes using a continuous two-phase anaerobic solids digestion system. Paper presented at ASAE/CSAE, Annual International Meeting. Ottawa, Ontario, Canada. ASAE Paper Number 047070.

24. Romano, R. T., & Zhang, R. (2007). Co-digestion of onion juice and wastewater sludge using an anaerobic mixed biofilm reactor. Bioresource Technol., 99, 631–637.

25. Roukas, T., & Liakopoulou-Kyriakides, M. (1999). Production of pullulan from beet molasses by *Aureobasidium pullulans* in a stirred tank fermentor. *J. Food. Eng.*, 40, 89–94.

26. Sagagi, B. S., Garba, B., & Usman, N. S. (2009). Studies on biogas production from fruits and vegetable waste. *Bajopas.*, 2, 115–118.

27. Sahayaraj, K., & Namasivayam, S. K. R. (2008). Mass production of entomopathogenic fungi using agricultural products and by products. Afr. J. Biotechnol., 7, 1907–1910.

28. Singh, A., Kuila, A., Adak, S., Bishai, M., & Bancrjcc, R. (2012a). Utilization of vegetable wastes for generation of bioenergy. *Agric. Res.*, 2, 1–10.

29. Singh, A., Kuila, A., Adak, S., Bishai, M., & Banerjee, R. (2012b). Use of fermentation technology on vegetable residues for value added product development: A concept of zero waste utilization. *Int. J. Food Fermentation Technol.*, 2, 173 –184.

30. Strati, I. F., Gogou, E., & Oreopoulou, V. (2015). Enzyme and high pressure assisted extraction of carotenoids from tomato waste. *Food Bioprod. Process.*, 94, 668–674.

31. Tincilley, A., Easwaramoorthy, G., & Santhanalakshmi, G. (2000). Attempts on mass production of *Nomuraea rileyi* on various agricultural products and byproducts. *J. Biol. Contr.*, 18, 33–40.

32. Trujillo, D., Perez, J. F., & Cebre, F. J. (1993). Energy recovery from wastes, anaerobic digestion of tomato plant mixed with rabbit wastes. *Bioresource Technol.,* 45, 81–83.

33. Vinson, J. A., Hao, Y., Su, X., & Zubik, L. (1998). Phenol antioxidant quantity and quality in foods, vegetables. *J. Agric. Food Chem.,* 46, 3630–3634.

34. Viswanath, P. (1992). Anaerobic digestion of fruit and vegetable processing wastes for biogas production. *Bioresource Technol.,* 40, 43–48.

35. Vrije, T., Budde, M. A. W., Lips, S. J., Bakker, R. R., Mars, A. E., & Claassen, P. A. M. (2010). Hydrogenproduction fromcarrotpulp by the extreme thermophiles *Caldicellulosiruptor saccharolyticus* and *Thermotoga neapolitana. Int. J. Hydrogen Energy.,* 35, 13026–13213.

36. Wang, Q., Li, B. B., Li, H., & Han, J. R. (2010). Yield, dry matter and polysaccharides content of the mushroom *Agaricus blazei* produced on asparagus straw substrate. *Sci. Hort.,* 125, 16–18.

37. Yoo, S. D., & Harcum, S. W. (1999). Xanthan gum production from waste sugar beet pulp. *Bioresource Technol.,* 70, 105–109.

CHAPTER 9

AN OVERVIEW OF BIOPROCESSING AND BIOREFINERY APPROACH FOR SUSTAINABLE FISHERIES

VEGNESHWARAN V. RAMAKRISHNAN, WINNY ROUTRAY, and DEEPIKA DAVE

CONTENTS

9.1   INTRODUCTION

Fish industry plays a major part in providing macro and micro-nutrients including highly beneficial proteins, carbohydrates, bioactive peptides, vitamins, enzymes, enzyme inhibitors, glycosoaminoglycans (GAG), collagen, gelatin, oils and omega-3 fatty acids [78, 110, 128]. The dietary consumption of different species of groundfish, finfish and shellfish is governed by various factors including popularity of various species as a food ingredient, health beneficial effects and ease of availability of the species as a food commodity in the markets [172]. Popularity of any food is decided by the sensory properties, which mainly depends on the bio-chemical composition decided by the climate of the breeding area, composition and quality of food available to the species. Health beneficial effects depend on the biochemical composition, their levels and the bioavailability of these compounds depending on physio-chemical composition of different species consumed as food. Availability of the different species depends on the rate of production varying with seasons, landscape, growth pattern, adaptability, migration, life cycle, biology of the different species and transportation and storage facility in the area.

Increasing globalization has led to change in the food market, food availability and food preferences. Presently, different varieties of fish are available world-wide which might not be locally grown or processed. This increases the nutrient accessibility and availability in other parts of the world with no access to various seafood and fisheries. This eventually creates demand leading to superfluous production in fisheries and aquaculture industries and helps nutritional food components economically available in other parts of the world. Also, the demands for processed forms of fish such as pickled, canned, dried and salted are increasing, which are either processed at the on sites or fish processing facility. However, different steps of unit operations lead to processing losses, which further increases with increase in the number of unit operation. In certain parts of the world, fish processing facilities are well equipped with all the technologies and equipment. However, in some parts with economic restrictions and climatic, geographical and technological constraints, most of the processing might depend on human labor and expertise, which lead to increase in fish processing waste.

Fish and shellfish are usually processed before being used for human consumption and generates unutilized byproducts up to 30–80% of the body weight of the processed fish and shellfish, which consists of shell, head, frames, fins, tails, skin and gut. Hence, fish production, handling and processing should be properly modulated or controlled to minimize the waste generation, proper storage facilities should be provided in different parts to slow down the bacteriological, chemical and physical degradation process. At present, there is limited capacity available to handle these unutilized resources in terms of composting, mink feed, rendering, ocean dumping and landfilling. Landfills are being rapidly filled and water resources are vulnerable to eutrophication and death of other aquatic animals due to oxygen deficiency in the water. This also leads to increased biological and chemical oxygen demand of the waste present in the water.. These fish by-products are also a rich source of essential high value nutraceuticals and pharmaceuticals with various biological properties including: collagen, gelatin, chitin, chitosan, carotenoids, glucosamine, protein, amino acids, bioactive peptides, omega-3 fatty acids, glycosoaminoglycans (GAGs), oil, calcium and enzymes.

Fish byproducts typically comprises of 58% proteins and 19% other extract or fat, in addition to minerals, such as manganese, phosphorous, magnesium, potassium, calcium, , sodium, iron, zinc, and copper. Therefore, unutilized fish and shellfish byproducts could be a great potential source for good quality nutraceuticals and pharmaceuticals that can be utilized for human intake rather than designate and dispose them as waste [3, 89, 99]. After the fish oil has been extracted, residual biomass can be used as a raw material for biodiesel, biomethane and bioethanol production via fermentation, chemical or enzymatic transesterification. These factors have necessitated the need for the development of biorefinery and bioprocessing approach for the production of various high value nutraceuticals and pharmaceuticals from fish and shellfish by-products thereby maximizing the sustainability and economic viability of the industry.

This chapter focus on the facts related to fish and shell fish industry and summarizes processing technologies for different categories of fish, food, feed and other products derived from different fishes and fishery byproducts; and the bioprocessing and biorefinery concept for complete utilization of fisheries waste resources for the sustainability and economic viability of the industry.

9.2 PROCESSING TECHNIQUES FOR DIFFERENT CATEGORIES OF FISH AND WASTE BY-PRODUCTS PRODUCED DURING PROCESSING

Fish can be classified into three main categories including: groundfish, pelagic and finfish and shellfish. Each of the three different categories has different techniques have distinctive processing parameters and therefore produce effluent with distinctive characteristics.

9.2.1 FISH PROCESSING

The phrase "fish processing" can include all types of groundfish and fin-fish processing. Most of the fish processing techniques include (a) stunning of fish, grading, (b) removal of slime, (c) scaling, (d) washing, (e) deheading, (f) gutting, (g) cutting of fins, (h) slicing into steaks, (g) fileting, (h) meat-bone separation, (i) packaging, (j) labeling and distribution [64]. Variations in the fish processing techniques are implemented according to different sites, however, the most of the processing flow in the plant remains constant. During any fish processing operation, there are two types of waste released, liquid waste and solid waste.

9.2.1.1 Groundfish

Groundfish category includes cod, haddock, redfish spp, flatfishes, greenland turbot, pollock, hake, cusk, catfish, skate, dogfish, etc., except halibut. Generally, halibut are stored whole or processed and washed before storage onboard Generally, these types of fish do not necessitate any treatment before subjected to fileting operation. Most groundfish processing plants use mechanized equipment, where the fish are washed in large wash tanks and passed on to the fileting machines. The skin is removed from the filet and inspected before plate freezing, packaging and storage. The typical groundfish fileting operation is described in the Figure 9.1 [7]. The solid waste released during groundfish processing includes skin, head, gut, frames, tails and fins. Solid waste is either processed and rendered for pet food or dumped into oceans and landfills. The liquid waste is wash water and approximate range for different properties of this water are shown in Table 9.1 [14].

9.2.1.2 Pelagic and Finfish

The pelagic and finfish includes herring, mackerel, swordfish, tuna, alewife, eel, salmon, smelt, silversides, shark, capelin, etc. Herring are round

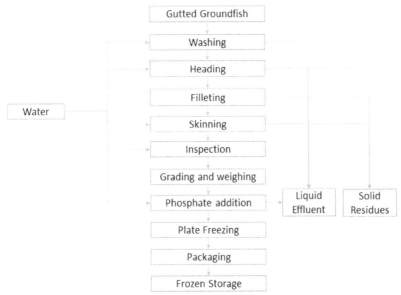

FIGURE 9.1 Groundfish processing operation [7]. (Modified from AMEC (2003). *Management of Wastes from Atlantic Seafood Processing Operations* p. 40, http://coinatlantic.ca/documents/aczisc_miscellaneous%20_documents/nparpt.pdf)

on site and delivered to processing plant for fileting and the heads, tails, fins and viscera are removed by automated machines.

The offal is later transported to the reduction plant for processing into fish meal or into pet food. Salmon processing involves removal of head and gutting of the fish prior to further processing. The manual salmon processing line usually consists of a specific task assigned for each worker including deheading, gutting and removal of internal organs. The processed fish is then cleaned with water to remove blood and remaining parts. The tails, fins and the collarbone attached to the head are not cut off. Prior to freezing and packaging, the salmon is subjected to a smooth coating of clear ice glaze [7].

Processing of the other types of pelagic and finfish is carried out in similar ways with slight variations in the processing techniques. Since, the pelagic fish have higher content of sarcoplasmic proteins compared to other fish, there is a rapid decrease in the pH after processing. Therefore, during washing addition of 0.5% sodium bicarbonate to water improves the quality of the fish mince and elevates the texture quality. The pelagic

TABLE 9.1   Inputs and Outputs of Fish Production Processing [14]

Process	Inputs		Outputs					
	Fish (kg)	Energy (kW h)	Wastewater (m³)	BOD (kg)	COD_5 (kg)	N (kg)	P (kg)	Solid waste (kg)
Deheading of white fish	1000	0.3–0.8	1	-	2–4	-	-	Head and debris: 270–320
De-icing and washing	1000	0.8–1.2	1		0.7–4.9			0–20
Filleting of deheaded white fish	1000	1.8	1–3	-	4–12	-	-	Frames and off cuts: 200–300
Filleting of ungutted oily fish	1000	0.7–2.2	1–2	-	7–15	-	-	Entrails, tails, heads and frames: 400
Freezing and storage	1000	10–14	-	-	-	-	-	-
Frozen fish thawing	1000	-	5	-	1–7	-	-	-
Grinding	1000	0.1–0.3	0.3–0.4	-	0.4–1.7	-	-	0–20
Handling and storage of fish	1000	10–12	-	-	130–140	-	-	-
Oily fish fileting	1000	Ice: 10–12 Freezing: 50–70 Filleting: 2–5	5–8	50	85	2.5	0.1–0.3	400–450
Packaging of filets	1000	5–7.5	-	-	-	-	-	-
Scaling of white fish	1000	0.1–0.3	10–15	-	-	-	-	Scales: 20–40

TABLE 9.1 (Continued)

Process	Inputs		Outputs					
	Fish (kg)	Energy (kW h)	Wastewater (m³)	BOD (kg)	COD₅ (kg)	N (kg)	P (kg)	Solid waste (kg)
Skinning oily fish	1000	0.2–0.4	0.2–0.9	-	3–5	-	-	Skin: 40
Skinning white fish	1000	0.4–0.9	0.2–0.6	-	1.7–5	-	-	Skin: 40
Trimming and cutting of white fish	1000	0.3–3	0.1	-	-	-	-	-
Unloading of fish	1000	3	2–5	-	27–34	-	-	-
White fish fileting	1000	Ice: 10–12	5–11	35	50	-	-	Skin: 40–50
		Freezing: 50–70						Heads: 210–250
		Fillet- ing: 5						Bones: 240–340

mince production process with the addition of sodium bicarbonate is described in the Figure 9.2 [97, 139].

9.2.1.3 Shellfish

Shellfish category includes clams, oyster, scallop, squid, mussels, lobster, shrimp, crab, whelks, cockles, sea cucumber, sea urchin, etc. Lobsters are kept alive until processing and they are marketed with their shells intact either live or cooked. The lobsters are steam cooked and water cooled before handling, packaging and shipping. The lobsters are sometimes butchered, in which the backs are removed and the remaining viscera are washed free. The wash-water contains significant amounts of solids and organic pollutants [7, 52].

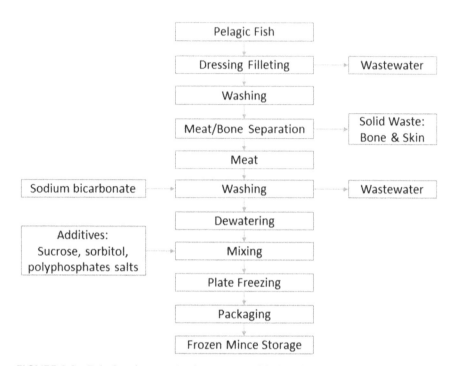

FIGURE 9.2   Pelagic mince production process with the addition of sodium bicarbonate [139]. (Modified from Pan, B. S. [1990]. Minced fish technology. In: *Seafood: Resources, Nutritional Composition, and Preservation*. pp. 190–210.)

The steps involved in shrimp processing is shown in Figure 9.3. The shrimp is processed with heads or without heads. The raw shrimp is usually held on ice for 2 days to allow internal enzymes and bacteria to degrade the tissue present between the shell and meat and to assist efficient removal of shell. During this process, the deterioration caused by the enzyme increases the water holding capacity and bacterial load on the shrimp tissues. The iced shrimp is later introduced to potable water to melt the ice and is sent into the pre-cooker conveyor. In the pre-cooking process, the shrimp is subjected to live steam to provide optimum peeling and recovery of meat and the microbial load is also significantly reduced.

The shrimp is then sent onto the oscillating peelers to remove the shell and meat. The water is sprayed and the waste shells are washed away. The shrimp is washed with potable water and immersed in salt solution and hand-packed into cans, vacuum-sealed, refrigerated or frozen. Shrimp heads and shells are usually discarded as waste from processing. The

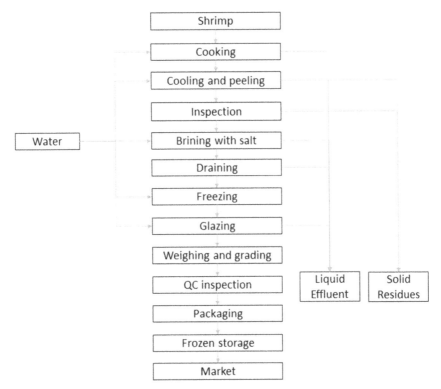

FIGURE 9.3 Typical shrimp processing [7]. (Modified from AMEC (2003). *Management of Wastes from Atlantic Seafood Processing Operations* p. 53, http://coinatlantic.ca/ documents/aczisc_miscellaneous%20_documents/nparpt.pdf)

crabs are usually cooked with salted water and cooled with fresh water. The crabmeat from body and claws is either manually or mechanically removed. The crab shell is discarded as waste [7, 52]. Other shellfish species are processed in similar ways with some variations, however they are not well documented as shrimp and crab, which are the most popularly consumed shellfishes.

9.2.2   CHARACTERISTICS OF FISH AND SHELLFISH PROCESSING WASTE

During any fish processing operation, there are two types of waste released including, liquid waste and solid waste.

9.2.2.1 Liquid Waste

During any fish processing operation, washing using fresh water is an integral part to remove accumulated bacteria on the fish surface. The quality of washing depends upon the fish to water ratio, water quality and kinetic energy of the water stream and most of the processing industries use fish to water ratio of 1:1 for washing. However, the amount of water used increases by two fold during processing. Generally, washing is carried out for 1–2 min in the vertical drum, horizontal drum and conveyor belt washers. These processors can be used to process whole fish, deheaded and gutted fish as well as fish filets, and does not cause any physical damage to the product. The processed water from processing is collected in the waste basins. The amount of wastewater produced during each step in the fish processing, canning and fish meal is shown in Tables 9.1–9.3 [14, 64]. During shellfish processing, lower BOD and nitrogen concentrations are expected due to the lack of offal, blood, slime and skin contents. The contaminant characteristics during shellfish processing is shown in Table 9.4 [7].

9.2.2.2 Solid Waste

The solid waste released during fish includes: head, tails, skin, gut, fins, and frames and shellfish processing includes: shell and head. For some fish species, the solid waste from fileting plants can generate 30–60% solid waste and shellfish processing plants can generate 75–80% solid waste. These by-products of the fish and shellfish processing industry can be a great source of value added products such as proteins and amino acids, collagen and gelatin, oil, minerals, pigments, chitin, chitosan and enzymes. Currently, the fish processing waste are being converted into fishmeal and fertilizers. However, most of these by-products are disposed of in landfills or disposed of in ocean under permit. These wastes contain proteins (58%), ether extract or fat (19%) and minerals. Also, monosaturated acids, palmitic acid and oleic acid are abundant in fish waste (22%) as shown in Table 9.5 [50]. The average dry proximate content of the shrimp shells, which is discarded as waste includes: ash (34%) consisting mainly of calcium carbonate ($CaCO_3$), protein (35–40%) and chitin (17–20%) [152].

TABLE 9.2 Inputs and Outputs of Fish Canning Process [14]

Process	Inputs		Outputs					
	Fish (kg)	Energy (kW h)	Wastewa- ter (m³)	BOD (kg)	COD₅ (kg)	N (kg)	P (kg)	Solid waste (kg)
Can sealing	1000	5–6	-	-	-	-	-	-
Canning	1000	150– 190	15	52	116	3	0.1– 0.4	Head: 250
								Bones: 100–150
Draining of cans containing precooked fish	1000	0.3	0.1–0.2	-	3–10	-	-	-
Nobbing and packing in cans	1000	0.4–1.5	0.2–0.9	-	7–15	-	-	Head and entrails: 150
								Bones and meat: 100–150
Precooking of fish to be canned	1000	0.3–11	0.07–0.27	-	-	-	-	Inedible parts: 150
Sauce filling	1000	-	-	-	-	-	-	Spillage of sauce and oil: varies
Sterilization of cans	1000	230	3–7	-	-	-	-	-
Unloading fish for can- ning	1000	3	2–5	-	27–34	-	-	-
Washing of cans	1000	7	0.04	-	-	-	-	-

TABLE 9.3   Inputs and Outputs of Fish Meal Production Process [14]

Process	Inputs		Outputs					
	Fish (kg)	Energy (kW h)	Wastewater (m³)	BOD (kg)	COD$_5$ (kg)	N (kg)	P (kg)	Solid waste (kg)
Cooking of fish	1000	90	-	-	-	-	-	-
Drying of press cake	1000	340.0	-	-	-	-	-	-
Fish meal and fish oil	1000	Electricity: 32	-	-	-	-	-	-
Fish oil polishing	1000	Hot water	0.05–0.1	-	5	-	-	-
Pressing the cooked fish	1000	-	750 kg water / 150 kg oil	-	-	-	-	Press cake: 100 dry matter
Stick water evaporation	1000	475	-	-	-	-	-	Concentrated stick water: 250 / Dry matter: 50

9.3   FISH INDUSTRY BASED FOOD PRODUCTS

Fish based products are available in different forms depending on the targeted market and the variety or type of fish being processed. The different fish based food products are summarized in the following subsections along with the summary of the extra unit operation steps followed for the purpose.

9.3.1   FRESH PRODUCTS

Most fishery products are preferred to be consumed in their fresh forms. However, in different parts of the world, the marketed form of different

TABLE 9.4 Contaminant Concentrations of Shellfish Processing Plant Effluents [7]

Species	BOD (mg/L)	COD (mg/L)	TS (mg/L)	TSS (mg/L)	TKN (mg/L)	NH$_3$-N (mg/L)
Blue crab	10000–14000	20000–25000	18000–25000	700–1000		200–250
Clam wash-water		637–3590	2528–3590		113–260	
Crab	4100	29000		95		
Crab & crab sections				210		
Crab meat				170		
Crab process	181–1281	320–2940	1040–1814	80–815 11429 (VSS)	23–166	6–13.6
Oyster	164–576	164–1000	240–400	50–284	224–91	20–10
Oyster	510		2280			
Oyster	310 (tot) 282 (flt)	407 (tot) 5–57 (flt)		12–11 (VSS)		
Scallop	580–1250	544–3184		31–1905		15.5–37.5
Scallop shucking		1965	9867	350	420	
Scallop shucking		1965	9867		420	
Shellfish	290–380 (flt), 280–1075 (tot)	250–738 (flt) 485–1623 (tot)	776–2000	125–825 120–81 (VSS)	36–45	6–15
Shrimp				2900		
Shrimp		3400–6500	1900–2000			
Shrimp canning	1070			550		
Shrimp packing	112–340	131–360	50–500	22–200	22.4–59.4	1.8–13.8
Shrimp processing	416–857		115–357			
Shrimp processing	530–1240 (tot.) 330–530 (sol.)		240–660			

Note: Flt – filtered; tot – total; sol – soluble; VSS – Volatile Suspended Solids; TS- Total solids; TSS- Total soluble solids; TKN-; NH$_3$-N. (Modified from AMEC (2003). *Management of Wastes from Atlantic Seafood Processing Operations* pp. 1–93. http://coinatlantic.ca/documents/aczisc_miscellaneous%20_documents/nparpt.pdf)

TABLE 9.5   Composition of Fish Waste [50]

Nutrient	Fish waste
Ash (%)	21.79 ± 3.52
Calcium (%)	5.80 ± 1.35
Copper (ppm)	1.00 ± 1.00
Crude fiber (%)	1.19 ± 1.21
Crude protein (%)	57.92 ± 5.26
Fat (%)	19.10 ± 6.06
Iron (ppm)	100.00 ± 42.00
Magnesium (%)	0.17 ± 0.04
Manganese (ppm)	6.00 ± 7.00
Phosphorous (%)	2.04 ± 0.64
Potassium (%)	0.68 ± 0.11
Sodium (%)	0.61 ± 0.08
Zinc (ppm)	62.00 ± 12.00

Note: Values in % or mg/kg (ppm) on a dry matter basis

fish sold by the traders varies depending on the preference of the local consumers. At some places, fish is sold in the live form, where they are caught and their metabolism is systematically slowed down after sorting according to the appropriate required/standard quality and transported to the places of consumption, where consumer can buy the live fish and process themselves [172]. However, at some other places, the fish are stunned and killed and kept on ice in their whole form to be sold in the retail market within next few days of the catch. Also in other parts of the world and in cases of larger finfish such as salmon, systematic processing of fish is conducted with separation of the head, guts and frames from the commercially consumed before releasing the product into the retail market. In certain cases, fish steaks are further skinned and cut into filets for ready to cook purposes. In case of molluscs and crustaceans too, depending on market demand, they are harvested and transported in their live forms and supplied to the customers to process further or processed and then released into the market.

9.3.2 SLICED AND MINCED PRODUCTS

Finfish is also sold in cut, sliced, minced or diced form. In many markets there are provision of appropriate small-scale machines for the cutting and processing of different fish products depending on the market opinion and demand. Large scale processing is also conducted with modulated processing lines, which consume a lot of water, human and machinery resources and generates a significant amount of biological waste as discussed in the previous sections. In most cases of food processing with the addition of any extra processing steps, the consumption of resources increases including storage, and packaging requirements and increases the possibility of degradation of the food material as it is more and more exposed to the environment. Fish which are minced or sliced are generally suggested to be done as soon as possible after the catch, and are suggested to be conducted in extremely hygienic facilities and under controlled conditions [1]. With appropriate packaging and storage conditions, the food quality can be maintained, however, it increases the possibility of creating a huge amount of biological waste in case of failure of any facility, machine or any other lapse in the quality control. Most of these sliced or products are vacuum packed and stored in the cold storage before being available on the market shelves.

9.3.3 PROCESSED PRODUCTS

Some of the cut/ diced products are also available in their freshly marinated forms to make it ready to cook, as preferred by many consumers. Processed fish can be further processed and sold in pickled, cured, dried, smoked, frozen or other preserved forms. Some of the products such as whelks and crabs are available in the market in their semi-cooked or cooked forms, which is the normal marketed form of these products. In most of the cases further processing is either done to increase the products variability in the market which increases options for the customers, make them cost effective or is done to increase the shelf life. Over the time processed products such as dried, pickled, smoked, or salted forms of fish products were developed to make available the nutrients during the seasons with no fish production.

Processed products such as surimi consist of myofibrillar proteins recovered by washing minced fish. This is a refined form of mechanically deboned fish meat and has distinct physio-chemical properties including gel-forming ability, water and oil binding properties. Surimi research projects are concentrating mainly on new sources of fish and underutilized fish with high nutritional value, which can be converted to optimized products with maximum food value [180]. Pacific pollock is one of the most applied fish variety for the purpose. This product was first developed in Japan, which is still one of the biggest consumer market for the product. Japan is also a significant market for other products such as boiled, grilled or fried fish patties [148], which have become popular products in other parts of the world too.

Processed fish products involve many unit operation steps including multiple steps of thawing, heating and cooking. These processes, when conducted imperfectly lead to increased amount of losses. Low temperature cooking or low concentrations of salt while processing should be avoided to prevent growth of different micro-organism leading to spoilage [1]. Furthermore, high temperature of processing and high salt concentration lead to degradation of the quality of product by increased liquid release and change in texture due to protein denaturation and increased oxidation of the fat molecules. Hence, researchers are still working on the new processing technologies or optimizing the established technologies depending on the biochemical and physio-chemical nature of different products.

High pressure processing is one of the latest recommended processing methods which has been successfully applied in case of several food products for inactivation of enzyme and microbes leading to increase in shelf-life, however high pressure application has been reported to increase lipid oxidation in muscle tissues [136]. High pressure processing has also been reported to have denaturation effects on the myofibrillar protein, degrees of which vary with variation in the pressure level. Effect of pressure on myfibrillar protein of the fish varies with temperature, pH, ionic strength, pressure and time levels. According to Chapleau et al. [35], aggregation of protein structure was observed at pressures higher than 300 MPa based on the study of protein conformation by quasielastic light scattering and gel filtration. Modification of the tertiary and quaternary structures of proteins are also expected to induce a molten globule state [35], which implies that application of this method for fish protein needs further optimization.

High pressure processing has also been applied for oyster shulking, and the study indicated that application of high pressure reduced the initial microbial load by 2 to 3 logs and counts remained lower throughout the storage period of 27 days, thus increasing the shelf-life [74]. In this study, high pressure treated oysters also received higher quality scores as compared to the control samples. Based on these observations, it can be concluded that effect of the processing method varies with the method, parameters of the method and the sample on which processing is applied.

Different processing technologies including extrusion processing [38] and thermal processing, which is also applied for canning [158] and is also used for production of processed food products from the processed fish filets and fish processing by-products suitable for human consumption. Canning is applied for the processing and preservation of different species most popularly for small pelagics, tuna, mackerel, and also for molluscs and crustacea. There are other technologies which have been developed and applied to increase the shelf-life of the fish products especially by decreasing the microbial content such as irradiation technology [61], high pressure thawing applicable for food products with high water content [55], and modified atmosphere packaging [73] for preservation of maximum quality.

Fish processing and different related techniques have been discussed in detail by George Hall in his book on fish processing [73]. Fish products are diverse biomaterials, hence researchers are continually working on developing and optimizing new methods and techniques to fulfill the increasing demand with minimum cost and possibly minimum loss of biological waste, however, biological waste is still increasing and will further increase with increase in production.

9.4   VALUE ADDED PRODUCTS FROM FISHERIES WASTE

9.4.1   FISH OIL

Fish processing by-products contain fish oil and the amount depends upon the fat content of the specific fish species. Generally, fish contains

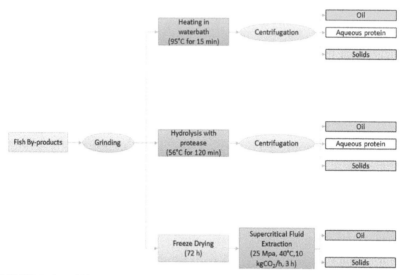

FIGURE 9.4   Different methods for the extraction of fish oil [156]. (Reprinted from Rubio-Rodríguez, N., Sara, M., Beltrán, S., Jaime, I., Sanz, M. T., & Rovira, J. (2012). Supercritical fluid extraction of fish oil from fish by-products: A comparison with other extraction methods. *Journal of Food Engineering*, 109(2) 238–248.   © 2002, with permission from Elsevier.)

2–30% fat [64]. Traditionally, fish oil is a by-product of processes that convert fish resources into fish meal which is also referred as the reduction process. There are many techniques being investigated to improve the quality of fish meal and fish oil by minimizing the temperature causing protein denaturation, heat exposure and oxidation, keeping the basic fish meal production still the same. Generally, the fish resources reduction process yields 20% fish meal and 5% fish oil. The fish meal reduction process depends on various industrial species including peruvian anchovy, mackerel, sand eel, capelin, menhaden, herring and pollock due to its high fat compared to other species [192].

Marine fish are commonly classified by the fat content of their filets as lean (<3%), medium (3–7%), and high fat (>7%). Generally, the lipids are present in the liver, muscle and gonads in lean fish which includes catfishes, tilapia, and carp, halibut, cod, hakes and pollock, whereas, the fat is present in the hypodermal portion in the fatty fish which include herring, mackerel, sardine, anchovy, salmon. In any fish, lipid content increases from tail to head, with higher level of fat deposition adjacent to the abdominal cavity [138].

9.4.1.1 Extraction Methods

There are different processes available for the efficient recovery of oil from fish tissues including: heat extraction, enzymatic extraction and supercritical fluid extraction as shown in Figure 9.4.

9.4.1.1.1 Heat extraction

The heat extraction of oil is carried out by rapidly heating the minced fish tissues for about 100 seconds and fed into the decanter and separator to recover fish oil, solids and stick water. The crude oil is added with food antioxidant until further processing and storage. The stick water are concentrated using an evaporator and mixed with the decanter and fed into a plate freezer. The plate freezer rapidly freezes the solids and concentrate into practical fish protein blocks. This method prolongs the storage of protein product and simplifies the handling of the product [104].

9.4.1.1.2 Enzymatic extraction

Enzymatic hydrolysis was developed to produce fish oil using commercial proteases. Enzymatic hydrolysis is carried out at low temperature with no waste by-product produced. It is used to extract oil for production of biodiesel, and/or omega-3 fatty acids. Several enzymes (alcalase, neutrase, lecitase ultra, protamex and flourzyme) can be used to extract fish oil. Several studies have reported that in the enzymatic extraction method, the fish samples are first minced and 50 g of fish mince is initially heated at 90°C for 5 min to deactivate endogenous enzymes and to the heated samples 50 mL of water or buffer is added. The temperature and pH of the fish mince is adjusted to 55°C and 8.0 for alcalase, 45°C and 6.5 for neutrase, 40°C and 6.8 for lecitase ultra, 60°C and 9.0 for protex and 50°C and 7.5 for protamex. The hydrolysis is started by adding 0.5–2% (w/w) enzyme to the mixture and continued for 1, 2, 3 or 4 h. After each hydrolysis time interval, the samples are heated at 90°C for 5 min to deactivate the added enzymes. The mixture is then centrifuged at 10,000 rpm for 20 min to separate oil fraction from the mixture [22, 60, 113, 121, 146, 175, 176].

During enzymatic extraction, increase in enzyme concentration increases the rate of hydrolysis, but may not significantly increase the oil yield due to the limitation of substrate availability for the enzyme to bind. Several studies used enzyme concentrations ranging between 0.05 and 2%, and indicated that increasing the enzyme concentration more than 1% was insignificant for the oil yield, and therefore, the enzyme concentration should not be higher than 2%. Therefore, it can be concluded that the concentration of 0.5% can be used for the oil extraction unless the enzyme is recycled or an immobilized reactor is used in order to reduce the cost associated with the enzyme [146]. The addition of buffer or water during the enzymatic production of oil has been observed to hinder process and also lead to reduction of oil yield due to the formation of emulsion. The formation of emulsion is not desirable and increase in the amount of emulsion decreases the amount of free oil produced [121].

The production time also plays an important role during the enzymatic extraction of oil. The increase in time will not improve oil yield, but will change the oil color to brown due to the formation of brown pigments from the reaction of carbonyls (aldehydes) produced from the oxidation of Polyunsaturated Fatty Acids (PUFA), with amino acids and proteins. A shorter reaction time will allow more throughputs, and/or reduce the volume of the reactor, thereby reducing the cost of oil extraction. Therefore, a 1–2 h hydrolysis time for oil extraction is recommended [121, 146]. The initial heating of the minced raw material to inactive the endogenous enzymes also plays an important role in the enzymatic oil production process. During heat inactivation, the proteins in the raw material are denatured and precipitated. Only a small portion of denatured proteins is solubilized and the remaining denatured protein forms a lipid-protein complex, which eventually reduces the release of lipids into the oil fraction. Ineffective handling of the raw material and processing parameters can be detrimental to the quality of oil, which can lead to dark colored oil due oxidation [64, 146].

9.4.1.1.3 *Supercritical fluid extraction*

Supercritical fluid extraction (SFE) is relatively new technology, which is currently tested to produce high quality fish oil. This methodology uses moderate temperatures, provides oxygen free environment during the extraction, which inhibits oxidation of polyunsaturated fatty acids and eliminate polar impurities including inorganic derivatives with heavy metals. It also efficiently extracts low polar lipid compounds in a selective manner. Rodriguez et al. [155] studied and optimized the SFE process during their study. In this process, the fish sample was minced and freezes dried before it was kept inside the extracting vessel that was pressurized to the extraction pressure using carbon dioxide. The circulation of solvent was started at the desired extraction temperature, for a specific time and desired flow of the solvent. The solvent used in the process was reused, the solute was removed through the separator. The pressure solvent in CO_2 was reduced to approximately 5 MPa with the temperature maintained lower than 40°C. The optimal extraction conditions reported in the study was: a pressure of 25 MPa, a temperature of 313 K (39.85°C) and the density of CO_2 maintained in the chamber was 13.75 Kg CO_2/h. The extraction yield obtained after 3h was 96.4%.

9.4.1.2 Quality Characteristics

Crude fish oil produced by any process may contain minor amounts of non-triglyceride substances and impurities including solid particles, gums, free fatty acids, pigments which can cause dark colored oil, foaming, precipitate when the oil is heated. Therefore, the quality of fish oil is determined by various physical and chemical properties as described in the Table 9.6 [26]. Allowed amount of additive in a food grade fish oil is shown in Table 9.7 [39].

9.4.1.3 Refining and Purification

On a large-scale fish oil processing for the feed and food industry, omega-3 concentration and biodiesel production, the fish oil is expected to be free of any impurities. Therefore, the fish oil is subjected to various refining steps including: degumming, neutralization, bleaching and deodorization. As a pretreatment step for production of edible oil and biodiesel, the fish oil is subjected to degumming. The degumming process is generally used

TABLE 9.6   Crude Fish Oil Quality Guidelines and Physical Characteristics [26]

Quality Guidelines	Permitted Limits
Moisture and impurities	0.5–1%
Free fatty acids	Range 1–7% but usually 2–5%
Peroxide value, meq/kg	3–20
Anisidine Number	4–60
Totox value	10–60
Iodine value	
Capelin	95–160
Herring	115–160
Menhaden	120–200
Sardine	160–200
Anchovy	180–200
Jack Mackerel	160–190
Sand Eel	150–190
Color, Gardner scale up to 14	
Iron, ppm	0.5–7.0
Copper, ppm	<0.3
Phosphorous, ppm	5–100
Physical Characteristics	Permitted Limits
Specific heat, cal/g	0.50–0.55
Heat of fusion, cal/g	About 54
Calorific value, cal/g	About 9,500
Slip melting point, °C	10–15
Flash point, °C	
As Triglycerides	±360
As fatty acid	±220
Boiling point, °C	>250
Specific gravity	
At 15°C	About 0.92
At 30°C	About 0.91
At 15°C	About 0.90
Viscosity, cp	
At 20°C	60–90
At 50°C	20–30
At 90°C	About 10

(From Bimbo, A. P. (1998). Guidelines for characterizing food-grade fish oil. *International News on Fats, Oils and Related Materials*, 9(5), 473–483. http://www.iffo.net/system/files/Guidelines%20Food%20grade%20fish%20oil%201998-3.pdf)

TABLE 9.7 Allowed Amount of Additive in a Food Grade Fish Oil [39]

Chemical	Functional class	Max level
Ascorbyl Esters	Antioxidant	500 mg/kg
Butylated Hydroxyanisole	Antioxidant	200 mg/kg
Beta-Carotenes	Color	1000 mg/kg
Carotenoids (beta-Carotenes, Carotenoic acid, ethyl ester	Color	25 mg/kg
Citric Acid	Acidity regulator, Antioxidant, Color retention agent, Sequestrant	GMP
Citric and fatty acid esters of glycerol	Antioxidant, Emulsifier, Flour treatment agent, Sequestrant, Stabilizer	100 mg/kg
Diacetyltartaric and fatty acid esters of glycerol	Emulsifier, Sequestrant, Stabilizer	10000 mg/kg
Fast Green FCF	Color	GMP
Guaiac Resin	Antioxidant	1000 mg/kg
Indigotine (Indigo carmine)	Color	300 mg/kg
Isopropyl citrates	Antioxidant, Preservative, Sequestrant	200 mg/kg
Polydimethylsiloxane	Anticaking agent, Antifoaming agent, Emulsifier	10 mg/kg
Polysorbates	Emulsifier, Stabilizer	5000 mg/kg
Propyl Gallate	Antioxidant	200 mg/kg
Propylene glycol esters of fatty acids	Emulsifier	10000 mg/kg
Stearyl Citrate	Antioxidant, Emulsifier, Sequestrant	GMP
Sunset yellow FCF	Color	300 mg/kg
Tertiary butylhydroquinone	Antioxidant	200 mg/kg
Thiodipropionates	Antioxidant	200 mg/kg

(Modified from Codex (2015). *General standard for food additives. Codex alimentarius international food standards*. pp. 1–396. http://www.fao.org/fao-who-codexalimentarius/sh-proxy/en/?lnk=1&url=https%253A%252F%252Fworkspace.fao.org%252Fsites%252Fcodex%252FStandards%252FCODEX%2BSTAN%2B192-1995%252FCXS_192e.pdf)

to remove gum-like material, which are mainly composed of carbohydrates, proteins and nitrogen compounds. The gums are divided into two types: hydratable and non-hydratable gums.

There are three types of process available for degumming including: water, chemical and enzyme. All the oils contain hydratable gums, which can be removed using water, because they absorb water and then become insoluble in oil. They can be later removed by centrifugation. Water degumming also has a major advantage being relatively simple, cheap and environmentally friendly process to remove hydratable gums. The non-hydratable gums are removed by acid hydrolysis most commonly using phosphoric acid. Enzymes can be used to remove both hydratable gums and non-hydratable gums. Enzymes such as phospholipase A (PLA) and phospholipase C (PLC) are specifically used to remove phospholipids in the oil. Neutralization is commonly carried out using caustic soda to purify crude fish oil. Once neutralization is complete, the fish oil is washed with water, centrifuged and dried in a vacuum dryer [103].

9.4.2   OMEGA-3 FATTY ACIDS

Fish oils are one of the most important sources for omega-3 fatty acids which mostly includes: cis-5,8,11,14,17-eicosapentaenoic acid (EPA) and cis-4,7,10,13,16,19-docosahexaenoic acid (DHA). The functional and biological properties of omega-3 fatty acids include: prevention of atherosclerosis, protection against arrhythmias, reduced blood pressure, beneficial to diabetic patients, fortification against manic-depressive illness, lowered symptoms in asthma patients, fortification against chronic obstructive pulmonary diseases, lighten symptoms of cystic fibrosis, recuperating survival of cancer patients, reduction in cardiovascular disease and improved learning ability. As per recommendation of The American Heart Association, at least two servings of fish every week can help in reducing the effect of cardiovascular diseases [64]. To be effectively applied for clinical and nutritional applications, the amount of omega-3 fatty acids present in the fish oil is not sufficient. Therefore, the concentrated form of these fatty acids is required to achieve sustained benefits. The concentration of omega-3 fatty acids is carried out using various processes including: urea complexation, low-temperature crystallization, supercritical fluid extraction, molecular distillation and enzymatic methods.

9.4.2.1   Urea Complexation

The concentration of omega-3 fatty acids from fish oil using urea complexation is one of the traditional and simplest technique. Initially, the triglycerides (TAG) present in the oil are converted into fatty acids esters via alkaline hydrolysis using potassium hydroxide or sodium hydroxide. The resulting fatty acids are mixed with an ethanolic solution of urea for a complex formation. In this method, the addition of urea initiates formation of complexes between urea and straight chain saturated fatty acids. Further, the urea molecules form solid-phase complexes with both saturated and monounsaturated fatty acids and crystallizes on cooling. The cooled crystals are removed by filtration and the filtered solution is highly enriched with omega-3 fatty acids [107, 166].

Gamez-Meza et al. [58] reported concentration of omega-3 fatty acids using urea complexation in which 15 g of sardine oil was mixed with 25 g of urea dissolved in 100 mL 95% ethanol and heated until the whole mixture turned into a clear homogeneous solution. The solution was rapidly cooled and stored in the refrigerator at 5°C for 8 h. The crystals were removed by centrifugation at 6000 g for 20 min at 5°C. The supernatant was stored at -30°C for 12 h and then centrifuged at 6000 g at –30°C for 20 min. Non-complexing supernatant containing polyunsaturated fatty acids was acidified at pH 4.0 and equal volumes of warm water (65°C) and hexane were added. The mixture was stirred for 30 min, the phases separated and the solvent was evaporated. The mass of oil was measured gravimetrically and the yield was calculated in comparison from the initial polyunsaturated fatty acids. The results from this study indicated, total reduction in saturated fatty acid concentration, significant decrease in the monounsaturated fatty acid concentrations and large increase in the polyunsaturated fatty acid concentration. The EPA was enriched from 14.5 to 34.1% and DHA from 12.5 to 39.4% and resulted in a total PUFA yield of 90.5%.

Estiasih et al. [51] studied the concentration of omega-3 fatty acids from the tuna canning processing oil by urea crystallization. The researchers optimized the crystallization reaction conditions including urea to fatty acid ratio (X1) and crystallization time (X2) using response surface methodology. The results from the study indicated the optimum conditions for producing ω-3 fatty acids concentrate were the urea to fatty acids ratio of 2.99:1 and the crystallization time of 23.64 h. The urea complexation process has a high selectivity towards saturated and mono-unsaturated long-chain fatty acids. However, handling of large volumes of inflammable solvents and water during the process can be an inhibitory factor. In addition, huge amounts of the urea-saturated fatty acid complexes have to be disposed, making this technology relatively expensive [107].

9.4.2.2   Low Temperature Crystallization

Low-temperature crystallization of lipids including triglycerides and fatty acids in organic solvents is also one of the very old techniques to concentrate omega-3 fatty acids. This technology separates triglycerides and fatty acids according to their melting points in various organic solvents including hexane, methanol, ethanol or acetone at very low temperature range, between $-50°C$ to $-70°C$ [90]. The melting point of the fatty acids depend upon the degree of saturation and therefore, the saturated fatty acids having the higher melting point crystallize out first, followed by the unsaturated fatty acids. During the process, the temperature of the solution increases, the mixture of saturated and unsaturated fatty acid decreases. The saturated fatty acids have higher melting point and therefore starts to crystallize out first and the liquid phase becomes enriched with unsaturated fatty acids. If the oil consists of a number of fatty acids, then repeated crystallization is required to obtain purified fractions [166].

Differences in the melting points of the triacylglycerols (TAG), crystallization process allow a great possibility of separation and concentration of PUFA. This process does not cause any damage to the fatty acids and provide more natural and better digestible omega-3 concentrates with good stability against oxidation. The low temperature crystallization can be carried out in the presence or absence of solvents. The low-temperature crystallization solvent process has to utilize large amounts of inflammable solvents, which presents a serious drawback for large-scale industrial applications. However, the low temperature crystallization in the absence of solvents is called as "Tritiaux Process" which involves slow cooling and slow agitation in the hydrophillization process including two main steps: crystallization that produces solid slurry in the liquid phase enriched with PUFA and separation.

The process parameters including: temperature, cooling rate and agitation influence the chemical and physical attributes of the PUFA concentrates [90, 107, 178]. Soleimanian et al. [178] investigated the effects of temperature, stages of crystallization, rate of cooling, agitation and addition of primary nucleus on separation efficiency of PUFA from Lantern fish oil. In this study, 10 g of fish oil was placed in a 15 mL falcon centrifuge tube and was cooled rapidly within 1 h from 25°C to the four different fractionation temperature of -5, 5, 0 and 7°C with holding time of 50–60 min. After each run, the samples were centrifuged at 12,000 rpm for 2 min to yield a liquid fraction containing low-melting-point fatty acids with higher PUFA concentration. The effects of crystallization procedures, slow cooling, effect of agitation and addition of primary nucleus were studied and the samples were analyzed using gas chromatograph. The optimized results indicated that a two-stage fractionation at the final temperatures of 5°C and 0°C under slow cooling and slow agitation and in the presence of primary nucleus resulted in 50 per cent increase in PUFA content over the original fish oil. There is no harmful organic solvents and carcinogenic carbamates, simple purification technique, reasonable cost of production and natural and digestible triglyceride with high oxidation stability makes this process more viable to concentrate omega-3 fatty acids.

9.4.2.3   Supercritical Fluid Fractionation

Supercritical fluid fractionation is a new and novel technology, which is being applied in the production of food and pharmaceutical products. In this technique, a specific compound or a fraction is extracted by changing the pressure and temperature without changing the extraction phase using supercritical fluid [166]. This technology works on the combination of two principle methods of distillation (separation is caused due to differences in component volatiles) and liquid extraction (separation is caused between the components that exhibit little difference in their relative volatilities or that are thermally labile). During this process, the substance existing as a supercritical fluid is determined by its critical pressure (Pc) and critical temperature (Pc).

The supercritical fluids have liquid like densities and values of viscosity and density between those typical for gases and liquids. The solubility of the solute in the supercritical fluid depends upon the density and it can be increased with increasing pressure. The supercritical fluid is highly compressible near critical pressure and a small increase in the pressure leads to proportional increase in fluid density. At critical pressures, a moderate temperature increase leads to significant decrease in the solute solubility, which is called retrograde behavior. At higher pressures, the fluid is less compressible and increase in temperature leads to decrease in density and increase the solubility of the solute, which is called non-retrograde behavior [200].

Carbon dioxide (CO_2) is used as the mobile phase in supercritical fluid extraction for edible applications due to various advantages including: moderate critical temperature and pressure, inexpensive, non-toxic, odorless, nonflammable, inert, environmentally friendly, readily available and safe. Carbon-di-oxide has a critical temperature of 31.1°C and critical pressure of 1070 PSIG, which allows the extraction to be carried out at temperatures below 100°C [107, 166, 200].

Homayooni et al. [80] studied the conditions for the optimized supercritical concentration of omega-3 fatty acids in case of sardine fish. The extractions were carried out at a static time of 20 min, temperatures of 40, 50 and 60°C, pressures of 150, 250 and 350 bar and a dynamic time of 40 min. The carbon dioxide flow rate was maintained at 0.3 mL/min. 0.5 g of lyophilized sardine fish samples were used during the extraction and the extracted oil was collected in 3 mL ethanol in a 5 mL volumetric flask. The results from this study indicated highest decrease in saturated (SFA) and monounsaturated (MUFA) fatty acids which were obtained at 50–60°C and 350 bar.

Pettinello et al. [141] investigated the concentrations of EPA-Ethyl Esters (EPA-EE) from fish oil on both bench scale and pilot scale using super critical CO_2 (SC-CO_2) as mobile phase and silica gel as stationary phase. The results from this study concluded that on bench scale extraction, using a column of 25 mm x 200 mm resulted in the highest EPA-EE purity of 90% and recovery of 49% at a temperature of 70°C, pressure between 180–220 bar and SC-CO_2 flow rate of 13.5 g/min. The pilot scale results used column 100 mm x 520 mm, at a temperature of 70°C and pressure between 180–220 bar, but a higher SC-CO_2 flow rate between 15–25 kg/h resulted in processing of 300 g of fish oil per cycle and yielded an EPA-EE purity of 93% and recovery of 24.6%. Rubio-Rodríguez et al. [154] reported that the supercritical processing technology have been installed at a production scale and companies such as KD-Pharma, Bex-bach, Germany and Solutex, Madrid, Spain produce a combination of omega-3 PUFA in a proportion over 90%, EPA up to a concentration of 95% and DHA up to 80% through application of this technology.

9.4.2.4 Molecular Distillation

Traditionally, distillation has been used as a partial separation of mixtures of fatty acid esters and utilized the existing boiling point and molecular weight differences of fatty acids under reduced pressure and high temperature ranging between 150–250°C. Technical improvements of the traditional distillation led to the modified process called molecular distillation. In this process, lower temperatures and short heating intervals were used for the enrichment of omega-3 fatty acids in the fish oil [200]. The molecular distillation is carried out using refined, bleached and deodorized fish oil. The refined fish oil is esterified using ethyl alcohol and this process is called interesterification. The molecular distillation is carried out in a two-stage distillation process with a degasser. The distillation process requires three passes and the first pass is through degasser which removes any moisture left after interesterification. The second pass through the distillation unit concentrates EPA and DHA and removes 20–50% of C_{10}–C_{18} fatty acid esters.

The third pass through the distillation unit concentrates EPA and DHA from 40–80% and separates the heavier fraction (5–10%) consisting of longer chain fatty acids. The operating temperature and pressure were reported as 170–190°C and 0.005–0.01 torr, respectively. The recovery yield was up to 70% and the DHA and EPA concentrations between 55–65 wt%. In this method vacuum is used to separate oil compounds based on weight, thereby leading superior reduction in impurities far below the industry specifications The molecular weight grouping also allows the process to concentrate specific fatty acids. Some of the disadvantages of this process include higher temperature requirement than urea complexation or enzymatic extraction process and the initial capital investment [166].

The molecular distillation process is an excellent technology to decontaminate, remove strip heavy metals, pesticides, polychlorinated biphenyl and cholesterol from fish oil triglycerides and concentrate omega-3 fatty acids. However, supercritical fluid extraction process is also capable of achieving the above mentioned results and also the molecular distillation process requires high investment, which is substantially lower than the investment required for an industrial supercritical fluid extraction plant and also operates at a much lower temperature than molecular distillation process [107].

Nordiac Naturals Inc., California, USA originally based in Norway is producing superior quality fish oil using a patented process using molecular distillation for omega-3 concentration. In this process, the fish oil is extracted by boiling and pressing the fish under oxygen-free processing environment. The fish oil is then subjected to purification to strip heavy metals, pesticides and polychlorinated biphenyl through molecular distillation. The purified fish oil is subjected to esterification to obtain ethyl esters of fish oil. The purified ethyl esters are subjected to a single-step molecular distillation process to concentrate omega-3 fatty acids as it avoids unnecessary reheating and transportation of the oil in order to finalize the concentration process. In the final step, after concentration, the purified oils are returned to their original natural triglyceride form using the process called re-esterification, hence the resulting product consists of 93% true triglycerides for better absorption by the human body [131].

9.4.2.5 Enzymatic Concentration

The enzymatic concentration of omega-3 fatty acids in the fish oil is carried out using various processes including: esterification, hydrolysis or exchange of fatty acids in the esters. This process can be selected based on the substrate and the reaction conditions. The major advantages of application of lipase include mild reaction conditions, less energy requirement, lower capital investment and selectivity of the catalyst [107, 200]. Microbial lipases extracted from *Aspergillus niger, Candida cylindracea (*now known as *Candida rugosa), Chromobacterium viscosum, Geotrichum candidum, Mucormiehei, Pseudomonas sp., Rhizopus oryzae,* and *Rhizopusniveus* have been used to concentrate omega-3 fatty acids.

Shahidi and Wanasundara [171] studied the concentration of omega-fatty acids in menhaden oil (MHO) in the form of acylglycerols using various microbial lipases and reported *Candida cylindracea* as the most suitable lipase for omega-3 concentration which results in the 44.4% concentrated fish oil containing 18.5% EPA, 3.62% Docosapentaenoic acid and 17.3% DHA. Carvalho et al. [33] also concentrated omega-3 fatty acids from sardine oil using four different lipases including: *Candida cylindracea lipase, Rhizopus delemar lipase, Aspergillus oryzae lipase* and *Chromobacterium viscosum lipase* and concluded *Candida cylindracea lipase* as the best lipase for omega-3 concentration.

Okada and Morrissey [137] investigated the enzymatic hydrolysis of concentrating omega-3 fatty acids from sardine oil using commercially available microbial lipases from: *Candida rugosa, Mucor javanicus,* and *Aspergillus niger* and found *Candida rugosa* to be more effective compared to other lipases. The extraction was carried out using 2 g sardine oil at 37°C for 1.5, 3, 6 and 9 h with the addition of 250 U of enzyme and 1.5 h was observed as the optimum time for producing omega-3 concentrates as the concentration of triglycerides did not significantly improve even after 9 h hydrolysis. The major factors that are involved in the enzymatic concentration of omega-3 fatty acids include enzyme concentration, reaction time, temperature and pH. These factors along with temperature and reaction time are significant in determining the economics of enzymatic concentration for large-scale production processes and determine the oxidative state of the omega-3 fatty acids [188].

9.4.3 PROTEIN

Fish are great source of protein and significant amount of muscle proteins are found in fish frames [198]. The proteins derived from fish are far superior in quality as compared to those of plant sources. The fish proteins are composed of balanced dietary essential amino acids compared to both plant and other animals sources, but are more heat sensitive than animal sourced protein [48, 56, 204]. Fish muscles can be divided into two categories: light and dark. The white fish such as cod and haddock are mostly composed of light muscles with small strips of dark muscles under the skin on both sides of the body. However, in fish with fatty acids such as herring and mackerel, the presence of dark muscles is higher and contains vitamins and fats. Light muscle is more abundant and contains about 18–23% proteins. The type of muscles of white and fatty fish is shown in Figure 9.5 [128].

Fish protein is composed of two types of protein including: structural and sarcoplasmic protein. The fish proteins are made up of 70–80% structural and 20–30% sarcoplasmic proteins with about 2–3% insoluble connective tissue proteins. Myofibrillar proteins are the primary food proteins and they make up about 66–77% of the total protein content in the fish meat. They are composed of 50–60% myosin and 15–30% actin [98, 181].

Currently, the fish and its waste by-products have been used as a source of producing protein in various products such as fish silage, fish meal and fish sauce. Fish silage is a liquid protein product which is produced from whole fish or fish waste that are liquefied using internal enzymes present in the fish along with addition of acid, addition of lactic acid bacteria or by addition of external enzymes. Several acids including hydrochloric acid, sulfuric acid and formic acid have been used for the production of fish silage. During fish silage preparation, the raw material is chopped into small pieces and 3% by weight solution of 98% formic acid is added and

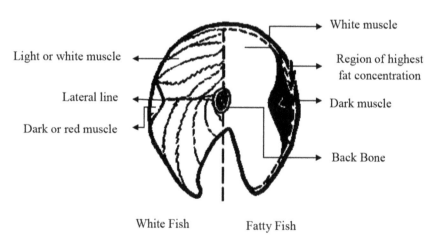

FIGURE 9.5   Types of muscles in white fish and fatty fish [128]. (From http://www.fao. org/wairdocs/tan/x5916e/x5916e00.htm)

mixed well and then stored for 48 days. The pH of the mixture is maintained less than 4 to prevent bacterial action [13, 64].

Fish meal is a dry powder prepared from whole fish or fish processing waste which are unacceptable for human consumption. Barlow et al. [20] explained the fish meal production process and reported that the production of fish meal is carried out in six steps (a) heating, (b) pressing, (c) separation, (d) evaporation, (e) drying and (f) grinding. During heating the fish, the protein coagulates and ruptures the fat deposits and liberates oil and water. During pressing step, the fish is pressed to remove large quantities of liquid from the fish raw material. The liquid collected is also known as stick water and it is evaporated resulting in a thick syrup containing 30–40% solids. Then it is subjected to drying using press-cake method to obtain a stable meal. This meal is ground to the desired particle size. Small pelagic fish or by-products can also be converted into fish sauce using salt fermentation. The fish by-products are mixed with salt in the ratio of 3:1 at 30°C for 6 months. The amber colored protein solution is drained from the bottom of the tank and can be used as a condiment on vegetable dishes and is nutritious due to the presence of essential amino acids [66].

Fish proteins are extracted from fish using chemical and enzymatic methods. Protein hydrolysates obtained from these processes can be applied for various industrial products development including milk replacers, protein supplements, stabilizers in beverages and flavor enhancers.

9.4.3.1 Chemical Extraction

Protein from whole fish or fish waste is traditionally extracted by using solvents or chemicals. The solvent extraction of fish protein is carried out in three step isopropanol extraction process at temperature between 20–75°C for 3.5 h as shown in Figure 9.6. During each extraction, the samples are centrifuged and the supernatant fraction is collected, dried, milled and screened to separate bone fragments. The disadvantages of this method include: poor functionality, off-flavors, high cost of production and traces of the solvent in the final product, making it commercially unsuccessful [173]. The chemical extraction of protein from fish and fish waste is usually carried out using sodium hydroxide (NaOH).

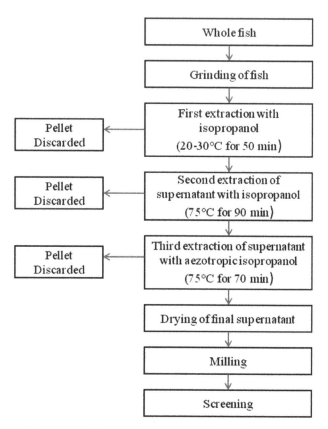

FIGURE 9.6   Extraction of fish protein using solvent [173]. (Reprinted by permission of the publisher [Taylor & Francis Ltd, http://www.tandfonline.com], from Sikorski, Z. E., Naczk, M., & Toledo, R. T. (1981). Modification of technological properties of fish protein concentrates. Critical Reviews in Food Science and Nutrition, 14(3), 201–230.)

Arnesen and Gildberg [12] reported extraction of protein from Atlantic cod using 3 M NaOH in a three step extraction process as shown in Figure 9.7. Two thousand grams of mined cod was mixed with 2000 mL in the ratio of 1:1 and the reaction was started by adjusting the pH to 11 with 62 mL of 3M NaOH. The samples were centrifuged at 4000 g for 15 min at 4°C and the supernatant was collected. The pellet extracted was subjected to same extraction process and centrifuged to collect supernatant. The pellet was again suspended in 2000 mL of water and the pH was adjusted to 2 with 3M HCl and the third extraction was carried out for 15 min and centrifugation was done to collect supernatant. The supernatants

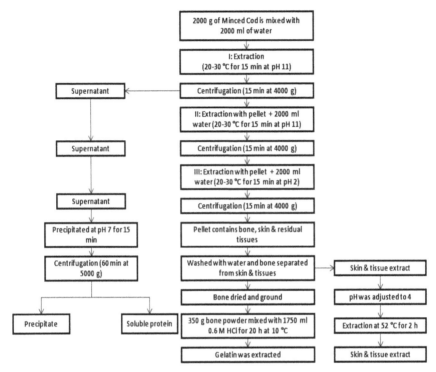

FIGURE 9.7 Extraction of muscle proteins and gelatin from Atlantic cod [12]. (Modified from Arnesen, J. A., & Gildberg, A. (2006). Extraction of muscle proteins and gelatine from cod head. *Process Biochemistry*, 41(3), 697–700.)

from the three extracts were pooled together and the pH was adjusted to 7 with 3 M NaOH. The samples were allowed to precipitate for 15 min at room temperature and the soluble protein was separated by centrifugation at 4°C for 60 min at 5000 g. Several other researchers including: Batista [21], Undeland et al. [194], Kelleher and Hultin [92] and Nurdiyana et al. [135] reported the use of 3M NaOH for the extraction of protein from fish and fish waste.

9.4.3.2 Enzymatic Extraction

Enzymatic extraction of protein is carried out by addition of commercially available proteases, carbohydrases and lipases under controlled extraction conditions without degrading their nutritional qualities for the acceptance in the food industry. Traditionally, the enzymatic extraction has been widely used in the food industry for the manufacture of broad spectrum of food products as special diets for babies and sick adults [169]. The products using enzymatic process is low in bitterness, osmotically balanced, hypoallergic and have good flavor. Most of these diets are composed of peptides and are rich in amino acids [68]. Kristinsson et al. [98] and Gildberg [65] reported that fish-processing waste is being underutilized as low quality animal feed or fertilizer. Most of the research studies conducted on the enzymatic processing of fish protein seems to be laboratory oriented or small scale, and have their limitations when scaled up to industrial scale due to the low yields, initial high cost of enzymes, inactivation of enzymes after hydrolysis either by heat or by pH and the inability to reuse enzymes.

Proteases are derived from animal, plant and microbial sources. Due to the inability of plant and animal proteases to meet current demand in the market, there is an increase in the demand for microbial proteases. Bacterial proteases are often used in the production of protein hydrolysate [98]. Alcalase is an alkaline enzyme, which is produced from Bacillus licheniformis, has been proven to be one of the best enzymes used to prepare fish protein hydrolysate. The protein extracted using alcalase enzyme was reported to have better functional properties, a high protein content with an excellent nitrogen yield, an amino acid composition comparable to that of muscle and a higher nutritional value than those produced by other enzymes such as Neutrase [23, 169, 187].

Ramakrishnan et al. [146] reported the extraction of protein hydrolysate from whole fish, head, fin, tail, skin and gut, and frames using alcalase at various concentrations (0.5, 1 and 2%) and time (1, 2, 3 and 4 h) and the corresponding extraction method is summarized in Figure 9.8. In this study, minced fish sample (50 g) was heated in a water bath at 90°C for 10 min before the extraction to deactivate the endogenous enzymes. Then, 50 mL of 1M potassium phosphate buffer (pH 7.5) was added to the fish in the ratio of 1:1 (fish: buffer). Once the samples reached the required temperature of 55°C, the enzymatic hydrolysis was started by adding 0.5% or 1% or 2% (by weight of raw material) alcalase and after 1, 2, 3 or 4 h of hydrolysis, the samples were heated to 90°C for 5 min to

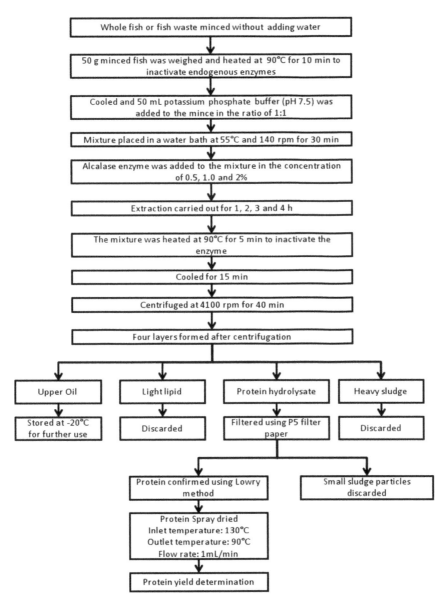

FIGURE 9.8 Enzymatic extraction of protein from whole fish [146]. (Reprinted and adapted from Ramakrishnan VV, Ghaly AE, Brooks MS, Budge SM (2013) Extraction of Oil from Mackerel Fish Processing Waste using Alcalase Enzyme. Enz Eng 2:115. doi:10.4172/2329-6674.1000115. With permission via the Creative Commons Attribution License.)

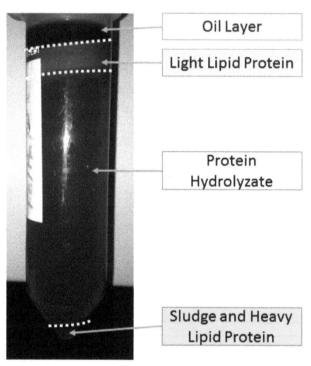

FIGURE 9.9 Four layers when recovering soluble fish protein hydrolysate [146]. (Reprinted and adapted from Ramakrishnan VV, Ghaly AE, Brooks MS, Budge SM (2013) Extraction of Oil from Mackerel Fish Processing Waste using Alcalase Enzyme. Enz Eng 2:115. doi:10.4172/2329-6674.1000115. With permission via the Creative Commons Attribution License.)

inactivate the enzymes. The mixture was then allowed to cool and centrifuged at 4100 rpm for 40 min. Four layers (Figure 9.9) containing: a light-lipid layer, a soluble clear protein layer and a bottom sludge layer containing the remaining fish tissues was obtained. The upper oil layer was carefully removed using a pipette and stored at −20°C. The soluble protein layer was filtered and spray dried to obtain a dried protein powder. Various researchers including − Kristinsson and Rasco [98], Shahidi et al. [169], Benjakul and Morrissey [23], Gildberg [65], Liaset et al.[111], Guerard et al. [71], Gbogouri et al. [59], Vieira et al. [199], Bhaskar and Mahendrakar [25], Bhaskar et al. [24] and Holanda and Netto [43] − have reported different methods of enzymatic extraction of protein from fish and fish processing waste.

9.4.4 ENZYMES

Fish are a rich source of enzymes including pepsin, trypsin, chymotrypsin and collagensae which exhibit high catalytic activities at relatively low concentrations and they possess better catalytic properties, good efficiency at lower temperatures, lower sensitivity to substrate concentrations and greater stability in a wide range of pH. These enzymes are currently extracted from fish viscera and commercially manufactured on large scale [64]. Pepsin is a proteolytic enzyme and is used in the extraction of collagen and gelatin from fish and other sources and can also be used as rennet substitute to digest proteins. The active conditions for effective functioning of pepsin include: pH 2–4 and temperature of 30°C.

Chymotrypsin occurs in two forms including: chymotrypsin A and chymotrypsin B is an endopeptidase which breaks the central backbone of peptide bonds and can be used to break complex amino acids and enrich their qualities. The active conditions for effective functioning of chymotrypsin include: pH 7–8 and temperature of 50°C. Trypsin is an exopeptidase which cleaves the peptide bond on the carboxyl side of arginine and lysine and the optimum conditions for trypsin is at pH 9 and temperature of 40°C. Fish collagenases are capable of hydrolyzing triple type I, II and III tropocollagen molecule and illustrate both trypsin and chymotrypsin like activities. They are most active in the pH range of 6.5–8.0 and at a temperature of 30°C [34, 91, 129, 130, 157, 170, 207, 208].

9.4.5 CHITIN AND CHITOSAN

Chitin is a linear polysaccharide composed of α-(1–4)-linked 2-acetamido-2-deoxy-D-glucose units which can be de-N-acetylated [102, 150]. A critical evaluation of potential sources of chitin and chitosan concluded that shrimp, prawn and crab waste were the principle source of them and would remain so for the immediate future. The renewable natural biopolymer chitin and its derivative chitosan, have gained an outstanding reputation with numerous applications in the fields of water engineering, cosmetics, paper engineering, textile engineering, food engineering, agriculture, photography, chromatographic separations, medical and pharmaceutical applications in the recent decades [85, 150]. Chitin and chitosan are generally non-toxic, non-soluble in water and most organic solvents. Physical methods using particle size reduction are promising as alternative means of chitin extraction, however obtaining high yields of purified chitin at low cost has been proved challenging. There has been limited industrial use of enzymatic methods for chitin and chitosan production due to the high costs associated with such methods on the industrial scale [185]. However, enzymatic hydrolysis produces a higher grade chitosan than chemical methods. Chemical methods may also produce toxic compounds rendering the chitosan unsuitable for biomedical applications, and is a source of environmental pollution [10]. However, a combined approach using physical, chemical and enzymatic methods is of interest as it may provide a cost effective and environmentally friendly compromise.

9.4.5.1   Extraction Methods

During a study by Percot et al. [140] on chitin production process, the shrimp shells were subjected to two major processing: demineralization and deproteinization. The shrimp shells were first subjected to mechanical washing to remove impurities and freeze dried. The dried shrimp shells were ground and subjected to demineralization using 1 M HCl for 24 h and deproteinization using 1 M NaOH for 1–72 h at temperatures ranging from 65 to 100°C. The resulting mixture was filtered to obtain chitin and filtrate. The filtrate was later washed and dried to obtain protein as a by-product.

Arbia et al. [10] reported different methods including: the chemical and biological method of chitin extraction from shrimp shells as shown in Figure 9.10. The traditional chemical method includes washing and crushing of shrimp shells, deproteinization of shells using alkali (1 M NaOH for 1–72 h at 65–100°C), demineralization of shells using acid (0.275–2 M HCl for 1–48 h at 0–100°C) and discoloration and bleaching using chloroform, methanol and water (1:2:4) at 25°C to obtain raw chitin. The biological method includes deproteinization of shells using protease-producing bacteria instead of alkali and demineralization of shells using organic acid producing bacteria.

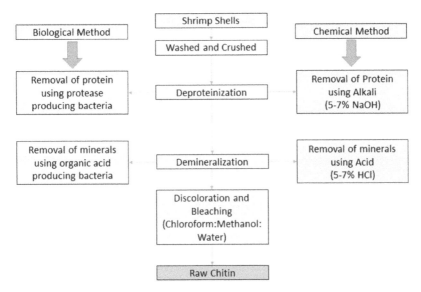

FIGURE 9.10 Chitin recovery by chemical and biological methods [10]. (Adapted from Arbia, W., Arbia, L., Adour, L., & Amrane, A. (2013). Chitin extraction from crustacean shells using biological methods-A review. *Food Technology and Biotechnology*, 51(1), 12.)

De Holanda and Netto [43] extracted chitin, protein, astaxanthin pigmented oil and lipids from minced shrimp processing waste as shown in Figure 9.11. In this study, the shrimp waste was subjected to deproteinization using enzymes (3% Alcalase at a temperature of 60°C, pH of 8.5 for 15 min and 1% Pancreatin at a temperature of 40°C, pH of 8.5 for 25 min) and alkali (using 1–5% NaOH and 1–5% KOH, at three different temperatures 50, 70 and 90°C for 1, 2 or 3 h) for comparative analysis. After the hydrolysis, the results indicated that 59.50 and 50.55% of protein can be recovered using Alcalase and Pancreatin, respectively. The alkaline hydrolysis using NaOH and KOH resulted in 80–86% protein recovery. The results suggested that the residual protein in the chitin after enzymatic and alkaline hydrolysis was 9.31 and 3.99%, respectively. They concluded that extracting chitin without residual is highly unlikely and incomplete deproteinization was due to the covalently bound muscle proteins to the chitin forming a stable complex.

Synowiecki and Al-Khateeb [189] extracted chitin from shrimp processing waste using chemo-enzymatic method. The shrimp waste was demineralized first using 10% HCl in the shell: water ratio of 1:20 (w/v)

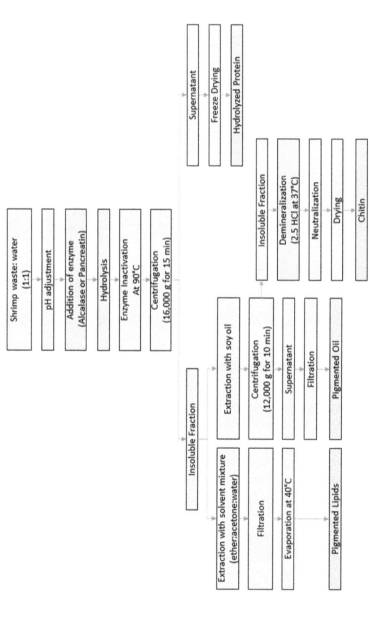

FIGURE 9.11   Extraction of chitin, protein, astaxanthin from shrimp waste [43]. (Modified from De Holanda, H. D., & Netto, F. M. (2006). Recovery of components from shrimp (*Xiphopenaeus kroyeri*) processing waste by enzymatic hydrolysis. *Journal of Food Science, 71*(5), C298–C303.)

while deproteinization was carried out as the second step with Alcalase concentration of 20 AU/kg. The results indicated the residual ash and protein content as 0.31–1.56% and 4.45–7.9%, respectively. They also suggested that these protein impurities in the chitin will not affect its use in various applications. Valdez-Pena et al. [195] used five food grade enzymes including Alcalase, Flavorzyme, Lysozyme, Papain and Trypsin VI for enzymatic deproteinization of shrimp shells. The deproteinized shrimps (0.5 g) were subjected to microwave assisted demineralization process using a treatment with 100 mL 1 M lactic acid at 121°C for 30 min. The results from this study indicated that the samples treated with Alcalase recorded the lowest protein content in insoluble solids and efficient protein removal from shrimp head waste. The report also indicated complete removal of proteins (275 mg/g residual protein) and shell waste had 17.85–56.58% of residual ash content. After microwave assisted demineralization process, the yield of chitin was 22% and the ash content was only 0.2%. This process was able to completely remove residual protein and ash from the final product.

Stanley [183] extracted chitin from three different species including red crab, Indian white shrimp and bivalve (Florida marsh clam using the traditional chemical process and the results indicated that the highest chitin yield can be obtained from bivalve (64%) and the lowest chitin yield can be obtained from red crab (31%) and Indian white shrimp (30%). Lertsutthiwong et al. [108] extracted chitin from shrimp shells using two types of chemical treatment process. In the first treatment process, the shrimp shells were first treated with NaOH (1, 2 and 4%) for deproteinization followed by demineralization using 4% HCl while, in the second treatment process, the demineralization was carried out first followed by deproteinization under the same conditions. The chitin yields from both the treatment process were compared and the results indicated that the chitin yield from the second treatment process (20–27%) was higher than first treatment process (15–20%). They concluded that the shells can lose the protective protein layer if the deproteinization carried out first and chitin in the shell can get exposed to the HCl during demineralization and further hydrolysis of chitin thereby leading to loss of final chitin fraction. In case of first demineralization step, the protein layer can be protected at the backbone of the shell thereby leading to lower loss of chitin during subsequent deproteinization using NaOH.

The schematic interpretation of organic matrix in shells structure is shown in Figure 9.12 [142]. It shows a thin layer of chitin sandwiched between two thicker layers of proteins. This complex is called a carrier protein (CP). The carrier protein is surrounded by the matrix of mineralized protein that is embedded in layers of $CaCO_3$. The protein in the mineralized layers and the protein in the carrier protein complex might protect the cleavage of the chitin chain against acid hydrolysis. $CaCO_3$ layers alternate with layers in which a chitin filament is covered with protein (CP) and is embedded in a mineralized proteinous matrix (MM). During deproteination, the protein is removed from CP; thereafter chitin is fully exposed to the chemical environment.

Manni et al. [119] extracted chitin from shrimp shell waste using enzymatic and chemical deproteinization and chemical demineralization process. The enzymatic deproteinization was carried out using SV1 enzyme (1 U/mg of protein) extracted from *Bacillus cereus* at pH 8.0, temperature 40°C for 3 h while the chemical deproteinization was carried out using 1.25 M NaOH at a shell:water ratio of 1:20 (w/v) for 4 h. The deproteinized shell waste was demineralized using 1.5 M HCl at a shell:water ratio of 1:10 (w/v) for 6 h at 50°C. They reported that the residual protein content present in the final chitin after enzymatic deproteinization was

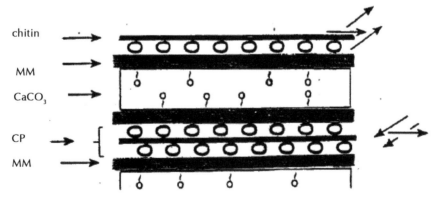

FIGURE 9.12 Structure of organic matrix in shells. MM: Mineralized Protein Matrix; CP: Carrier Protein [108]. (From Lertsutthiwong, P., How, N. C., Chandrkrachang, S., & Stevens, W. F. (2002). Effect of chemical treatment on the characteristics of shrimp chitosan. *Journal of Metals, Materials and Minerals*, 12(1), 11–18. Used with permission via the Creative Commons License.)

higher (11.2%) than chemical deproteinization (7.4%). However, enzymatic deproteinization can avoid heavy metal residues, over-hydrolysis and breakdown of chitin. The ash content was reduced to 0.4% during chitin extraction. The chitin recovery yield from shrimp waste after enzymatic and chemical deproteinization was 16.5 and 20%, respectively.

Younes et al. [205] extracted chitin from shrimp shells using enzymatic deproteinization and chemical demineralization process. The deproteinization was carried out using microbial proteases from six sources including *Bacillus mojavensis* A21, *Bacillus subtilis* A26, *Bacillus licheniformis* NH1, *Bacillus licheniformis* MP1, *Vibrio metschnikovii* J1 and *Aspergillus clavatus* ES1. The protein removal using A26, J1 and MP1 was 76% while 65 and 59% was obtained using NH1 and ES1, respectively. *Bacillus mojavensis* A21 protease at an enzyme/substrate ratio of 7.75 U/mg, at a temperature of 60°C and incubation time of 6 h was able to remove up to 88% of the proteins from the shrimp shells. The minerals were completely removed by chemical demineralization process. The chitin recovery yield was 18.5% of its initial dry mass as water insoluble white fibrous material.

Kjartansson et al. [93] extracted chitin from fresh water prawns (FWP) which were washed, boiled in 4% NaCl, dried in freeze dryer and milled to get a FWP powder. The FWP powders were demineralized using 0.25 M HCl in shell: water ratio of 1:40 w/v at 40°C for 4 h. The demineralized FWP powder was deproteinized using 0.2 M NaOH in a shell: water ratio of 1:40 w/v at 40°C for 4 h. The deproteinized samples were subjected to additional deproteinization using 1 M NaOH in a shell: water ratio of 1:40 w/v at 90°C for 2 h. The samples were then dried and milled to obtain chitin powder. The samples were sonicated for 0, 1 and 4 h during each step of chitin extraction process and the results indicated that the samples sonicated for 0 h resulted in highest chitin yield of 8.28 g/100 g of raw material. The samples sonicated for 1 and 4 h resulted in 7.55 and 5.03 g of chitin per 100 g of raw material, respectively.

In general, during shrimp chitin extraction, the deproteinization efficiency varied from 54–97% and the demineralization efficiency varied from 0–98%. The chitin recovery yield from shrimp shells varied from 16.5–64% [55].

The conversion of chitin to chitosan is carried out using a process called deacetylation. In this process, the chitin are subjected to 50% NaOH in the ratio of 1:1 at 80–110°C for 2 h. Stanley [183] obtained chitosan from chitin extracted from four different species including red crab, Indian white shrimp, *Abra alba* (white furrow shell) and *Cyrenoida floridana* (Florida marsh clam). The extraction was carried out using the traditional chemical process using 50% NaOH and the results indicated that the highest chitosan conversion yield was obtained from Abra alba (white furrow shell) (97%), the lowest chitosan conversion yield was obtained from red crab (70%), Indian white shrimp yielded 92% chitosan and *Cyrenoida floridana* (Florida marsh clam) yielded 60% chitosan yield.

Gildberg and Stenberg [67] compared the chitosan extraction efficiency from shrimp waste using the newly modified enzymatic method and traditional method. In the enzymatic method, 2 kg of shrimp waste was treated with 20 mL of Alcalase at 40°C for 2 h and the treated samples were manually pressed with a nylon net screen (1.5 mm pore size) to remove the crude hydrolyzate. The press cake (904 g) was stored in a plastic bag at –20°C and used for further chitosan production process. The crude hydrolyzate was centrifuged at 10000 g for 40 min to obtained crude protein hydrolyzate. The press cake was demineralized by adding 4 L of water and 0.5 L of 37% HCl for 12 h at 17–18°C. The demineralized shells were deproteinized by adding 5 kg of shell and 0.55 L of 50% (w/w) NaOH. The deproteinized shells were subjected to deacetylation by adding 1 kg of shells with 1 kg of solid NaOH and heated for 4 h at 80°C. The chitosan obtained was washed and dried in freeze dryer. The traditional process was carried out in the same way except the pre-enzymatic treatment was eliminated. The results indicated that in traditional process 12.8% of nitrogen and 8.8% of dry matter was obtained whereas using the enzymatic pretreatment 68.5% of nitrogen and 33.2% of dry matter can be achieved. The total weight recovery of chitosan using traditional and enzymatic process was 2.37 and 2.32%, respectively. The study indicated that, it was possible to recover about 70% of the total nitrogen retained in valuable products as compared to less than 15% in the traditional chitosan production process.

Puvvada et al. [144] extracted chitin and chitosan from shrimp shells, in which 1.5 kg of shrimp shells were crushed and subjected to deproteinization using 4% NaOH for 1 h to remove proteins from the shells. The deproteinized shells were demineralized using 1% HCl in the ratio of 1:4

for 24 h to remove minerals. The samples were again treated with 50 mL of 2% NaOH to decompose albumen into water soluble amino acids and the extracted chitin was washed and dried for chitosan production process. The dried chitin was subjected to deacetylation using 50% NaOH at 100°C for 2 h. The chitosan samples were washed and dried at 110°C for 6 h. The chitosan obtained was purified by removing insoluble fraction using filtration, re-precipitation with 1 N NaOH and demetallization using sodium dodecyl sulfate (SDS) and ethylenediaminetetraacetic acid (EDTA) precipitation. The chitosan yield was 34% after purification of the total raw material. The average molecular weight was 1,599,558.029, degree of deacetylation was 89.79%, viscosity was 304 cps, pH was 8.5, ash value was 0.25%, loss on drying was 9.34% and heavy metals were >10 ppm.

9.4.5.2 Properties and Applications

The important characteristics determining the quality of chitosan are its molecular weight, viscosity, solubility and degree of acetylation. Chitosan with high molecular weight has a linear unbranched structure and can be used to enhance viscosity in an acidic environment. Chitosan also behaves as a pseudo-plastic material with decreased viscosity at increasing rates of shear. The increase in chitosan concentration, decrease in temperature and increase in degree of deacetylation causes increase in viscosity of chitosan. Increase in viscosity or the degree of deacetylation of chitosan increased the digestibility of fats [174].

The degree of acetylation is an important characteristic that influences the performance of chitosan in many of its applications. At 20% degree of deacetylation, chitosan exhibits highest structural charge density and polyelectrolyte behavior during long distance intra and intermolecular electrostatic interactions. These interactions are responsible for expansion of chains in chitosan, increased solubility and ionic condensation of chitosan. At 20–50% degree of deacetylation, the hydrophobic and hydrophilic interactions are progressively counterbalanced in chitosan chains. Above 50% degree of deacetylation, electrostatic interactions in the chitosan are short distance interactions and therefore more applicable for biomedical applications. Majority of the biological properties of chitosan are based on cationic behavior due to the presence of free amino groups in the polymeric chain and depends significantly on the degree of deacetylation [101].

Biological properties of chitosan includes: biodegradability, biocompatibility, haemostatic, anticholesterolemic and antioxidative. Chitosan with lower deacetylation degree induces acute inflammatory response while the chitosan with higher deacetylation degree induces minimal inflammatory response due to the lower degradation rate. The toxicity of the chitosan also depends upon the degree of deacetylation and chitosan with <35% deacetylation caused dose dependent toxicity while chitosan with >35% deacetylation had lower toxicity. The cell adhesion, mucoadhesion, permeation, anticholesterolemic and antioxidative properties of chitosan are directly proportional to the degree of deacetylation [9].

Biomedical applications of chitosan include: wound healing, drug delivery and tissue regeneration systems. Chitosan activates immunocytes and inflammatory cells (macrophages, fibroblasts and angio-epithelial cells) and stimulates cell proliferation and histoarchitectural tissue organization. Chitosan is a natural haemostat, which can naturally assist in blood clotting and can also block nerve endings to reduce pain. Chitosan also depolymerizes itself to release N-acetyl-β-D-glucosamine to initiate fibroblast proliferation and to stimulate increased level of natural hyaluronic acid synthesis at the wound site. This helps in faster wound healing and scar prevention. These biomedical effects of chitosan depend upon the deacetylation degree. The wound break strength and collagenase activity of the chitosan increases with increase in deacetylation degree [9, 28, 81, 85, 193].

Naturally chitosan is insoluble at an alkaline and neutral pH. However, chitosan is soluble in water and the solubility depends on the degree of deacetylation. At 40% degree of deacetylation, chitosan is soluble up to pH 9 whereas at higher degree of deacetylation chitosan is soluble in solution with pH up to 6.5 [174].

Molecular weight of chitosan and deacetylation degree plays an important role in the drug delivery systems using microspheres. Microspheres prepared with high molecular weight chitosan gelled faster than those prepared with low molecular weight chitosan. Low molecular weight chitosan has higher activation energy, require more interaction time with other chains and to gelate with glutaraldehyde. Higher deacetylation degree increases the covalent cross-linking, compactness and hydrophobicity and decreases the size, surface roughness, swelling, loading capacity and burst release of the microspheres. Higher molecular weight increases the sphericity, morphology, homogeneity, and cross-linking and decrease the swelling, release rate and diffusion coefficient [9, 72, 101].

9.4.6 BIODIESEL

The continuously increasing demand for energy has been translated into increased cost of crude oils, shortage of fossil fuels and intensified emission of greenhouse gases worldwide. If the utilization of fossil fuels is continued at the present rate, local air quality will deteriorate severely and global warming will increase beyond repairable extent [4, 57]. Renewable energy resources of biological origin (biofuels) have smaller net greenhouse gas emissions. Currently, biodiesel and bioethanol production are gaining momentum all across the globe due to the shrinking supply of oil reserves, security of source, cost of production and the impending threat of global warming [44]. However, sustainable production of biofuels will require a resourceful biomass conversion process.

Biodiesel is a biofuel that is obtained from plant and marine oils or animal fats. Biodiesel, as a diesel-equivalent, has a potential share among biofuels of about three quarter of all refinery distillate fuel oils. In comparison to petroleum diesel, biodiesel significantly reduces emissions of carbon dioxide (about 50–60%), sulfur dioxide and harmful air pollutants. Green house gas emissions can be reduced by 10–20% and 40–90% with the use of at least 20% (B20) and 100% (B100) biodiesel blends, respectively [77, 182, 186].

One of the biggest challenges in biodiesel production is the availability of feed-stock. There is a concern about using plant-derived oils and fats since the crops used for biodiesel production are also needed for food, feed and oleochemical industries. Biodiesel factories must compete with food, cosmetic, chemical and livestock feed demands. An increased demand for vegetable oils will increase the use of fertilizers in the crop fields, which will contribute to greenhouse gas emissions. In fact, biodiesel production depending on heavily fertilized crops would lead to 70% increase (from the current value) in greenhouse gas emissions [86, 122, 147].

Fish oil can be converted into biodiesel through a process called transesterification. The transesterification process can be carried out in two different methods: chemical transesterification and enzymatic transesterification. Transesterification is the process of reaction of a fat or oil (triacylglycerol-TAG) with an alcohol to yield biodiesel (fatty acid methyl ester-FAME) and glycerol in the presence of a catalyst (chemicals or enzymes) [4, 147]. It has been shown that the transesterification seems to be the best practical option (compared to dilution, micro-emulsification or pyrolysis) since this process can significantly reduce the high viscosity of oils and produces a biodiesel with the same physical properties as petroleum diesel fuel [116].

9.4.6.1   Chemical Transesterification

In chemical transesterification process, either KOH or NaOH are used as alkali catalyst together with an alcohol (methanol or ethanol) to convert oil into biodiesel. The schematic principle of the transesterification reaction is shown in Figure 9.13. The simplified form of its chemical reaction is presented in Eq. (1) in which R_1, R_2, R_3 are long-chain hydrocarbons. Usually, vegetable oils and animal oils consists of five chains including: palmitic, stearic, oleic, linoleic, and linolenic. During the transesterification process, the triglyceride is converted stepwise to diglyceride, monoglyceride, and finally to glycerol and 1 mol of fatty ester is liberated at each step as shown in Figure 9.14 [116].

The chemical transesterification has been widely used in the industries for its high conversion rate (>90%) in short period of time (4–10 h). However, there are several disadvantages including: product separation, soap

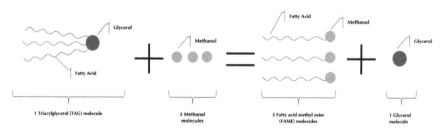

FIGURE 9.13   Schematic of the transesterification reaction.

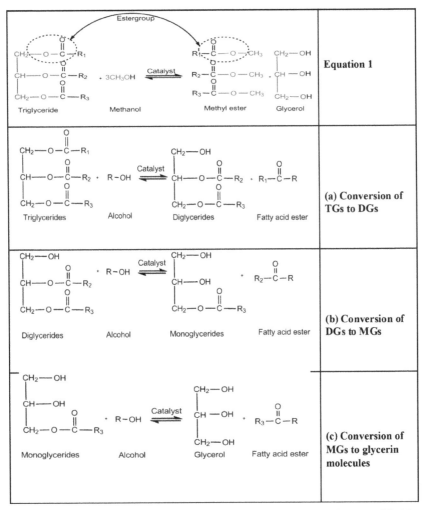

FIGURE 9.14 Three continuous stages of the transesterification reaction. (Modified from Ma, F., & Hanna, M. A. (1999). Biodiesel production: A review. *Bioresource Technology*, 70(1), 1–15.)

formation and negative environmental impacts such as greenhouse gas, CO, hydrocarbons, NO_x and particles in exhaust emissions.

There are several factors which affect the rate at which chemical transesterification proceeds and the ultimate yield of biodiesel. These include: (a) type of alcohol, (b) alcohol to oil molar ratio, (c) reaction temperature and time, (d) catalyst type and concentration, (e) water activity, and (f) mixing intensity.

9.4.6.1.1   Type of alcohol

The most frequently used alcohol (acyl-acceptors) that can be used for transesterification include: methanol, ethanol, propanol, butanol, amyl alcohol, octanol, and branched alcohols [57]. Methanol is most widely used for FAME production, and the reaction is known as methanolysis. In addition, ethanol is also used but it is relatively expensive, less volatile and less reactive, renewable and eco-friendly as it is produced from agricultural products as compared to methanol [16].

9.4.6.1.2   Alcohol to oil molar ratio

The optimum molar alcohol to oil ratio is based on the reaction system, feedstock, catalyst and type of alcohol used. Theoretically, for each molecule of triglyceride, three molecules of alcohol are needed to produce three molecules of FAME. Alcohol is used in excess to shift the equilibrium towards the formation of esters [151]. A molar ratio of 6:1 is normally used in industrial process because the yield of FAME is higher than 98% by weight. The molar ratio has no effect on acid, peroxide, saponification and iodine value of methyl esters. Depending on the feedstock oil, optimal molar ratios of methanol to oil have been reported between 3:1 and 24:1 [120].

9.4.6.1.3   Reaction to temperature and time

Various research studies revealed that reaction temperature and time have a significant effect on the rate of transesterification. However, the reaction can be conducted at room temperature if sufficient time is provided [182]. However, a longer reaction time does not necessarily increase the conversion but favors the backward reaction (hydrolysis of esters) which results in a reduction of product yield [109]. It is already established that in most cases, the reaction temperature is kept close to the boiling point of methanol (60–70°C) at atmospheric pressure for a given time. Such mild reaction conditions require pre-treatment (degumming or esterification) of oil for the removal of free phospholipids and fatty acids. Simultaneous esterification and transesterification can take place under high pressure (9000 kPa) and high temperature (240°C) so a pretreatment step can be eliminated if the reaction is carried out under these conditions [18].

9.4.6.1.4 Catalyst type and concentration

Alkali, acid, enzyme or heterogeneous catalysts are used in transesterification reaction, among which alkali catalysts like NaOH and KOH are more effective and used industrially due to their availability, low cost and large continuous-flow production processes. Advantages of using alkali catalyst are (a) their ability to neutralize free fatty acids and water to mediate the reaction (b) their higher yield (c) a reduced reaction time (d) and, a lower concentration of alcohol needed for the reaction to continue to completion [117]. While using NaOH as a catalyst, it is very essential to maintain its anhydrous form and store it in a dry place as presence of moisture leads to the production of soap during transesterification. Concentrations ranging from 0.5 to 1.0% (w/w) have been found to yield conversion rates of 94 to 99% [160].

9.4.6.1.5 Water activity

The water/moisture content of the reactants play a vital role in the transesterification reaction as water content above 1500 ppm in oil significantly increases soap formation during transesterification. This produced soap also increases the viscosity of the reaction mixture, sometimes causing gel formation, thereby making the separation of glycerol from ester difficult [202].

9.4.6.1.6 Mixing intensity

Mixing plays a significant role in the transesterification reaction, to produce efficient reactivity between oil and alkali (NaOH)-methanol solution and reduce immiscibility in the solution. The stirring is no longer needed once the two phases are mixed properly. The rate of stirring ranges between 180–600 rpm depending upon the type of substrate, alcohol and catalyst. More gentle mixing methods should be implemented to prevent formation of tiny glycerol bubbles (emulsion). Reaction time is also an important factor for determining the yield of methyl esters [116].

9.4.6.1.7   Separation and final treatment of FAME (Biodiesel)

An important step in the production of biodiesel is the final treatment of the raw fatty acid alkyl ester. Once the transesterification reaction is completed, two major products exist: esters (biodiesel) and glycerol. Due to the higher density of glycerol compared to biodiesel phase it settles at the bottom of the reaction vessel, allowing an easy separation from the biodiesel phase. The separation of phase during a transesterification reaction can be observed within 10 min and might be completed after several hours of settling. The initial separation of biodiesel and glycerol can be achieved after prolonged settling time, or a centrifuge can be used to efficient and faster separation of phases [196].

After transesterification, the obtained biodiesel and glycerol are contaminated with unreacted catalyst, alcohol, oil and soap. Although the glycerol contains more contaminants than the biodiesel, yet it also affects the quality of biodiesel. Therefore, crude biodiesel needs to be purified before use [164].

The unreacted alcohol present in the biodiesel is removed by distillation and the process is called alcohol stripping. The biodiesel is then purified using three different types of approaches including: water washing, dry washing and membrane extraction [20, 24, 61, 80].

9.4.6.1.8   Storage stability

Biodiesel is safe to store and the properties of biodiesel should remain stable even after long-term storage. There are several key factors that need to be considered for the storage of biodiesel, including exposure temperature, oxidative stability, fuel solvency, and material compatibility [127]. Viscosity, peroxide value and more dramatically, Rancimat Induction Period have shown changes compared to other properties of biodiesel during storage [123]. Most pure biodiesel are generally stored at 7 and 10°C to avoid the formation of crystals, which can plug fuel lines and fuel filters. Antioxidant additive is necessary to maintain acid value and fuel viscosity and avoid formation of gums and sediments. Additionally, biocides need to add for prevention of biological growth in the fuel. Storage tanks made of aluminum, steel, Teflon®, and fluorinated polyethylene or polypropylene should be selected for biodiesel storage purposes [125].

9.4.6.2 Enzymatic Transesterification

To overcome the problems associated with chemical transesterification, enzymes can be used as catalysts which have several advantages including less downstream processing with no difficulties with product separation, no alkaline wastewater generation and a high degree of product purity especially with the use of lipases as catalysts [62, 86, 116, 133]. The enzymatic transesterification also have several factors affecting the process including oil: alcohol molar ratio, temperature, alcohol, organic solvents, and reaction time. The enzymatic process also has several disadvantages including high reaction time and requirement of highly concentrated catalyst, which is required to complete the reaction leading to very high cost of production. In order to reduce the cost, immobilized enzymes have to be used. However, the activity of the enzymes is reduced significantly within 100 days of application [54].

9.4.6.2.1 Enzyme choice

During the enzymatic transesterification process, the choice of enzyme plays an important role in the quality and yield of biodiesel. Depending on the source of lipase and alcohol used, some of the lipases are capable of converting more than 90% of raw material into biodiesel under different reaction temperatures of 30–50°C and reaction time of 8–90 h. There are 38 unique bacteria which produce the lipase that can be used for biodiesel production including: *Aspergillus niger, Bacillus thermoleovorans, Burkholderia cepacia, Candida antarctica, Candida cylindracea, Candida rugosa, Chromobacterium viscosum, Fusarium heterosporum, Fusarium oxysporum, Getrichum candidum Humicola lanuginose, Oospora lactis, Penicillium cyclopium, Penicillium roqueforti, Pseudomonas aeruginosa, Pseudomonas cepacia, Pseudomonas fluorescens, Pseudomonas putida, Rhizomucor miehei, Rhizopus arrhizus, Rhizopus chinensis, Rhizopus circinans, Rhizopus delemr, Rhizopus fusiformis, Rhizopus japonicus NR400, Rhizopus oryzae, Rhizopus stolonifer NRRL1478, Rhodotorula rubra, Saccharomyces cerevisiae, Staphylococcus hyicus* and *Thermomyces lanuginose* [54, 62]. However, the most effective lipases that are produced for transesterification processes have been obtained from *Candida antarctica, Candida rugosa, Pseudomonas cepacia, Pseudomonas fluorescens, Rhizomucor miehei, Rhizopus chinensis, Rhizopus oryzae* and *Thermomyces lanuginose* [197]. Several studies on enzymatic transesterification of biodiesel from different sources are shown in Table 9.8.

9.4.6.2.2   *Alcohol type*

The presence of an acyl acceptor in the reaction system plays an important role in the enzymatic transesterification process. The methyl and ethyl acetate are the most efficient acyl acceptors but they are more expensive compared to the commonly used alcohols. Therefore commonly used alcohols including: methanol, ethanol, propanol, iso-propanol, 2-propanol, n-butanol and iso-butanol [82, 197]. Breivik et al. [31] reported use of ethanol to perform enzymatic transesterification on fish oil with Novozyme 435 as the enzyme catalyst and reported a 100% conversion yield of fatty acid methyl esters. Salis et al. [161] reported the use of alcohols such as methanol, 2-butanol, ethanol, propanol, 2-methyl-1-propanol using *Pseudomonas cepacia* lipase in triolein and conversion yield of fatty acid esters, which were 40, 83, 93, 99 and 99%, respectively. The study also reported that the lipases are deactivated due to the addition of insoluble methanol to the reaction system, which also affects the conversion yield.

Chen and Wu [37] reported methanol and ethanol as the two most economically feasible alcohols, which can be utilized for the enzymatic transesterification process. However, they both have the property to inhibit and deactivate the enzymes. Especially methanol was found to be the most deactivating alcohol. Transesterification carried out using ethanol as acyl

TABLE 9.8 Enzymatic Production of Biodiesel Using Immobilized Lipases

Feed stock	Lipase	Acyl-acceptor	Alcohol: Substrate	Solvent	Temp. (°C)	Other conditions	Yield (%)	Authors
Cottonseed oil	Candida antarctica	Methanol	-	tert-butanol	-	-	97	Royon et al. [153]
Ethanol	Candida antarctica	Ethanol	-	None	-	-	82	Mittelbach [124]
Fish oil	Novozyme 435	Ethanol		Solvent free	20°C	22 h	100	Breivik et al. [31]
Jatropha oil	Chromobacterium viscosum	Ethanol	4:1	None	40	8 hrs, 200 rpm, addition of 0.5% (w v-1) water	92	Shah et al. [168]
Jatropha oil	Candida antarctica	2-propanol	4:1	hexane	50°C	8 hrs, 150 rpm	92.8–93.4	Modi et al. [126]
Jatropha oil	Pseudomonas cepacia	Ethanol	4:1 molar ratio	-	50	8 hrs, 200 rpm	98	Shah and Gupta [167]
Jatropha oil	Whole cell Rhizopusoryzae	Methanol	3:1	-	30	60 hrs, glutaraldehyde treatment	80	Tamalampudi et al. [190]
Mahua oil	Pseudomonas cepacia	Ethanol	4:1 molar ratio	-	40	6 hrs, 200 rpm	96	Kumari et al. [100]
Palm kernel Oil	Pseudomonas cepacia	Ethanol	-	None	-	-	72	Abigor et al. [2]

TABLE 9.8 (Continued)

Feed stock	Lipase	Acyl-acceptor	Alcohol: Substrate	Solvent	Temp. (°C)	Other conditions	Yield (%)	Authors
Palm kernel	Pseudomonas cepacia	t-butanol	-	None	-	-	62	Abigor et al. [2]
Rapeseed Oil	Candida antarctica	Methanol	-	tert-butanol	-	-	95	Li et al. [110]
Rapeseed oil	Candida sp. 99–125	Methanol	3:1 molar ratio added in three steps	petroleum ether	40	36 hrs, 180 rpm, batch stirred reactor	83	Deng et al. [45]; Nie et al. [132]
Rapeseed oil	Thermomyces lanuginose	Methanol	4:1	tert butanol	35	12 hrs130 rpm	95	Li et al. [110]
Salad oil	Candida sp. 99–125	Methanol	-	n-hexane	40	30 hrs, 180 rpm, batch stirred reactor	95	Deng et al. [45]; Nie et al. [132]
Soybean oil	Candida antarctica	Methanol	-	-	-	preincubated in ethyl oleate for 0.5 hours	97	Samukawa et al. [162]
Soybean oil	Candida antarctica	Methanol	-	none	-	Stepwise addition of methanol	93.80	Watanabe et al. [201]

Soybean oil	Candida antarctica	Methyl acetate	12:1	none	40	14 hrs, 150 rpm	92	Du et al. [47]
Soybean oil	Rhizopus oryzae	Methanol	-	none	-	-	80–90	Kaieda et al. [89]
Soybean oil	Rhizopus oryzae	Methanol	-	-	-	Stepwise addition of methanol, glutaraldehyde treatment	90	Ban et al. [19]
Soybean oil	Pseudomonas cepacia	Methanol	-	None	-	-	67	Noureddini et al. [134]
Soybean oil	Pseudomonas cepacia	Ethanol	-	None	-	-	65	Noureddini et al. [134]
Sunflower Oil	Rhizomucor miehei	Methanol	3:1 molar ratio added in three steps	n-hexane	40	30 hrs, 200 rpm	>80	Soumanou and Born-scheuer [179]
Sunflower Oil	Rhizomucor miehei	Ethanol	3:1 molar ratio added in four steps	n-hexane	40	24 hrs	79.10	Soumanou and Born-scheuer [179]
Sunflower Oil	Thermomyces lanuginose	Methanol	3:1 molar ratio added in three steps	n-hexane	40	30 hrs, 200 rpm	>60	Soumanou and Born-scheuer [179]

TABLE 9.8 (Continued)

Feed stock	Lipase	Acyl-acceptor	Alcohol: Substrate	Solvent	Temp. (°C)	Other conditions	Yield (%)	Authors
Sunflower Oil	Thermomyces lanuginose	1-propanol	3:1 molar ratio added in four steps	-	40	24 hrs	89.80	Deng et al. [45]
Sunflower Oil	Thermomyces lanuginose	2-propanol	3:1 molar ratio added in four steps	-	40	24 hrs	72.80	Deng et al. [45]
Sunflower Oil	Pseudomonas fluoresces	Methanol	4.5:1 molar ratio added in three steps	None	40	24 hours, 200 rpm	>95	Soumanou and Born-scheuer [179]
Sunflower Oil	Pseudomonas fluoresces	Isobutanol	3:1 molar ratio added in four steps	-	40	24 hrs	45.30	Deng et al. [45]
Sunflower oil	Candida antarctica	Methanol	-	None	-	-	3	Mittelbach [124]

Sunflower oil	Candida antarctica	Methanol	3:1 molar ratio added in four steps	Propanol	-	24 hrs	93.20	Deng et al. [45]
Sunflower oil	Candida antarctica	Methyl acetate	12:1	None	40	10 hrs	92	Xu and Wu [203]
Sunflower oil	Mucor miehei	Ethanol	3:1	None	30	5 hrs	83	Selmi and Thomas [165]
Tallow	Candida antarctica	Methanol	3:1	-	30°C	72 hrs, 200 rpm	74	Lee et al. [106]
Vegetable Oil	Candida sp. 99–125	Methanol	-	Petroleum ether	40	30 hours, 180 rpm, batch stirred reactor	96	Deng et al. [45]; Nie et al. [132]
Waste oil	Candida sp. 99–125	Methanol	-	Petroleum ether	40	22 hrs, 180 rpm, three packed bed reactors	92	Deng et al. [45]; Nie et al. [132]

acceptor has been reported to be comparatively better than in methanol because the rate of transesterification increases when the length of carbon chain increases. In addition, ethanol is currently produced from a renewable source and thereby leaves very less environmental footprint and makes the production of biodiesel more economically feasible. Watanabe et al. [201] reported that the inhibiting effects of the lower chain alcohols like methanol and ethanol can be overcome by either the stepwise addition of alcohols or the use of solvents in the reaction medium.

9.4.6.2.3   Presence of solvents in the reaction system

Generally, the solvents are added to the transesterification process to protect the enzyme from inhibition and denaturation by the alcohol, as the solvents increase the solubility of the alcohol. The presence of solvents in the reaction system increases the rate of reaction and ensures a homogenous mixture compared to the solvent-free system. However, the solvents may also lead to glycerol solubility, which can form a coating over the enzyme causing reduction in enzyme performance [99, 197]. Most commonly used solvents in the transesterification process include cyclohexane, hexane, n-heptane, isooctane, petroleum ether, 2-butanol and tert-butanol.

The most efficient solvent among all is tert-butanol, because the solvent is slightly polar, has an enzyme- stabilizing effect and is not influenced by any reactants or products and solvents based on their polarity [42, 57, 117, 143, 192]. The presence of solvents for biodiesel production process has a great advantage when lower chain alcohols such as methanol or ethanol are used, because they are effective to decrease the inhibitory effect of the alcohol. However, the addition of solvent increases the reaction volume, which has to be removed from the final product as they are volatile and hazardous, and this also increases the production cost of biodiesel.

9.4.6.2.4   Alcohol to oil molar ratio

The alcohol to oil molar ratio in a reaction system is determined based on the reaction system, feedstock, enzyme and alcohol used. In general, during the transesterification process, three moles of alcohol are reacted with one mole of triglyceride to convert into three mole of fatty acid methyl esters and one mole of glycerol [8]. The presence of the higher amounts of alcohol induces glycerin (by-product) solubility thereby interfering with the activity of the enzyme.

The excess alcohol also forms droplets that could potentially deactivate enzymes [49]. Stepwise addition of alcohol is more efficient because it minimizes the inhibition level in a solvent free system. On the other hand, using ethanol in biodiesel production resulted in lower inhibition of enzymes by using a higher molar ratio of alcohol to oil [94].

9.4.6.2.5 *Reaction temperature*

The reaction temperature of the transesterification process depends upon the alcohol to oil molar ratio, enzyme stability and the type of solvents. In general, the optimal temperatures for most of the lipases are between 30 and 60°C [54]. Jeong and Park [87] investigated enzymatic transesterification using *Candida antarctica* lipase with methanol at a temperature in the range of 25–55°C and the optimum temperature was found to be 40°C. Lee et al. [105] investigated the biodiesel production process using a combination of two lipases obtained from *Rhizopus oryzae* and *Candida rugose*, and the optimal temperature was found to be 45°C in the methanolysis reaction. Qin et al. [145] reported the optimal temperature of lipase obtained from *Rhizopus chinensis* using methanol to be 30°C. Salis et al. [161] reported the optimum temperature of *Pseudomonas cepacia* lipase tested in butanol as 50°C after 1 h and 40°C after 2 h.

Though there are few studies on application of fish oil as a source of biodiesel, it can be considered as a potential resource. Based on the observations obtained from the other bioresources, the processing technology can be specifically designed for fish oils, optimized and applied, which will be beneficial to the aquaculture industry.

9.5 BIOREFINERY AND BIOPROCESSING

Bioprocessing includes all the different methods and technologies applied for the complete utilization of biological cells and the biomaterials. As evident, from the previous sections significant amount of by-products as a waste are produced during fish and shellfish processing and fish based products manufacture, which will increase with increased production and application of various extra processing steps.

In the current scenario, most of the fish byproducts are utilized as fish feed, pet food or are dumped in the landfill or oceans. Different machines and various optimized unit operations are applied to obtain the minced products and other processed food products. The composition of food products derived from fish are variable and the machines and methods developed over time for processing of food cannot be applied for the by-products, as in most cases the textural and biochemical compositions of by-products are different from the fish derived food products. Water content is a major part of all the biomaterials and varies in different cases depending on sources of various biomaterials. By-products can be a mix of protein and fatty acids as in case of salmon by-products or it can be a mix of protein, chitin and caretonoid as in case of shrimp shells. Depending on the nature of the by-products and the targeted component to be extracted or produced by chemical conversions, different unit-operations, technologies and machineries are decided, aligned and/or optimized.

In many cases, for the further application of the targeted compound present in a complex biological matrix, with high affinity for the bio-matrix, or present in a complex form, the compound is first isolated from the bio-matrix and extracted.

Extraction methods applied for different components varies with the targeted compound and the source matrix. For example, standard methods of total lipid extraction include soxhlet extraction, Roese-Gottlieb extraction, Bligh and Dyer, and Modified Bligh and Dyer extraction methods [27, 118, 177], which are traditional solvent extraction methods. There are several studies and reviews which have analyzed and summarized the application of different solvents. Some of the regularly used solvents include hexane, methanol, acetone, propanol, cyclohexane, petroleum ether, chloroform and a combination of these solvents [118].

Selection of solvents differs based on the nature and solubility of target compounds in the solvent, physico-chemical properties of solvent (boiling point, solubility, miscibility with water/ oil) and the source of the compounds (the biomaterial), which has been discussed in detail for different targeted samples in the previous sections. However, in case of the application of components as a food additive or nutraceutical, solvents which can be applied are limited. Other extraction methods such as supercritical extraction (in case of small scale production or targeted extraction of omega-3/6 fatty acids) and enzymatic extraction (for large scale application) methods have been also applied. Enzymes have been also applied for selective hydrolysis, esterification and transesterification as summarized during various reviews [159, 171] and in the previous sections. Different studies have reported the application of proteases for extraction of oil from the fish matrix, and different lipases for selective hydrolysis and esterification of omega-3 fatty acids [113, 114].

Other extraction methods including ultrasonic extraction and microwave extraction can also be applied at lab-scale, however for the commercialization of these methods, further analyzes and research is still required. These methods are eco-friendly as compared to traditional solvent extraction methods, less time consuming and energy efficient. Detailed discussion about the application of these methods is beyond the scope of this chapter, however specific applications for the fish-byproducts have been discussed in detail in previous sections.

Combination of different extraction and purification methods have been also applied for the isolation of different components from complex matrix. Combination of various methods which can be successfully applied for various fish-byproducts components have been discussed in the previous sections. Another component is astaxanthin, which is a complex carotenoid and is categorized as xanthophyll pigment which can be potentially extracted from the shrimp shells. Astaxanthin is also extracted from the green algae *Haematococcus pluvialis* and the red yeast *Phaffia rhodozyma* [75]. Some other microorganisms which have been reported as an efficient natural source of this compound include microalgae *Chlorella vulgaris* [69, 70], *Chlorella zofingiensis*, and *Chlorococcum* [6] and different fungi [32] and yeasts [6].

Astaxanthin has been extracted from other aquaculture waste such as blue crab (*Callinectes sapidus*) shell waste [53] and underutilized species including krill [5]. Methods of extraction include different solvent extraction methods [5, 6, 36] (solvents include acetone, hexane, ethanol or combination of various solvents, edible oils) and supercritical extraction [163] (with co-solvents including ethanol and edible oil [96]). Some pre-treatment steps can also be added to the extraction of astaxanthin, including fermentation and freeze-drying [11]. Depending on the nature (polarity) of the astaxanthin to be extracted, a step-wise combination of all these methods can be applied for the efficient extraction and storage of astaxanthin compounds. Solvent extraction using organic solvents with different polarities has been extensively applied for astaxanthin extraction, however the major drawback of extraction of astaxanthin from the shrimp shells is the hardness of the shells for which digestion is extremely required.

During past decades, digestion with enzymes and fermentation of the shells has been applied which were observed to increase the extractability of astaxanthin from the shrimp shells [11]. Hence, fermentation or enzymatic digestion followed by solvent extraction can be an effective and economic way of extraction of astaxanthin. Extraction of astaxanthin with supercritical extraction has also been extensively applied for astaxanthin extraction or extraction of oil rich in astaxanthin. However, necessity of freeze-dried raw materials and high cost of supercritical extraction are the major drawbacks of the method and complete extraction of astaxanthin from the shrimp shells with oil might not be possible. However, after the solvent extraction, transfer of astaxanthin into the oil and further concentration of the astaxanthin content in the oil using supercritical fluid extraction might be a viable and cost effective solution and should be further explored. Also transfer of astaxanthin from eco-friendly organic solvent into the oil can lead to astaxanthin rich oil which can be a potential value-added product with different applications.

In certain cases, extraction methods are developed focused on concurrent extraction of several compounds from a biomaterial. Chitin and astaxanthin were simultaneously extracted by Auerswald and Gade [15] from lobster *Jasus lalandii* waste, and it was reported that astaxanthin extracted from this source could be used as an aquacultural feed additive. Similarly, from fish waste, protein and oil can be extracted simultaneously as mentioned in previous sections and the process can be further optimized. During a study, after the optimized super-critical CO_2 extraction of fish oil, the residue was observed to be a potential source of fish protein with intact functionality [191].

To increase the stability or bioavailability of the compounds, the extraction step is followed by additional steps of *purification and encapsulation* of the extracted compound. Chitosan has been reported as a potential stabilizer for astaxanthin, and it was observed that chitosan encapsulated astaxanthin did not undergo isomerization nor chemical degradation under the investigated storage conditions of 25, 35 and 45°C where samples were in desiccators sealed with parafilm and covered with aluminum foil [76]. Oxidative stability, structure, and physical characteristics of micro-capsules formed by spray drying of fish oil with protein and dextrin wall materials were studied by Kagami et al. [88].

The properties of the micro-capsule are affected by the type of stabilizers or encapsulating agent, proportion of the components, method of dispersion of the solute into substrate or solvent or medium, and micro-encapsulation technique. Nano-particle encapsulation of fish oil was conducted by spray drying after preparing sub-micron emulsions via high energy emulsifying techniques, including microfluidization and ultrasonication [83]. During this study, microfluidization was reported as an efficient emulsification technique resulting in fish oil encapsulated powder with the lowest unencapsulated oil at the surface of particles, attributed mainly to its ability to produce emulsions at the nano-range (d43 of 210–280 nm). The encapsulation methods and variables affect the stability and the bioavailability of the target compound also [46, 76, 78].

Manufacture of certain other products involves many unit operations including extraction, bioprocesses including deproteination, demineralization and neutralization (in case of astaxanthin extraction [43]). For production of films, from the materials, including fish myofibrillar protein [41], fish gelatin mixed with the texturizers gellan and κ-carrageenan [143], water soluble fish protein collected from surimi wastewater [29], fish gelatin combined with nanoclay composite film [17] and processes such as sonication (for mixing the sample) [17], casting [41], extrusion or other methods might be more useful. However, studies in this regard are in their preliminary phase which requires further optimization. Continuous research in this area has led to development of many different products and there are several methods which are applied for analyzes of the extracts or products developed during these research projects, which analyzes the composition (spectroscopy [79, 95], liquid chromatography [30, 115], gas chromatography [84, 206], mass spectrometry), quality (microbial count, sensory evaluation) and possible performance (low NO_x emissions, high Cetane number and oxidative stability [3]) during their application.

Based on the discussion throughout this chapter, it can be deduced that through the combination of different bioprocessing methods, fish industry by-products can be converted to food, feed, nutraceuticals, pharmaceuticals and energy based products. This implies that with successful and planned application of *biorefinery concepts*, complete utilization of fishery by-products leading to minimization of waste and new job and business opportunities in this sector. The biorefinery concept for the production of various value added products from fish and shellfish waste is demonstrated in Figures 9.15 and 9.16.

The digestive enzymes present in the fish lead to autolysis causing solubilization of fish-byproduct containing both protein and oil. Also other enzymes such as lipases can cause rapid deterioration leading to loss in quality and yield of fish oil [63]. Separation of these enzymes will slow down the spoilage significantly. Oxygen and microbial spoilage also leads to rapid deterioration of fish tissues and therefore require

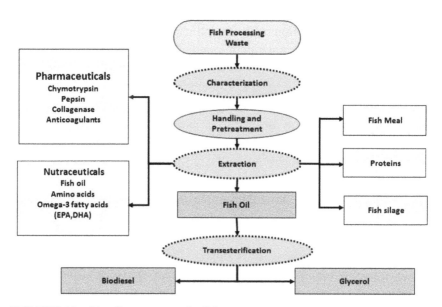

FIGURE 9.15   Biorefinery concept for fish waste.

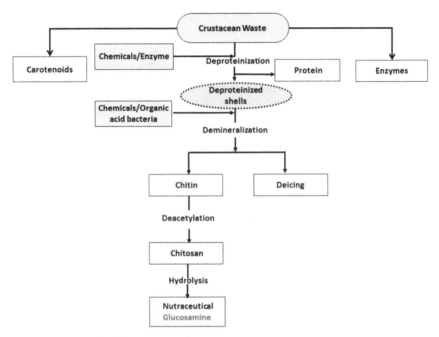

FIGURE 9.16 Biorefinery concept for crustacean waste.

special handling and pretreatment to minimize spoilage. Several studies are going on to improve the handling methods to increase the retention of initial quality of the product. Then the sample can be homogenized for further processing. The homogenized sample can be subjected to thermal/ enzymatic hydrolysis.

With optimization of the different variables and factors, maximum quantity of targeted product with high quality can be obtained. After hydrolysis, the samples can be centrifuged and separated into oil and protein rich fractions which can be further analyzed and applied in appropriate sector. Value added products including: enzymes, omega-3/6/9 fatty acids, biodiesel, protein based films and relatively cheaper products including silage and fish meal (which can be used as animal feed) can be produced simultaneously, by careful step-wise application of technology and various bioprocesses. However, the quality of all these products depends on the quality of the starting sample, its freshness, composition and also on availability of technology in the local area.

Similarly, the crustacean byproducts produced from the shellfish processing industry can be utilized for production of different compounds including astaxanthin, protein hydrolysates and chitin. Shellfish processing industry produces huge amount of wastewater and shrimp shell. Wastewater can be utilized as a source of nutrient for cultivation of microorganisms [42, 149], for production of shrimp flavor compounds using membrane separation methods [112]. Wastewater is also rich in protein content and astaxanthin, however research focused on extraction of these two components from the wastewater is not significantly available, which is currently being further explored by several groups around the world [184]. Also, crustacean heads have been used as fish feeds after fermentation [40]. Crustacean shells can be deproteinated, where protein fraction can be further utilized, and the carotenoid fraction can also be isolated. The remaining fraction can be further utilized for the production of chitin, which can be directly used or can be further converted to chitosan and glucosamine [119].

9.6 CONCLUSIONS

Fish industry is expanding throughout the world and disposal of the processing wastes and by-products is going to be one of the major concerns of the future fish based food market. Fish and shellfish waste resources are a rich source of essential high value nutraceuticals and pharmaceuticals with various biological properties including: collagen, gelatin, chitin, chitosan, carotenoids, glucosamine, protein, amino acids, bioactive peptides, omega-3 fatty acids, glycosoaminoglycans (GAGs), oil, biodiesel, bioethanol, biomethane, calcium and enzymes. Fish byproducts typically comprises of 58% proteins and 19% other extract or fat, in addition to minerals, such as calcium, phosphorous, potassium, sodium, magnesium, iron, zinc, manganese and $_{copper}$. However, systematic application of biorefinery concept is still new to a very large part of the world and further research is going on world-wide. Hence, with application of appropriate bioprocesses and development and application of different technologies and biorefinery concepts can be beneficial to the aquaculture and fishery industries, thus creating new jobs and potentially decreasing the biological waste.

9.7 SUMMARY

Fishery food products including fin-fish and shell fish are considered important part of daily food recipes or delicacies worldwide. Hence, the ever-expanding fisheries industry generates lot of revenue and is a major source of income for a large part of the world. As fin-fish and shell-fish are major source of various essential and health beneficial proteins, amino-acids, polysaccharides, oils, fatty acids, vitamins and minerals and other nutritional elements, large amount of research and extension work is oriented towards the increase in the production and maintenance of high quality of the product obtained.

Food habits vary with different cultures, available resources, climatic conditions and food preferences around the world. Hence, the varieties and species of fish available and consumed in different parts of the world vary. However, in case of all the seafood or inland fish industry, processing of these biomaterials generates massive quantities of waste/ by-products, which will further increase in the future with increased demand and production. In most parts of the world, the waste produced is generally dumped into the water bodies leading to significant increase in biochemical oxygen demand and chemical oxygen demand; or in landfills, which are filling very fast. However, to prevent such a situation and for the sustainable future of fish industry, biorefineries concepts are encouraged and being investigated worldwide, where the underutilized species and processing industry wastes, such as crab and shrimp shells, fish processing waste (guts, head and frame), clam shells and other bio-materials are converted to value-added products.

In past few decades, increasing number of studies have reported the beneficial components of different processing wastes which can be extracted and applied as nutraceuticals or pharmaceuticals. Hence, with effective and careful research and pilot scale application, the fisheries waste resources can be converted to value-added nutraceuticals, pharmaceuticals, cosmetics, functional foods and applied in energy sector, thus generating extra source of income and leading to a sustainable fish processing industry.

This chapter summarizes the processing of different biomaterials collected from the seafood and inland fisheries and their different extractable components. This chapter also highlights some of the latest value-added products and the process technology involved in their development, thus justifying the relevance of using of these biomaterials and the future scope of this industry boosting the blue economy thus leading to the bio-sustainability.

ACKNOWLEDGEMENTS

Authors would like to acknowledge Research and Development Corporation of Newfoundland for their funding to carry out research on fish waste resources and the editors of this book series Dr. Murlidhar Meghwal and Dr. Megh R. Goyal for giving us this great opportunity and for their continued support throughout the process.

KEYWORDS

- astaxanthin
- biocompatibility
- biodiesel
- bioprocessing
- biorefinery
- chemical extraction
- chitin
- chitosan
- crab
- crustacean shells
- deproteinization
- DHA
- enzymatic concentration
- enzymatic extraction
- enzyme
- EPA

- extraction
- fish meal
- fish oil
- fish processing
- fish silage
- fish waste
- fishery
- groundfish
- heat extraction
- high pressure processing
- nutraceutical
- omega-3 fatty acids
- protein
- protein hydrolysates
- purification
- shellfish
- shrimp
- spoilage
- supercritical extraction
- sustainability
- sustainable fisheries
- transesterification
- value-added products
- waste

REFERENCES

1. Aberoumand, A. (2010). Estimation of microbiological variations in minced lean fish products. *World Journal of Fish and Marine Sciences*, 2(3), 204–207.
2. Abigor, R. D., Uadia, P. O., Foglia, T. A., Haas, M. J., Jones, K. C., Okpefa, E., Obibuzor, J. U., & Bafor, M. E. (2000). Lipase-catalyzed production of biodiesel fuel from some Nigerian lauric oils. *Biochemical Society Transactions*, 28(6), 979–981.

3. Adewale, P., Dumont, M.-J., & Ngadi, M. (2015). Recent trends of biodiesel production from animal fat wastes and associated production techniques. *Renewable and Sustainable Energy Reviews*, 45, 574–588.

4. Akoh, C. C., Chang, S.-W., Lee, G.-C., & Shaw, J.-F. (2007). Enzymatic approach to biodiesel production. *Journal of Agricultural and Food Chemistry*, 55(22), 8995–9005.

5. Ali-Nehari, A., Kim, S.-B., Lee, Y.-B., Lee, H.-Y., & Chun, B.-S. (2012). Characterization of oil including astaxanthin extracted from krill (Euphausia superba) using supercritical carbon dioxide and organic solvent as comparative method. *Korean Journal of Chemical Engineering*, 29(3), 329–336.

6. Ambati, R. R., Phang, S.-M., Ravi, S., & Aswathanarayana, R. G. (2014). Astaxanthin: Sources, extraction, stability, biological activities and its commercial applications—A review. *Marine Drugs*, 12(1), 128–152.

7. AMEC (2003). *Management of Wastes from Atlantic Seafood Processing Operations* pp. 1–93.

8. Antczak, M. S., Kubiak, A., Antczak, T., & Bielecki, S. (2009). Enzymatic biodiesel synthesis–key factors affecting efficiency of the process. *Renewable Energy*, 34(5), 1185–1194.

9. Aranaz, I., Mengíbar, M., Harris, R., Paños, I., Miralles, B., Acosta, N., Galed, G., & Heras, Á. (2009). Functional characterization of chitin and chitosan. *Current Chemical Biology*, 3(2), 203–230.

10. Arbia, W., Arbia, L., Adour, L., & Amrane, A. (2013). Chitin extraction from crustacean shells using biological methods-A review. *Food Technology and Biotechnology*, 51(1), 12.

11. Armenta, R. E., & Guerrero-Legarreta, I. (2009). Stability Studies on Astaxanthin Extracted from Fermented Shrimp Byproducts. *Journal of Agricultural and Food Chemistry*, 57(14), 6095–6100.

12. Arnesen, J. A., & Gildberg, A. (2006). Extraction of muscle proteins and gelatine from cod head. *Process Biochemistry*, 41(3), 697–700.

13. Arruda, L. F. d., Borghesi, R., & Oetterer, M. (2007). Use of fish waste as silage: A review. *Brazilian Archives of Biology and Technology*, 50(5), 879–886.

14. Arvanitoyannis, I. S., & Kassaveti, A. (2008). Fish industry waste: treatments, environmental impacts, current and potential uses. *International Journal of Food Science and Technology*, 43(4), 726–745.

15. Auerswald, L., & Gäde, G. (2008). Simultaneous extraction of chitin and astaxanthin from waste of lobsters Jasus lalandii, and use of astaxanthin as an aquacultural feed additive. *African Journal of Marine Science*, 30(1), 35–44.

16. Bacovsky, D., Körbitz, W., Mittelbach, M., & Wörgetter, M. (2007). *Biodiesel Production: Technologies and European Providers*. IEA, Task 39 Report T39-B6. p. 104, Graz, Austria.

17. Bae, H. J., Darby, D. O., Kimmel, R. M., Park, H. J., & Whiteside, W. S. (2009). Effects of transglutaminase-induced cross-linking on properties of fish gelatin–nanoclay composite film. *Food Chemistry*, 114(1), 180–189.

18. Bailey, A. E., & Hui, Y. (1996). Industrial oil and fat products. *Bailey's Industrial Oil and Fat Products*. Wiley Interscience, New York.

19. Ban, K., Kaieda, M., Matsumoto, T., Kondo, A., & Fukuda, H. (2001). Whole cell biocatalyst for biodiesel fuel production utilizing Rhizopus oryzae cells immobilized within biomass support particles. *Biochemical Engineering Journal*, 8(1), 39–43.

20. Barlow, S. M., & Windsor, M. L. (1984). Fishery by-products. *International Association of Fish Meal Manufacturers*, 1–23.

21. Batista, I. (1999). Recovery of proteins from fish waste products by alkaline extraction. *European Food Research and Technology*, 210(2), 84–89.

22. Batista, I., Ramos, C., Mendonça, R., & Nunes, M. L. (2009). Enzymatic hydrolysis of sardine (*Sardina pilchardus*) by-products and lipid recovery. *Journal of Aquatic Food Product Technology*, 18(1–2), 120–134.

23. Benjakul, S., & Morrissey, M. T. (1997). Protein hydrolysates from Pacific whiting solid wastes. *Journal of Agricultural and Food Chemistry*, 45(9), 3423–3430.

24. Bhaskar, N., Benila, T., Radha, C., & Lalitha, R. G. (2008). Optimization of enzymatic hydrolysis of visceral waste proteins of Catla (*Catla catla*) for preparing protein hydrolysate using a commercial protease. *Bioresource Technology*, 99(2), 335–343.

25. Bhaskar, N., & Mahendrakar, N. S. (2008). Protein hydrolysate from visceral waste proteins of Catla (*Catla catla*): Optimization of hydrolysis conditions for a commercial neutral protease. *Bioresource Technology*, 99(10), 4105–4111.

26. Bimbo, A. P. (1998). Guidelines for characterizing food-grade fish oil. *International News on Fats, Oils and Related Materials*, 9(5), 473–483.

27. Bligh, E. G., & Dyer, W. J. (1959). A rapid method of total lipid extraction and purification. *Canadian Journal of Biochemistry and Physiology*, 37(8), 911–917.

28. Boateng, J. S., Matthews, K. H., Stevens, H. N. E., & Eccleston, G. M. (2008). Wound healing dressings and drug delivery systems: A review. *Journal of Pharmaceutical Sciences*, 97(8), 2892–2923.

29. Bourtoom, T., Chinnan, M. S., Jantawat, P., & Sanguandeekul, R. (2006). Effect of select parameters on the properties of edible film from water-soluble fish proteins in surimi wash-water. *LWT – Food Science and Technology*, 39(4), 406–419.

30. Breithaupt, D. E. (2004). Identification and quantification of astaxanthin esters in shrimp *(Pandalus borealis)* and in a microalga (*Haematococcus pluvialis*) by liquid chromatography-mass spectrometry using negative ion atmospheric pressure chemical ionization. *Journal of Agricultural and Food Chemistry*, 52(12), 3870–3875.

31. Breivik, H., Haraldsson, G. G., & Kristinsson, B. (1997). Preparation of highly purified concentrates of eicosapentaenoic acid and docosahexaenoic acid. *Journal of the American Oil Chemists' Society*, 74(11), 1425–1429.

32. Campbell, R., Morris, W. L., Mortimer, C. L., Misawa, N., Ducreux, L. J., Morris, J. A., Hedley, P. E., Fraser, P. D., & Taylor, M. A. (2015). Optimizing ketocarotenoid production in potato tubers: Effect of genetic background, transgene combinations and environment. *Plant Science*, 234(27–37.

33. Carvalho, P. O., Campos, P. R. B., Noffs, M. D., Bastos, D. H. M., & Oliveira, J. G. D. (2002). Enzymic enhancement of ω3 polyunsaturated fatty acids content in Brazilian sardine oil. *Acta Farmaceutica Bonaerense*, 21(2), 85–88.

34. Castillo-Yánez, F. J., Pacheco-Aguilar, R., García-Carreño, F. L., de los Ángeles Navarrete-Del, M., & López, M. F. (2006). Purification and biochemical characterization of chymotrypsin from the viscera of Monterey sardine (*Sardinops sagax caeruleus*). *Food Chemistry*, 99(2), 252–259.

35. Chapleau, N., Mangavel, C., Compoint, J.-P., & de Lamballerie-Anton, M. (2004). Effect of high-pressure processing on myofibrillar protein structure. *Journal of the Science of Food and Agriculture*, 84(1), 66–74.

36. Chen, H.-M., & Meyers, S. P. (1982). Extraction of astaxanthin pigment from crawfish waste using a soy oil process. *Journal of Food Science*, 47(3), 892–896.

37. Chen, J.-W., & Wu, W.-T. (2003). Regeneration of immobilized Candida antarctica lipase for transesterification. *Journal of Bioscience and Bioengineering*, 95(5), 466–469.

38. Choudhury, G. S., & Gogoi, B. K. (1996). Extrusion processing of fish muscle: a review. *Journal of aquatic food product technology*, 4(4), 37–67.

39. Codex (2015). *General standard for food additives. Codex alimentarius international food standards*. pp. 1–396

40. Coward-Kelly, G., Agbogbo, F. K., & Holtzapple, M. T. (2006). Lime treatment of shrimp head waste for the generation of highly digestible animal feed. *Bioresource Technology*, 97(13), 1515–1520.

41. Cuq, B., Aymard, C., Cuq, J.-L., & Guilbert, S. (1995). Edible Packaging Films Based on Fish Myofibrillar Proteins: Formulation and Functional Properties. *Journal of Food Science*, 60(6), 1369–1374.

42. da Silva Copertino, M., Tormena, T., & Seeliger, U. (2009). Biofiltering efficiency, uptake and assimilation rates of *Ulva clathrata* (Roth) J. Agardh (*Clorophyceae*) cultivated in shrimp aquaculture waste water. *Journal of Applied Phycology*, 21(1), 31–45.

43. De Holanda, H. D., & Netto, F. M. (2006). Recovery of components from shrimp (*Xiphopenaeus kroyeri*) processing waste by enzymatic hydrolysis. *Journal of Food Science*, 71(5), C298–C303.

44. Demirbas, A. (2007). Progress and recent trends in biofuels. *Progress in Energy and Combustion Science*, 33(1), 1–18.

45. Deng, L., Xu, X., Haraldsson, G. G., Tan, T., & Wang, F. (2005). Enzymatic production of alkyl esters through alcoholysis: A critical evaluation of lipases and alcohols. *Journal of the American Oil Chemists' Society*, 82(5), 341–347.

46. Djordjevic, D., McClements, D. J., & Decker, E. A. (2004). Oxidative Stability of Whey Protein-stabilized Oil-in-water Emulsions at pH 3: Potential ω-3 Fatty Acid Delivery Systems (Part B). *Journal of Food Science*, 69(5), C356-C362.

47. Du, W., Xu, Y., Liu, D., & Zeng, J. (2004). Comparative study on lipase-catalyzed transformation of soybean oil for biodiesel production with different acyl acceptors. *Journal of Molecular Catalysis B: Enzymatic*, 30(3), 125–129.

48. Dunajski, E. (1980). Texture of fish muscle. *Journal of Texture Studies*, 10(4), 301–318.

49. Encinar, J. M., Gonzalez, J. F., Rodriguez, J. J., & Tejedor, A. (2002). Biodiesel fuels from vegetable oils: transesterification of *Cynara c ardunculus* L. oils with ethanol. *Energy and Fuels*, 16(2), 443–450.

50. Esteban, M. B., Garcia, A. J., Ramos, P., & Marquez, M. C. (2007). Evaluation of fruit–vegetable and fish wastes as alternative feedstuffs in pig diets. *Waste Management*, 27(2), 193–200.

51. Estiasih, T., Ahmadi, K., & Nisa, F. C. (2013). Optimizing conditions for the purification of omega-3 fatty acids from the by-product of tuna canning processing. *Journal of Food Science and Technology*, 5(5), 522–529.

52. FAO, Handling and processing shrimp. 2001. http://www.fao.org/wairdocs/tan/x5931e/x5931e00.htm#Contents, Department of Trade and Industry Torry Research Station.

53. Felix-Valenzuela, L., Higuera-Ciapara, I., Goycoolea-Valencia, F., & Arguelles-Monal, W. (2001). Supercritical CO_2/ethanol extraction of astaxanthin from blue crab (*Callinectes sapidus*) shell waste. *Journal of Food Process Engineering*, 24(2), 101–112.

54. Fjerbaek, L., Christensen, K. V., & Norddahl, B. (2009). A review of the current state of biodiesel production using enzymatic transesterification. *Biotechnology and Bioengineering*, 102(5), 1298–1315.

55. Flick, G. J. (2003). *Novel Applications of High Pressure Processing*. Commercial fish and shellfish technology fact sheet, 6(3), Virginia cooperative extension, Virginia Tech.

56. Friedman, M. (1996). Nutritional value of proteins from different food sources. A review. *Journal of Agricultural and Food Chemistry*, 44(1), 6–29.

57. Fukuda, H., Kondo, A., & Noda, H. (2001). Biodiesel fuel production by transesterification of oils. *Journal of Bioscience and Bioengineering*, 92(5), 405–416.

58. Gamez-Meza, N., Noriega-Rodrıguez, J. A., Medina-Juárez, L. A., Ortega-Garcıa, J., Monroy-Rivera, J., Toro-Vázquez, F. J., Garcıa, H. S., & Angulo-Guerrero, O. (2003). Concentration of eicosapentaenoic acid and docosahexaenoic acid from fish oil by hydrolysis and urea complexation. *Food Research International*, 36(7), 721–727.

59. Gbogouri, G. A., Linder, M., Fanni, J., & Parmentier, M. (2004). Influence of hydrolysis degree on the functional properties of salmon byproducts hydrolysates. *Journal of Food Science*, 69(8), C615–C622.

60. Gbogouri, G. A., Linder, M., Fanni, J., & Parmentier, M. (2006). Analysis of lipids extracted from salmon (*Salmo salar*) heads by commercial proteolytic enzymes. *European Journal of Lipid Science and Technology*, 108(9), 766–775.

61. Gen, M. Y. B. K. (2001). Irradiation processing of fish and shellfish products. *Food Irradiation: Principles and Applications*, 193.

62. Ghaly, A. E., Dave, D., Brooks, M. S., & Budge, S. (2010). Production of biodiesel by enzymatic transesterification: review. *American Journal of Biochemistry and Biotechnology*, 6(2), 54–76.

63. Ghaly, A. E., Dave, D., Budge, S., & Brooks, M. S. (2010). Fish spoilage mechanisms and preservation techniques: Review. *American Journal of Applied Sciences*, 7(7), 859.

64. Ghaly, A. E., Ramakrishnan, V. V., Brooks, M. S., Budge, S. M., & Dave, D. (2013). Fish processing wastes as a potential source of proteins, amino acids and oils: A critical review. *Journal of Microbial and Biochemical Technology*, 5, 107–129.

65. Gildberg, A. (1993). Enzymic processing of marine raw materials. *Process Biochemistry*, 28(1), 1–15.

66. Gildberg, A. (2004). Enzymes and bioactive peptides from fish waste related to fish silage, fish feed and fish sauce production. *Journal of Aquatic Food Product Technology*, 13(2), 3–11.

67. Gildberg, A., & Stenberg, E. (2001). A new process for advanced utilization of shrimp waste. *Process Biochemistry*, 36(8), 809–812.

68. Gonzàlez-Tello, P., Camacho, F., Jurado, E., Paez, M. P., & Guadix, E. M. (1994). Enzymatic hydrolysis of whey proteins: I. Kinetic models. *Biotechnology and Bioengineering*, 44(4), 523–528.

69. Gouveia, L., Choubert, G., Pereira, N., Santinha, J., Empis, J., & Gomes, E. (2002). Pigmentation of gilthead seabream, Sparus aurata (L. 1875), using Chlorella vulgaris (Chlorophyta, Volvocales) microalga. *Aquaculture Research*, 33(12), 987–993.

70. Gouveia, L., Gomes, E., & Empis, J. (1996). Potential use of a microalga (Chlorella vulgaris) in the pigmentation of rainbow trout (Oncorhynchus mykiss) muscle. *Zeitschrift für Lebensmittel-Untersuchung und Forschung*, 202(1), 75–79.

71. Guerard, F., Dufosse, L., De La Broise, D., & Binet, A. (2001). Enzymatic hydrolysis of proteins from yellowfin tuna (*Thunnus albacares*) wastes using Alcalase. *Journal of Molecular Catalysis B: Enzymatic*, 11(4), 1051–1059.

72. Gupta, K., & Jabrail, F. H. (2007). Glutaraldehyde cross-linked chitosan microspheres for controlled release of centchroman. *Carbohydrate Research*, 342(15), 2244–2252.

73. Hall, G. M. (2012). *Fish Processing Technology*. Springer Science and Business Media.

74. He, H., Adams, R. M., Farkas, D. F., & Morrissey, M. T. (2002). Use of high-pressure processing for oyster shucking and shelf-life extension. *Journal of Food Science*, 67(2), 640–645.

75. Higuera-Ciapara, I., Felix-Valenzuela, L., & Goycoolea, F. (2006). Astaxanthin: a review of its chemistry and applications. *Critical Reviews in Food Science and Nutrition*, 46(2), 185–196.

76. Higuera-Ciapara, I., Felix-Valenzuela, L., Goycoolea, F., & Argüelles-Monal, W. (2004). Microencapsulation of astaxanthin in a chitosan matrix. *Carbohydrate Polymers*, 56(1), 41–45.

77. Hinton, D., Masterson, N., Lippert, A., & King, R. (1999). Petroleum An Energy Profile. EIA Publication DOE/EIA-0545, Pittsburgh, PA.

78. Hogan, S. A., O'riordan, E. D., & O'sullivan, M. (2003). Microencapsulation and oxidative stability of spray-dried fish oil emulsions. *Journal of Microencapsulation*, 20(5), 675–688.

79. Holtin, K., Kuehnle, M., Rehbein, J., Schuler, P., Nicholson, G., & Albert, K. (2009). Determination of astaxanthin and astaxanthin esters in the microalgae Haematococcus pluvialis by LC-(APCI) MS and characterization of predominant carotenoid isomers by NMR spectroscopy. *Analytical and Bioanalytical Chemistry*, 395(6), 1613–1622.

80. Homayooni, B., Sahari, M., & Barzegar, M. (2014). Concentrations of omega-3 fatty acids from rainbow sardine fish oil by various methods. *International Food Research Journal*, 21(2), 743–748.

81. Ishihara, M., Nakanishi, K., Ono, K., Sato, M., Kikuchi, M., Saito, Y., Yura, H., Matsui, T., Hattori, H., & Uenoyama, M. (2002). Photocross-linkable chitosan as a dressing for wound occlusion and accelerator in healing process. *Biomaterials*, 23(3), 833–840.

82. Iso, M., Chen, B., Eguchi, M., Kudo, T., & Shrestha, S. (2001). Production of biodiesel fuel from triglycerides and alcohol using immobilized lipase. *Journal of Molecular Catalysis B: Enzymatic*, 16(1), 53–58.

83. Jafari, S. M., Assadpoor, E., Bhandari, B., & He, Y. (2008). Nano-particle encapsulation of fish oil by spray drying. *Food Research International*, 41(2), 172–183.

84. Jalali-Heravi, M., & Vosough, M. (2004). Characterization and determination of fatty acids in fish oil using gas chromatography–mass spectrometry coupled with chemometric resolution techniques. *Journal of Chromatography A*, 1024(1), 165–176.

85. Jayakumar, R., Prabaharan, M., Kumar, P. T. S., Nair, S. V., & Tamura, H. (2011). Biomaterials based on chitin and chitosan in wound dressing applications. *Biotechnology Advances*, 29(3), 322–337.

86. Jegannathan, K. R., Abang, S., Poncelet, D., Chan, E. S., & Ravindra, P. (2008). Production of biodiesel using immobilized lipase—A critical review. *Critical Reviews in Biotechnology*, 28(4), 253–264.

87. Jeong, G.-T., & Park, D.-H. (2008).Lipase-catalyzed transesterification of rapeseed oil for biodiesel production with tert-butanol. In: *Biotechnology for Fuels and Chemicals* pp. 649–657, Springer

88. Kagami, Y., Sugimura, S., Fujishima, N., Matsuda, K., Kometani, T., & Matsumura, Y. (2003). Oxidative stability, structure, and physical characteristics of microcapsules formed by spray drying of fish oil with protein and dextrin wall materials. *Journal of Food Science*, 68(7), 2248–2255.

89. Kaieda, M., Samukawa, T., Kondo, A., & Fukuda, H. (2001). Effect of Methanol and water contents on production of biodiesel fuel from plant oil catalyzed by various lipases in a solvent-free system. *Journal of Bioscience and Bioengineering*, 91(1), 12–15.

90. Kapoor, R., & Patil, U. K. (2011). Importance and production of omega-3 fatty acids from natural sources. *International Food Research Journal*, 18(493–499.

91. Karim, A. A., & Bhat, R. (2009). Fish gelatin: properties, challenges, and prospects as an alternative to mammalian gelatins. *Food Hydrocolloids*, 23(3), 563–576.

92. Kelleher, S. D., & Hultin, H. O. (1991). Lithium chloride as a preferred extractant of fish muscle proteins. *Journal of Food Science*, 56(2), 315–317.

93. Kjartansson, G. T., Zivanovic, S., Kristbergsson, K., & Weiss, J. (2006). Sonication-assisted extraction of chitin from shells of fresh water prawns (*Macrobrachium rosenbergii*). *Journal of Agricultural and Food Chemistry*, 54(9), 3317–3323.

94. Köse, Ö., Tüter, M., & Aksoy, H. A. (2002). Immobilized *Candida antarctica* lipase-catalyzed alcoholysis of cotton seed oil in a solvent-free medium. *Bioresource Technology*, 83(2), 125–129.

95. Krawczyk, S., & Britton, G. (2001). A study of protein–carotenoid interactions in the astaxanthin-protein crustacyanin by absorption and Stark spectroscopy; evidence for the presence of three spectrally distinct species. *Biochimica et Biophysica Acta (BBA)-Protein Structure and Molecular Enzymology*, 1544(1), 301–310.

96. Krichnavaruk, S., Shotipruk, A., Goto, M., & Pavasant, P. (2008). Supercritical carbon dioxide extraction of astaxanthin from Haematococcus pluvialis with vegetable oils as co-solvent. *Bioresource Technology*, 99(13), 5556–5560.

97. Kristinsson, H. G., Lanier, T. C., Halldorsdottir, S. M., Geirsdottir, M., & Park, J. W. (2013). Fish protein isolate by pH shift. In: *Surimi and Surimi Seafood*. Park, J. W., ed., CRC Press.

98. Kristinsson, H. G., & Rasco, B. A. (2000). Fish protein hydrolysates: production, biochemical, and functional properties. *Critical Reviews in Food Science and Nutrition*, 40(1), 43–81.

99. Kumari, A., Mahapatra, P., Garlapati, V. K., & Banerjee, R. (2009). Enzymatic transesterification of Jatropha oil. *Biotechnol Biofuels*, 2(1), 1–6.

100. Kumari, V., Shah, S., & Gupta, M. N. (2007). Preparation of biodiesel by lipase-catalyzed transesterification of high free fatty acid containing oil from Madhuca indica. *Energy and Fuels*, 21(1), 368–372.

101. Kumirska, J., Weinhold, M. X., Thöming, J., & Stepnowski, P. (2011). Biomedical activity of chitin/chitosan based materials—influence of physicochemical properties apart from molecular weight and degree of N-acetylation. *Polymers*, 3(4), 1875–1901.

102. Kurita, K. (2006). Chitin and chitosan: functional biopolymers from marine crustaceans. *Marine Biotechnology*, 8(3), 203–226.

103. Laval, A. (2010). *Multiple choice for fats and oils refining.* Alfa Laval degumming and neutralization solutions, http://local.alfalaval.com/de-de/wichtige-industrien/lebensmittel-molkerei-getraenke/oele/Documents/Pflanzliche%20%C3%96le.pdf. pp. 1–20

104. Laval, A. (2015). *A continuous process for high-grade fish oil and fish protein.* http://local.alfalaval.com/de-de/wichtige-industrien/lebensmittel-molkerei-getraenke/oele/Tierische-Fette/Documents/Fischverarbeitung.pdf. pp. 1–2

105. Lee, J. H., Lee, D. H., Lim, J. S., Um, B.-H., Park, C., Kang, S. W., & Kim, S. W. (2008). Optimization of the process for biodiesel production using a mixture of immobilized Rhizopus oryzae and Candida rugosa lipases. *Journal of Microbiology and Biotechnology*, 18(12), 1927–1931.

106. Lee, K.-T., Foglia, T. A., & Chang, K.-S. (2002). Production of alkyl ester as biodiesel from fractionated lard and restaurant grease. *Journal of the American Oil Chemists' Society*, 79(2), 191–195.

107. Lembke, P. (2013). Production techniques for omega-3 concentrates. In *Omega-6/3 Fatty Acids* pp. 353–364, Springer.

108. Lertsutthiwong, P., How, N. C., Chandrkrachang, S., & Stevens, W. F. (2002). Effect of chemical treatment on the characteristics of shrimp chitosan. *Journal of Metals, Materials and Minerals*, 12(1), 11–18.

109. Leung, D. Y. C., & Guo, Y. (2006). Transesterification of neat and used frying oil: optimization for biodiesel production. *Fuel Processing Technology*, 87(10), 883–890.

110. Li, L., Du, W., Liu, D., Wang, L., & Li, Z. (2006). Lipase-catalyzed transesterification of rapeseed oils for biodiesel production with a novel organic solvent as the reaction medium. *Journal of Molecular Catalysis B: Enzymatic*, 43(1), 58–62.

111. Liaset, B., Lied, E., & Espe, M. (2000). Enzymatic hydrolysis of by-products from the fish-fileting industry; chemical characterisation and nutritional evaluation. *Journal of the Science of Food and Agriculture*, 80(5), 581–589.

112. Lin, C. Y., & Chiang, B. H. (1993). Desalting and recovery of flavor compounds from salted shrimp processing waste water by membrane processes. *International Journal of Food Science & Technology*, 28(5), 453–460.

113. Linder, M., Fanni, J., & Parmentier, M. (2005). Proteolytic extraction of salmon oil and PUFA concentration by lipases. *Marine Biotechnology*, 7(1), 70–76.

114. Linder, M., Matouba, E., Fanni, J., & Parmentier, M. (2002). Enrichment of salmon oil with n-3 PUFA by lipolysis, filtration and enzymatic re-esterification. *European Journal of Lipid Science and Technology*, 104(8), 455–462.

115. López-Cervantes, J., Sánchez-Machado, D. I., Gutiérrez-Coronado, M. A., & Ríos-Vázquez, N. J. (2006). Quantification of astaxanthin in shrimp waste hydrolysate by HPLC. *Biomedical Chromatography*, 20(10), 981–984.

116. Ma, F., & Hanna, M. A. (1999). Biodiesel production: A review. *Bioresource Technology*, 70(1), 1–15.

117. Machek, J., & Skopal, F. (2001). Biodiesel from rapeseed oil, methanol and KOH 3. Analysis of composition of actual reaction mixture. *European Journal of Lipid Science and Technology*, 103(363–371.

118. Manirakiza, P., Covaci, A., & Schepens, P. (2001). Comparative Study on Total Lipid Determination using Soxhlet, Roese-Gottlieb, Bligh & Dyer, and Modified Bligh & Dyer Extraction Methods. *Journal of Food Composition and Analysis*, 14(1), 93–100.

119. Manni, L., Ghorbel-Bellaaj, O., Jellouli, K., Younes, I., & Nasri, M. (2010). Extraction and characterization of chitin, chitosan, and protein hydrolysates prepared from shrimp waste by treatment with crude protease from Bacillus cereus SV1. *Applied Biochemistry and Biotechnology*, 162(2), 345–357.

120. Matassoli, A. L. F., Corrêa, I. N. S., Portilho, M. F., Veloso, C. O., & Langone, M. A. P. (2009). Enzymatic synthesis of biodiesel via alcoholysis of palm oil. *Applied Biochemistry and Biotechnology*, 155(1–3), 44–52.

121. Mbatia, B., Adlercreutz, D., Adlercreutz, P., Mahadhy, A., Mulaa, F., & Mattiasson, B. (2010). Enzymatic oil extraction and positional analysis of ω-3 fatty acids in Nile perch and salmon heads. *Process Biochemistry*, 45(5), 815–819.

122. McNeff, C. V., McNeff, L. C., Yan, B., Nowlan, D. T., Rasmussen, M., Gyberg, A. E., Krohn, B. J., Fedie, R. L., & Hoye, T. R. (2008). A continuous catalytic system for biodiesel production. *Applied Catalysis A: General*, 343(1), 39–48.

123. Meher, L., Sagar, D. V., & Naik, S. (2006). Technical aspects of biodiesel production by transesterification—a review. *Renewable and Sustainable Energy Reviews*, 10(3), 248–268.

124. Mittelbach, M. (1990). Lipase catalyzed alcoholysis of sunflower oil. *Journal of the American Oil Chemists'Society*, 67(3), 168–170.

125. Mittelbach, M., & Gangl, S. (2001). Long storage stability of biodiesel made from rapeseed and used frying oil. *Journal of the American Oil Chemists'Society*, 78(6), 573–577.

126. Modi, M. K., Reddy, J. R. C., Rao, B. V. S. K., & Prasad, R. B. N. (2006). Lipase-mediated transformation of vegetable oils into biodiesel using propan-2-ol as acyl acceptor. *Biotechnology Letters*, 28(9), 637–640.

127. Mudge, S. M., & Pereira, G. (1999). Stimulating the biodegradation of crude oil with biodiesel preliminary results. *Spill Science and Technology Bulletin*, 5(5), 353–355.

128. Murray, J., & Burt, J. R. (2001). *The composition of fish. Ministry of Technology*. Torry Advisory Note No. 38), Torry Research Station.

129. Nalinanon, S., Benjakul, S., Visessanguan, W., & Kishimura, H. (2007). Use of pepsin for collagen extraction from the skin of bigeye snapper (*Priacanthus tayenus*). *Food Chemistry*, 104(2), 593–601.

130. Nalinanon, S., Benjakul, S., Visessanguan, W., & Kishimura, H. (2008). Tuna pepsin: characteristics and its use for collagen extraction from the skin of threadfin bream (*Nemipterus* spp.). *Journal of Food Science*, 73(5), C413-C419.

131. Naturals, N. (2011). *Distilling fact from distillation: Fish oil manufacturing clarified, USA*. https://www.nordicnaturals.com/en/About_Us/Why_Omega-3s/601.

132. Nie, K., Xie, F., Wang, F., & Tan, T. (2006). Lipase catalyzed methanolysis to produce biodiesel: optimization of the biodiesel production. *Journal of Molecular Catalysis B: Enzymatic*, 43(1), 142–147.

133. Nielsen, P. M., Brask, J., & Fjerbaek, L. (2008). Enzymatic biodiesel production: technical and economical considerations. *European Journal of Lipid Science and Technology*, 110(8), 692–700.

134. Noureddini, H., Gao, X., & Philkana, R. (2005). Immobilized *Pseudomonas cepacia* lipase for biodiesel fuel production from soybean oil. *Bioresource Technology*, 96(7), 769–777.

135. Nurdiyana, H., Siti Mazlina, M. K., & Siti Nor Fadhilah, M. (2008). Optimization of protein extraction from freeze dried fish waste using response surface methodology (RSM). *International Journal of Engineering and Technology*, 5(1), 48–56.

136. Ohshima, T., Ushio, H., & Koizumi, C. (1993). High-pressure processing of fish and fish products. *Trends in Food Science and Technology*, 4(11), 370–375.

137. Okada, T., & Morrissey, M. T. (2007). Production of n–3 polyunsaturated fatty acid concentrate from sardine oil by lipase-catalyzed hydrolysis. *Food Chemistry*, 103(4), 1411–1419.

138. Ovissipour, M., Rasco, B., & Bledsoe, G. Aquatic Food Products. In: *Food processing: Principles and applications, Second Edition,* pp. 501–534.

139. Pan, B. S. (1990). Minced fish technology. In: *Seafood: Resources, Nutritional Composition, and Preservation.* pp. 190–210.

140. Percot, A., Viton, C., & Domard, A. (2003). Optimization of chitin extraction from shrimp shells. *Biomacromolecules*, 4(1), 12–18.

141. Pettinello, G., Bertucco, A., Pallado, P., & Stassi, A. (2000). Production of EPA enriched mixtures by supercritical fluid chromatography: from the laboratory scale to the pilot plant. *The Journal of Supercritical Fluids*, 19(1), 51–60.

142. Poulicek, M., Voss-Foucart, M. F., & Jeuniaux, C. (1986).Chitinoproteic complexes and mineralization in mollusk skeletal structures. In: *Chitin in Nature and Technology.* pp. 7–12, Springer.

143. Pranoto, Y., Lee, C. M., & Park, H. J. (2007). Characterizations of fish gelatin films added with gellan and κ-carrageenan. *LWT – Food Science and Technology*, 40(5), 766–774.

144. Puvvada, Y. S., Vankayalapati, S., & Sukhavasi, S. (2012). Extraction of chitin from chitosan from exoskeleton of shrimp for application in the pharmaceutical industry. *International Current Pharmaceutical Journal*, 1(9), 258–263.

145. Qin, H., Yan, X., Yun, T., & Dong, W. (2008). Biodiesel production catalyzed by whole-cell lipase from *Rhizopus chinensis*. *Chinese Journal of Catalysis*, 29(1), 41–46.

146. Ramakrishnan, V. V., Ghaly, A. E., Brooks, M. S., & Budge, S. M. (2013). Extraction of oil from mackerel fish processing waste using Alcalase enzyme. *Enzyme Engineering*, 2(1–10).

147. Ranganathan, S. V., Narasimhan, S. L., & Muthukumar, K. (2008). An overview of enzymatic production of biodiesel. *Bioresource Technology*, 99(10), 3975–3981.

148. Rasco, B., & Bledsoe, G. (2006). Surimi and surimi analog products. In *Handbook of food science, technology, and engineering*

149. Rattanakit, N., Plikomol, A., Yano, S., Wakayama, M., & Tachiki, T. (2002). Utilization of shrimp shellfish waste as a substrate for solid-state cultivation of Aspergillus sp. S1–13: evaluation of a culture based on chitinase formation which is necessary for chitin-assimilation. *Journal of Bioscience and Bioengineering*, 93(6), 550–556.

150. Rinaudo, M. (2006). Chitin and chitosan: properties and applications. *Progress in Polymer Science*, 31(7), 603–632.

151. Robles-Medina, A., González-Moreno, P., Esteban-Cerdán, L., & Molina-Grima, E. (2009). Biocatalysis: towards ever greener biodiesel production. *Biotechnology Advances*, 27(4), 398–408.

152. Rødde, R. H., Einbu, A., & Vårum, K. M. (2008). A seasonal study of the chemical composition and chitin quality of shrimp shells obtained from northern shrimp (*Pandalus borealis*). *Carbohydrate Polymers*, 71(3), 388–393.

153. Royon, D., Daz, M., Ellenrieder, G., & Locatelli, S. (2007). Enzymatic production of biodiesel from cotton seed oil using t-butanol as a solvent. *Bioresource Technology*, 98(3), 648–653.

154. Rubio-Rodríguez, N., Beltrán, S., Jaime, I., Sara, M., Sanz, M. T., & Carballido, J. R. (2010). Production of omega-3 polyunsaturated fatty acid concentrates: a review. *Innovative Food Science and Emerging Technologies*, 11(1), 1–12.

155. Rubio-Rodríguez, N., Sara, M., Beltrán, S., Jaime, I., Sanz, M. T., & Rovira, J. (2008). Supercritical fluid extraction of the omega-3 rich oil contained in hake (*Merluccius capensis–Merluccius paradoxus*) by-products: study of the influence of process parameters on the extraction yield and oil quality. *The Journal of Supercritical Fluids*, 47(2), 215–226.

156. Rubio-Rodríguez, N., Sara, M., Beltrán, S., Jaime, I., Sanz, M. T., & Rovira, J. (2012). Supercritical fluid extraction of fish oil from fish by-products: A comparison with other extraction methods. *Journal of Food Engineering*, 109(2), 238–248.

157. Sabapathy, U., & Teo, L.-H. (1995). Some properties of the intestinal proteases of the rabbitfish, *Siganus canaliculatus* (Park). *Fish Physiology and Biochemistry*, 14(3), 215–221.

158. Sacchi, R., Paduano, A., Fiore, F., Della Medaglia, D., Ambrosino, M. L., & Medina, I. (2002). Partition behavior of virgin olive oil phenolic compounds in oil-brine mixtures during thermal processing for fish canning. *Journal of Agricultural and Food Chemistry*, 50(10), 2830–2835.

159. Sahena, F., Zaidul, I. S. M., Jinap, S., Karim, A. A., Abbas, K. A., Norulaini, N. A. N., & Omar, A. K. M. (2009). Application of supercritical CO_2 in lipid extraction – A review. *Journal of Food Engineering*, 95(2), 240–253.

160. Saka, S., & Kusdiana, D. (2001). Biodiesel fuel from rapeseed oil as prepared in supercritical methanol. *Fuel*, 80(2), 225–231.

161. Salis, A., Pinna, M., Monduzzi, M., & Solinas, V. (2005). Biodiesel production from triolein and short chain alcohols through biocatalysis. *Journal of Biotechnology*, 119(3), 291–299.

162. Samukawa, T., Kaieda, M., Matsumoto, T., Ban, K., Kondo, A., Shimada, Y., Noda, H., & Fukuda, H. (2000). Pretreatment of immobilized Candida antarctica lipase for

biodiesel fuel production from plant oil. *Journal of Bioscience and Bioengineering*, 90(2), 180–183.

163. Sánchez-Camargo, A. P., Martinez-Correa, H. A., Paviani, L. C., & Cabral, F. A. (2011). Supercritical CO 2 extraction of lipids and astaxanthin from Brazilian redspotted shrimp waste (Farfantepenaeus paulensis). *The Journal of Supercritical Fluids*, 56(2), 164–173.

164. Schumacher, J. (2007). Small scale biodiesel production: An overview. Agricultural marketing policy. http://www.ampc.montana.edu/documents/policypaper/policy22.pdf

165. Selmi, B., & Thomas, D. (1998). Immobilized lipase-catalyzed ethanolysis of sunflower oil in a solvent-free medium. *Journal of the American Oil Chemists' Society*, 75(6), 691–695.

166. Senanayake, S. P. J. N., & Rizvi, S. S. H. (2010). Methods of concentration and purification of omega-3 fatty acids. In: *Separation, Extraction and Concentration Processes in the Food, Beverage and Nutraceutical Industries*. pp. 483–505.

167. Shah, S., & Gupta, M. N. (2007). Lipase catalyzed preparation of biodiesel from Jatropha oil in a solvent free system. *Process Biochemistry*, 42(3), 409–414.

168. Shah, S., Sharma, S., & Gupta, M. N. (2004). Biodiesel preparation by lipase-catalyzed transesterification of Jatropha oil. *Energy and Fuels*, 18(1), 154–159.

169. Shahidi, F., Han, X.-Q., & Synowiecki, J. (1995). Production and characteristics of protein hydrolysates from capelin (*Mallotus villosus*). *Food Chemistry*, 53(3), 285–293.

170. Shahidi, F., & Kamil, Y. J. (2001). Enzymes from fish and aquatic invertebrates and their application in the food industry. *Trends in Food Science and Technology*, 12(12), 435–464.

171. Shahidi, F., & Wanasundara, U. N. (1998). Omega-3 fatty acid concentrates: nutritional aspects and production technologies. *Trends in Food Science and Technology*, 9(6), 230–240.

172. Sharma, S. K., Mulvaney, S. J., & Rizvi, S. S. (2000). Food process engineering: theory and laboratory experiments. Wiley-Interscience, New York.

173. Sikorski, Z. E., Naczk, M., & Toledo, R. T. (1981). Modification of technological properties of fish protein concentrates. *Critical Reviews in Food Science and Nutrition*, 14(3), 201–230.

174. Singla, A. K., & Chawla, M. (2001). Chitosan: Some pharmaceutical and biological aspects-an update. *Journal of Pharmacy and Pharmacology*, 53(8), 1047–1067.

175. Šližyte, R., Daukšas, E., Falch, E., Storrø, I., & Rustad, T. (2005). Yield and composition of different fractions obtained after enzymatic hydrolysis of cod (*Gadus morhua*) by-products. *Process Biochemistry*, 40(3), 1415–1424.

176. Šližytė, R., Rustad, T., & Storrø, I. (2005). Enzymatic hydrolysis of cod (Gadus morhua) by-products: Optimization of yield and properties of lipid and protein fractions. *Process Biochemistry*, 40(12), 3680–3692.

177. Smedes, F., & k Askland, T. (1999). Revisiting the development of the Bligh and Dyer total lipid determination method. *Marine Pollution Bulletin*, 38(3), 193–201.

178. Soleimanian, Y., Sahari, M. A., & Barzegar, M. (2015). Influence of processing parameters on physicochemical properties of fractionated fish oil at low temperature crystallization. *Nutrition and Food Science*, 45(1), 2–19.

179. Soumanou, M. M., & Bornscheuer, U. T. (2003). Improvement in lipase-catalyzed synthesis of fatty acid methyl esters from sunflower oil. *Enzyme and Microbial Technology*, 33(1), 97–103.
180. Spencer, K., & Tung, M. (1994).Surimi processing from fatty fish. In: *Seafoods: Chemistry, Processing Technology and Quality.* pp. 288–319, Springer.
181. Spinelli, J., & Dassow, J. A. (1982). Fish proteins: their modification and potential uses in the food industry. In: *Chemistry and Biochemistry of Marine Food Products* Martin, R. E., ed. pp. 13–25, AVI Publishing Company, Westport, Connecticut.
182. Srivastava, A., & Prasad, R. (2000). Triglycerides-based diesel fuels. *Renewable and Sustainable Energy Reviews*, 4(2), 111–133.
183. Stanley, S. A. (2013). Studies on the extraction of chitin and chitosan from different aquatic organisms. *Studies on the Extraction of Chitin and Chitosan from Different Aquatic Organisms*, 12(12), 12–15.
184. Stepnowski, P., Olafsson, G., Helgason, H., & Jastorff, B. (2004). Recovery of astaxanthin from seafood wastewater utilizing fish scales waste. *Chemosphere*, 54(3), 413–417.
185. Stevens, W. F. (2000). Production of chtin and chitosan: refinement and sustainability of chemical and biological processing. *Chitin, Chitosan Korea Society*, 5(3), 183–183.
186. Subramanian, K. A., Singal, S. K., Saxena, M., & Singhal, S. (2005). Utilization of liquid biofuels in automotive diesel engines: an Indian perspective. *Biomass and Bioenergy*, 29(1), 65–72.
187. Sugiyama, K., Egawa, M., Onzuka, H., & Oba, K. (1991). Characteristics of sardine muscle hydrolysates prepared by various enzymic treatments. *Bulletin of the Japanese Society of Scientific Fisheries (Japan)*.
188. Sun, T., Pigott, G. M., & Herwig, R. P. (2002). Lipase-assisted concentration of n-3 polyunsaturated fatty acids from viscera of farmed atlantic salmon (*Salmo salar* L.). *Journal of Food Science*, 67(1), 130–136.
189. Synowiecki, J., & Al-Khateeb, N. A. A. Q. (2000). The recovery of protein hydrolysate during enzymatic isolation of chitin from shrimp Crangon processing discards. *Food chemistry*, 68(2), 147–152.
190. Tamalampudi, S., Talukder, M. R., Hama, S., Numata, T., Kondo, A., & Fukuda, H. (2008). Enzymatic production of biodiesel from Jatropha oil: a comparative study of immobilized-whole cell and commercial lipases as a biocatalyst. *Biochemical Engineering Journal*, 39(1), 185–189.
191. Temelli, F., LeBlanc, E., & Fu, L. (1995). Supercritical CO_2 Extraction of Oil from Atlantic Mackerel (*Scomber scombrus*) and Protein Functionality. *Journal of Food Science*, 60(4), 703–706.
192. Turchini, G. M., Ng, W.-K., & Tocher, D. R. (2010). Fish oil replacement and alternative lipid sources in aquaculture feeds. CRC Press.
193. Ueno, H., Mori, T., & Fujinaga, T. (2001). Topical formulations and wound healing applications of chitosan. *Advanced Drug Delivery Reviews*, 52(2), 105–115.
194. Undeland, I., Kelleher, S. D., & Hultin, H. O. (2002). Recovery of functional proteins from herring (*Clupea harengus*) light muscle by an acid or alkaline solubilization process. *Journal of Agricultural and Food Chemistry*, 50(25), 7371–7379.

195. Valdez-Peña, A. U., Espinoza-Perez, J. D., Sandoval-Fabian, G. C., Balagurusamy, N., Hernandez-Rivera, A., De-la-Garza-Rodriguez, I. M., & Contreras-Esquivel, J. C. (2010). Screening of industrial enzymes for deproteinization of shrimp head for chitin recovery. *Food Science and Biotechnology*, 19(2), 553–557.

196. Van Gerpen, J., Shanks, B., Pruszko, R., Clements, D., & Knothe, G. (2004). *Biodiesel Production Technology: August 2002–January 2004. Other information: PBD: 1.* p. 110, Iowa State University, Renewable Products Development Laboratory, and USDA/NCAUR

197. Vasudevan, P. T., & Briggs, M. (2008). Biodiesel production—current state of the art and challenges. *Journal of Industrial Microbiology and Biotechnology*, 35(5), 421–430.

198. Venugopal, V., Chawla, S. P., & Nair, P. M. (1996). Spray dried protein powder from threadfin bream: preparation, properties and comparison with fpc type-b. *Journal of Muscle Foods*, 7(1), 55–71.

199. Vieira, G. H. F., Martin, A. M., Saker-Sampaiao, S., Omar, S., & Goncalves, R. C. F. (1995). Studies on the enzymatic hydrolysis of Brazilian lobster (*Panulirus* spp) processing wastes. *Journal of the Science of Food and Agriculture*, 69(1), 61–65.

200. Wanasundara, U. N., Wanasundara, P. K. J. P. D., & Shahidi, F. (2005).Novel separation techniques for isolation and purification of fatty acids and oil by-products. In *Bailey's Industrial Oil and Fat Products* pp. 585–621, John Wley and Son's Inc.

201. Watanabe, Y., Shimada, Y., Sugihara, A., & Tominaga, Y. (2002). Conversion of degummed soybean oil to biodiesel fuel with immobilized *Candida antarctica* lipase. *Journal of Molecular Catalysis B: Enzymatic*, 17(3), 151–155.

202. Wright, H., Segur, J., Clark, H., Coburn, S., Langdon, E., & DuPuis, R. (1944). A report on ester interchange. *Oil and Soap*, 21(5), 145–148.

203. Xu, G., & Wu, G.-y. (2003). The investigation of blending properties of biodiesel and No. 0 diesel fuel. *Journal-Jiangsu Polytechnic University*, 15(2), 16–18.

204. Yanez, E., Ballester, D., Monckeberg, F., Heimlich, W., & Rutman, M. (1976). Enzymatic fish protein hydrolyzate: chemical composition, nutritive value and use as a supplement to cereal protein. *Journal of Food Science*, 41(6), 1289–1292.

205. Younes, I., Ghorbel-Bellaaj, O., Nasri, R., Chaabouni, M., Rinaudo, M., & Nasri, M. (2012). Chitin and chitosan preparation from shrimp shells using optimized enzymatic deproteinization. *Process Biochemistry*, 47(12), 2032–2039.

206. Yu, Q. T., Liu, B. N., Zhang, J. Y., & Huang, Z. H. (1989). Location of double bonds in fatty acids of fish oil and rat testis lipids. Gas chromatography-mass spectrometry of the oxazoline derivatives. *Lipids*, 24(1), 79–83.

207. Zhao. L., Budge, S. M., Ghaly, A. E., Brooks, M. S., & D. D. (2011). Extraction, purification and characterization of fish pepsin: A critical review. *Journal of Food Processing and Technology*.

208. Zhou, L., Budge, S. M., Ghaly, A. E., Brooks, M. S., & Dave, D. (2011). Extraction, purification and characterization of fish chymotrypsin: A Review. *American Journal of Biochemistry and Biotechnology*, 7, 104–123.

PART IV

FOODS FOR BETTER HUMAN HEALTH

CHAPTER 10

FORTIFICATION OF PADDY AND RICE CEREALS

SHRUTI PANDEY and A. JAYADEEP

CONTENTS

10.1 INTRODUCTION

Fortification is the practice of adding essential vitamins and minerals (e.g., Iron, vitamin A, folic acid, iodine) to staple foods to improve their value and nutritional content. Fortification is a safe, effective mode to improve public well-being that has been used around the world since the 1920s. Normally fortified foods include staple products such as salt, maize flour, wheat flour, sugar, vegetable oil, and rice. Many diets, especially those of the poor, contain inadequate amounts of vitamins and minerals, which leads to micronutrient deficiency. Since most populations in resource-poor settings do not have access to adequate amounts of food, food fortification is a useful and inexpensive option. One of the most fundamental decisions underlying food fortification scheme is selecting suitable foods to be fortified with the essential micronutrients, deficient in a population's diet. Criteria to identify potential food fortification vehicles generally include selecting a food that is commonly eaten by the target groups, is affordable and accessible all year long, and is processed in such a manner that fortification is technically feasible and can be done economically. Staple foods such as wheat flour, rice and sugar have been popular foods to fortify to address micronutrient deficiencies in several developing countries.

This chapter focuses on the fortification of popular foods with special emphasis on rice, success stories about the fortification of rice, advantages, disadvantages and limitations of fortification.

10.2   MICRONUTRIENT DEFICIENCY DISORDERS

Micronutrient deficiency situation of nutritional insecurity are extensive among an estimated 2 billion people in both developing and developed countries. These "silent epidemics" of vitamins and minerals deficiency affects people of all genders and ages, as well as sure risk groups, most prominently for women, children and the elderly. Nutrient deficiency is a global problem due to poverty in developing nations and hidden hunger or micronutrient deficiency in the developed world. In 2010, Georg addressed combating under-nutrition in developing countries, through fortification of staple foods placing great emphasis on the role of the food industry. This stress on the private sector is mirrored in the workshop [23]. Vitamin deficiency in children can lead to the corneal lesion that can result in blindness. Mild vitamin A and Zinc deficiency lead to low immunity to infections. Iron efficiency anemia results in lethargy, infections, cognitive problems, etc. Iodine deficiency is also a matter of concern as it is required for normal growth, fetal development, and mental activity. Thiamine and riboflavin deficiency and general weakness and lack of resistance to diseases are also rampant. Globally as of 2014,

82 countries authorize fortification: 81 countries plus Punjab province in Pakistan have legislation to fortify wheat flour; of these 12 countries have legislation for fortification of maize products, and 6 countries made rice fortification mandatory [57].

10.3   FOOD FORTIFICATION IN PRACTICE

Food fortification has a long history of being used in industrialized countries for the flourishing control of deficiency of vitamins A and D, several B vitamins (thiamine, riboflavin, and niacin), iodine and iron. Food fortification is a demonstrated and effectual tool for tackling nutritional deficiencies among populations, particularly among 'emergent deficiencies' that were not previously considered a problem [60]. Salt iodization was introduced in the early 1920s in both Switzerland and the United States of America and has since expanded progressively all over the world and is mandatory in some countries [9]. From the early 1940s onwards, the fortification of cereal products with thiamine, riboflavin and niacin became a common practice. Foods for young children were fortified with iron, a practice that has substantially reduced the risk of iron-deficiency anemia in this age group. In more recent years, folic acid fortification of wheat has become widespread in America, a strategy adopted by Canada and the United States and about 20 Latin American countries. Assortment of micronutrients depends not only on their legal status, price, expected bioavailability, stability, and sensory acceptability but also on the product forms decent to the applied fortification technology. In some applications, water-soluble forms might be more appropriate, and in others water insoluble or even oily forms might be ideal [61]. In the less industrialized countries, fortification has become an increasingly attractive option in recent years. Given the success of the relatively long-running program to fortify sugar with vitamin A in Central America, where the prevalence of vitamin A deficiency has been reduced considerably, similar initiatives are being attempted in other parts of the world.

As outlined by the FAO, the most common fortified foods are:
- Milk and Milk products
- Fats and oils

- Cereals and cereal based products
- Accessory food items
- Tea and other beverages
- Infant formulas

10.3.1   CEREAL FORTIFICATION

Cereals are widely consumed by the majority of the world population and hence whole cereals, flour, products made from flour are fortified with variety of fortificants so as to reduce micronutrient malnutrition. The addition of vitamins B1 and B2, iron, niacin, and calcium to wheat flour is a common practice in many developed countries. The addition of other vitamins and minerals is also technologically feasible [14]. Two premixes to fortify wheat flour have been used; the first comprised vitamin A, pyridoxine, folic acid, tocopherol acetate, thiamine, riboflavin, niacin and iron while the second contained calcium, magnesium and zinc. The form of vitamin A most commonly used in the fortification of flour was dry stabilized vitamin A palmitate (type 250-sd) powder form. The water-soluble vitamins (thiamin, riboflavin, niacin, pyridoxine, folate and calcium pantothenate) are used in a crystalline form. The mononitrate salt of thiamine is preferred for this use. Iron is normally used in its reduced elemental form. Careful selection of the physical characteristics of the fortificant compound is important so as to ensure adequate mixing and to minimize segregation on storage in mixing dry fortificants with dry foods, [12]. Wheat Flour enrichment is mandatory in many countries, and the native level in 70% extraction flour (11–12 mg/kg) is enriched up to 44 mg/kg, which is the approximate content of whole-wheat grains. In the United States, corn meal, corn grits, and pasta products also have federal standards for voluntary iron enrichment, and these commodities are mostly enriched by manufacturers similarly to other baked goods such as crackers, rolls, cookies, and doughnuts.

The contribution of fortified iron-to-iron intake is highest in the United States, where it accounts for 20–25% of total iron intake [41]. The contribution of fortified iron to iron intake in the United Kingdom is much lower, around 6% [30–40]. Technology also exists for fortifying whole grains such as rice. This can be done by coating, infusing, or by using extruded grain analogs. The fortified grains are then mixed 1:100 or 1:200 with the normal grains. A sophisticated method of preparing fortified rice grains by first infusing B vitamins and then adding iron, calcium, and vitamin E in separate layers of coating material has been developed [29]. Other commonly fortified foods are breakfast cereals and infant cereals. In industrialized countries, breakfast cereals can potentially provide a significant amount of iron, particularly for children and adolescents. In the United Kingdom, for instance, they can provide up to 15% of total iron intake in 11–12-year-olds. The contribution of fortified iron from infant cereals is potentially much greater because they often provide the major source of iron at a critical time in a child's growth and brain development [29–50]. There are two major disadvantages to being used cereal products as vehicles for iron fortification. First, they contain high levels of phytic acid, a potential inhibitor of iron absorption—up to 1% in whole grains and about 100 mg/ 100 g in high-extraction flours. Second, they are extremely sensitive to fat oxidation during storage when highly bioavailable iron compounds such as ferrous sulfate are added [31].

For organoleptic reasons, cereal flours such as wheat and maize are usually fortified with poorly absorbed elemental iron powders, and rice with ferric orthophosphate or ferric pyrophosphate. Only bread, wheat flour (storage three months), and pasta products, because of their low moisture content, can be fortified with the highest available ferrous sulfate [29]. However, even with these foods, iron absorption will be inhibited by the presence of phytic acid unless an absorption enhancer is present. Although NaFeEDTA would appear to be ideally suited to the fortification of cereal flours and perhaps even pasta products. The usefulness of the fortification of these cereal foods can, therefore, be questioned, because rather low levels of poorly absorbed iron compounds are added without absorption enhancers to products containing phytic acid [4]. Breakfast cereals are similarly fortified with reduced elemental iron and in the absence of vitamin C, the usefulness of this fortification is also doubtful. Infant cereals, by contrast, are fortified with much higher levels of iron (200–500 mg/kg) in the presence of large amounts of vitamin C. More bioavailable iron compounds such as ferrous fumarate are also often used, and even with the electrolytic form of elemental iron, the efficiency of infant cereals to provide a nutritionally useful source of iron has been demonstrated [4, 63].

10.3.2   MILK AND MILK PRODUCTS

The production of a large proportion of milk in the market involves the removal of the cream. Along with the cream, much of the fat-soluble vitamins are also removed. The fortification of milk commonly involves the addition of vitamins A and D.

The possibility exists to add vitamins in their oily form or use water dispersal forms of these since milk is an oil-in-water emulsion. Ease of mixing has been identified as an advantage of using dry, water dispersal forms of fortificants. However, the disadvantage of this was that vitamins were less stable in this form after adding to milk as the protective coating dissolved leaving the vitamin susceptible to degradation [52]. The addition of oily vitamin preparations was recommended after dilution and pre-homogenization with a suitable quantity of milk. The addition of the vitamin mixture prior to the homogenization of the bulk supply facilitates uniform mixing. Fortification of milk with iron has also been attempted, however the relatively low iron bioavailability from milk products can be assumed to be due to the presence of two inhibitory factors, calcium [24] and the milk protein casein [32].

In a series of fortification trials in Chile in which iron-fortified formulas were fed to infants, the improvement of iron status was only modestly in the absence of vitamin C but improved considerably when it was added to the formula. The widespread consumption of iron-fortified (and vitamin C-fortified) formulas by infants in the United States is regarded as the reason for the dramatic fall in the prevalence of anemia over the last 30 years [66]. Whole milk could also be considered as a vehicle for iron fortification, but because of the presence of calcium and casein, an absorption enhancer should be added to improve absorption. Unfortunately, it is difficult to add vitamin C to fluid milk, and it has been reported to degrade rapidly to di keto gluconic acid leading to changes in flavor [25]. Many soluble iron compounds rapidly produce off-flavors when added to milk, owing to the promotion of lipolytic rancidity, oxidative rancidity by the oxidation of free fatty acids, and the partial or complete loss of vitamins A, C, and ß-carotene [11]. After evaluation of a series of compounds, the addition of ferric ammonium citrate has been proposed for liquid milk and skim milk, skim milk concentrate, and dry milk powder [18–64].

The usefulness of milk as a vehicle for iron fortification has been demonstrated in a Mexican school-feeding program [39]. The hemoglobin level of children fed 200 mL milk containing 20 mg iron as ferrous chloride improved by 1 g/dL in 3 months. The study demonstrated that with high levels of added iron, the addition of vitamin C was not essential. As with iron-fortified sugar, when iron-fortified milk is added to tea, coffee, or cocoa, the beverages undergo unacceptable color changes. Iron-fortified milk-based chocolate drinks are also food products that can be usefully targeted to children and adolescents. A variety of products is commercially available, although the phenolic compounds present in cocoa powder readily undergo color changes with soluble iron and also bind iron in the gut and inhibit its absorption. Compounds such as ferrous fumarate, ferrous succinate, ferric saccharate, and ferric pyrophosphate have shown acceptable organoleptic properties, with fumarate showing the highest absorption. The addition of vitamin C would presumably be necessary to overcome the inhibitory factors in the cocoa and milk [34].

Calcium fortification of milk and milk-based beverages has been carried out. Calcium fortificant preparations, including stabilizers and emulsifiers, have been used for this purpose to maintain calcium in suspension so as to improve mouth feel and appearance of products [2]. In Germany a milk-based fruit beverage has been marketed which is fortified with calcium, phosphorous as well as vitamins A, E, B and C.

10.3.3   FATS AND OILS

Margarine is a spread that has been commonly used interchangeably with butter. For this reason, in many countries fortification of this spread with vitamins A and D is practiced, since the food is that it replaces a good source of these vitamins. The vitamin A requirement is met using β-carotene as well as oil soluble vitamin A esters [5]. The oil soluble vitamins add in the required amounts to a portion of warm oil, which is then added to the bulk prior to homogenization. Particularly in the case of margarine with a high content of polyunsaturated fatty acids, vitamin E has also been added. Due to the mild processing conditions only small averages are required to compensate for processing losses: 10 % of vitamins A and D and between 5–15% for vitamin E [52].

Fortification of oil with vitamin A in the form of retinyl palmitate has been attempted in Brazil [51]. Storage studies demonstrated that after 18 months of storage in dark sealed containers losses of more than half of the vitamin content were experienced. When storage was not carried out in the dark, most of the vitamin content was lost after six months. Packaging of the fortified oil in opaque containers was therefore confirmed to be a critical consideration.

Vitamin A fortified oil showed good vitamin retention after five months of storage in sealed metal containers at high temperature and humidity [12]. Partially hydrogenated vegetable oil was used in the study and the fortificant was all-trans-retinyl palmitate, at a level of 491 mg, 10 g of BHA and BHT were used as antioxidants.

10.3.4 ACCESSORY FOOD ITEMS

Although staple foods are generally used as vehicles in food fortification programs, at times when none can be identified which have all the required characteristics, it has been necessary to find other options. In outlining fortification policy in the Philippines, it was stated that condiments may be fortified, but only with nutrients that are deficient in the diet and provided that such food is an appropriate vehicle for the micronutrient and is widely consumed by the general population or is intended for intervention programs to address a deficiency in a specific target population [53].

10.3.4.1 Salt

Salt iodisation began in 1922 in Switzerland and has been implemented in many countries as the major mechanism for eliminating iodine deficiency. Today, IDD remains a problem in many countries. WHO and UNICEF have established the goal of 'Universal Salt Iodization' to be achieved by the end of 1995. Salt has been favored as a carrier for iodine due to its widespread coverage, effectiveness, the simple technology involved and low cost [43]. Based on the suitability of salt as a widely used and low-cost vehicle, fortification of salt with other nutrients has also been attempted [51–49]. Almost all of the development work for the fortification of salt with iron has been conducted in India.

Color changes during storage have been the main problem because the salt in India is relatively crude and contains up to 4% moisture. All soluble iron compounds and vitamin C caused unacceptable color changes. Fortification was possible only with insoluble iron compounds, and ferric orthophosphate was recommended at 1 mg iron per gram salt, so as to provide about 15 mg extra iron per day. When NaHSO4 was added as an absorption promoter, absorption was reported to be 80% that of ferrous sulfate. A small-scale fortification trial in which the fortified salt was included in a school-feeding program demonstrated an improvement in iron status [48–59]. The salt that contains fewer impurities would undoubtedly be easier to fortify, but the extra cost to the consumer is always a major consideration in developing countries. Also, there is always the possibility that the iron-fortified salt will cause unacceptable color reactions if added to vegetables in a meal.

According to the Codex standard for food grade salt, use can be made of potassium or sodium iodides and iodates [19]. The iodates have been found to be more stable than the iodides under a wide range of conditions. Stability studies of iodized salt using potassium iodate as the fortificant demonstrated that there was no significant loss of iodine on storage in polyethylene bags for up to two years, and that boiling of the salt solutions led to the negligible iodine loss.

There have been four major technologies used in the addition of iodine to salt. These are dry mixing, drip feed addition, spray mixing and submersion: Fortification of Iron with salt has also been tried out [51]. It has been reported that use of a fortificant mixture containing 40 ppm of potassium iodate, and 1000 ppm of iron as ferrous sulfate and 10,000 ppm of a permitted stabilizer which rendered good bioavailability of both iodine and iron after prolonged storage.

Fortification of salt with vitamin A has been attempted under laboratory conditions [51]. The fortificant used was dry vitamin A palmitate type

250 SD protected by a lipid. The fortificant was found to be unstable at moisture contents above 2%, since salt is hygroscopic, packaging material with an adequate moisture barrier is desirable. Impurities in the salt were also found to destabilize the vitamin A. The particle size and shape must be such that uniform mixing could be achieved, and segregation does not occur on storage.

10.3.4.2   Monosodium Glutamate

MSG is a condiment that is widely used in many Asian countries. It has GRAS status in Codex Alimentarius as a flavor enhancer. Field studies on MSG fortification with vitamin A have been conducted in the Philippines and Indonesia [5–51]. In Philippines dry vitamin, A type 250 SD was used. This is vitamin A palmitate stabilized in an acacia-lactose matrix. The MSG was first ground to 100 mesh to facilitate mixing with the fortificant. Problems were encountered with segregation of fortificant and carrier, loss of vitamin A activity and color deterioration of the fortified product. Vitamin A acetate 325 L, a granular fortificant stabilized in a gelatin – sucrose matrix, was used in place of the finely powdered vitamin A palmitate 250 SD. Report also says that, under conditions of high humidity, there was a hardening of the gelatin coat with associated loss of vitamin potency.

In the Indonesian study, dry vitamin A palmitate type 250 CWS dispersed in an edible carbohydrate, stabilized with antioxidants and coated with a white protective layer, were used. This product was found to be both hot and cold water miscible. This fortificant was demonstrated to retain half of its potency when stored at 25°C for 18 months. Under moist, dark conditions, half of the activity is retained after seven months. Whereas when subjected to light the fortificant is destabilized more rapidly, the uncoated form of the vitamin is less stable under these conditions. The fortificant mixture was added to MSG at the rate of 0.171 wt. %. The vitamin A was aggregated into clusters so as to minimize the problem of segregation during mixing and storage [58].

Iron fortification of MSG has also been attempted using micronized ferric orthophosphate and zinc stearate coated ferrous sulfate [5]. The coated ferrous sulfate had reduced bioavailability about the uncoated form, but the fortified product was judged to have acceptable color, taste, bioavailability and particle size properties. Preliminary investigations indicated that the inclusion of dry stabilized vitamin A (type 250 SD) into the iron-fortified mixture might be technically feasible [22].

10.3.4.3 Sugar

Sugar has been found to be a suitable vehicle for nutrients in fortification programs in Latin America and the Caribbean. Sugar is an alternative vehicle for iron fortification in regions of the world where it is produced, such as the Caribbean and Central America, but in other developing countries refined sugar consumption is more common in the Middle and upper socioeconomic segments of the population. In the vitamin A fortification of sugar, vitamin A 250-CWS was proven to be the most effective fortificant.

Fortification of sugar with iron has also been attempted [5]. It has been reported that using sodium ferric EDTA as the fortificant, gave a promising result. Segregation of the fortificant and the carrier was not a problem as the iron compound became stuck to the sugar crystals at moisture contents exceeding 1% [13].

Iron from fortified sugar would be expected to be well absorbed if consumed with citrus drinks but poorly absorbed from coffee and tea owing to phenolic compounds or if added to cereal products, owing to phytate. As with salt, the main technical problem is to select a bioavailable iron compound that does not cause unwanted color changes in less pure sugar products. In Guatemala, this was overcome by adding NaFeEDTA [62]. Commercial white cane sugar would appear easier to fortify. The successful addition of several different ferric and ferrous compounds (100–200 mg iron/kg) together with vitamin C has been demonstrated [17]. There were, however, unacceptable color reactions when added to coffee and tea or to certain maize products. A successful fortification trial was reported in Guatemala, where NaFeEDTA added to sugar at 13 mg iron/kg to provide an extra 4 mg iron/day per person increased iron stores in all population groups receiving the fortified product [62].

10.3.4.4   Sauces

Iron fortification of strongly flavored or dark colored food products is often simplified because less care is required in consideration of the avoidance of off-flavors and off-colors. Largely for this reason iron fortification of curry powder demonstrated no technical difficulties [51]. It was reported that NaFeEDTA was used as fortificant in the enrichment of curry powder in South Africa.

Condiments that are traditionally used in developing countries, such as monosodium glutamate, fish sauce, curry powder, and bouillon cubes, could be useful fortification vehicles. Monosodium glutamate is widely used as a flavor enhancer in Asia and has been successfully fortified with ferric orthophosphate and ferrous sulfate encapsulated in zinc stearate [67]. The latter compound had 70% of the relative bioavailability of ferrous sulfate in rodents and the capsule had a melting point of 122°C. Pilot fortification trials with iron-fortified fish sauce or curry powder both fortified with NaFeEDTA, resulted in significant improvement in iron status in the population consuming the fortified products [21, 22]. [21–3]. The success of fortified condiments presumably depends both on the absence of adverse color reactions and on the addition of an absorption enhancer, such as EDTA.

10.3.5   TEA AND OTHER BEVERAGES

Successful attempts to fortify tea have been reported [8]. Finely powdered vitamin A palmitate 250 SD was used by dry mixing with the tea dust. An emulsion of vitamin A palmitate, diluted with 50% sucrose solution was sprayed onto tea leaves. The added vitamin showed excellent storage stability for periods up to 6 months and showed 100% recovery after brewing.

10.3.5.1   Fruit Juices and Drinks

In a majority of the cases, the pH of fruit drinks and juices is below 4.5 and the heat treatment is required for pasteurization. Some loss of heat-labile vitamins, thiamin, folic acid and ascorbic acid, occur as a cause of the thermal treatment. The acidity of these drinks causes problems in stability with vitamin A, folic acid, and calcium pantothenate. Carbonation of these beverages, with the resulting exclusion of oxygen, improves the stability of vitamins. The presence of sulfur dioxide in the fruit juices used in the manufacture of these beverages has been shown to have a detrimental effect on thiamine content [52].

Precautions that should be taken in the production of fortified liquid beverage have been outlined [16]. These included the use of stainless equipment, the addition of vitamins at the latest possible stage, avoidance of excessive aeration or even de-aeration, rapid cooling after thermal treatment and the avoidance of readily auto-oxidisable ingredients. Data reported from a storage stability test of pasteurized multivitamin orange juice demonstrated 50% loss of vitamin A after six months of storage, whereas B-carotene showed only about 6% loss. Riboflavin and Thiamin showed good stability, but vitamin C showed about a 23% loss [52].

10.3.5.2   Coffee

In some populations, coffee is consumed by most adults as well as some children, and it is technically and economically feasible to fortify coffee with iron. It was reported that use of ferrous fumarate in roasted and ground coffee, in which one cup (200 mL) provided 1 mg added iron. The addition of iron to soluble coffee is also relatively easy. It has been reported that the addition of a range of soluble ferrous and ferric compounds was possible [38]. Flavor and color changes are potential problems, and coffee, like tea and cocoa, contains phenolic compounds that strongly inhibit iron absorption [45].

10.3.6   INFANT FORMULAS

An adapted formula is designed to supply the total energy and nutrient requirements of healthy full-term infants during the first year of life. A follow-up formula designed to be a part of a mixed feeding regimen can sometimes be introduced after 4–6 months. Other specialized formulas have been produced which take into account specific nutritional problems of infants. In a study of iron fortification of infant cereals, it was proposed the use of ferrous fumarate and ferrous succinate as they gave rise to no objectionable flavors odors or colors on storage [33]. Ferrous sulfate coated with hydrogenated fats, mono- or di-glycerides and ethyl cellulose caused discoloration on reconstitution with hot milk and hot water.

10.4 FORTIFICATION: PADDY/RICE

Rice is the most popular cereal worldwide, serving as a staple food for 38–39 countries and half of the world's population [37]. Globally rice accounts for 22% of total energy intake [7]. For populations living in many developing countries, rice contributes the greatest percentage of calories and protein. Milling of rice is different from other cereals since the objective is to produce a maximum yield of milled grains rather than flour as with most other cereal grains. Processing the un-hulled rice grain, also called paddy or rough rice, involves cleaning, milling to remove the hull, germ, and bran layers, and sizing to produce white uncoated rice.

The B vitamins and iron are found primarily in the germ and bran layers and are removed in the milling process. Milling of brown rice to white rice removes approximately 80 percent of the thiamin. Other nutrients contained in the bran layer are also lost, including niacin, iron, and riboflavin. Nutrients can be preserved in the rice grain by parboiling. Fortification with Fe during parboiling resulted in 15 to 50 folds increase in grain Fe concentration, depending on the grain properties among different rice varieties. Though, the broken rice of Fe-fortified parboiled rice contained 5 to 7 times the Fe concentration of the full grain, which is often bought and consumed by people in the low-income category [57].

Parboiling results in movement of the nutrients contained in the bran layer in the inner endosperm layer prior to milling and removal of the bran. The parboiling process involves soaking of rough rice and applying heat and then drying and milling. Parboiled rice is produced in India, Bangladesh, Burma, Thailand, Sri Lanka, and other Asian countries by both traditional and modern parboiling processes [54].

Rice enrichment processes in commercial use are: powder and whole grain enrichment. Powder enrichment uses a pre-blended powder mixture of B vitamins (thiamin, riboflavin, niacin, or niacinamide), and iron (ferric orthophosphate–white iron; ferric sulfate– yellow iron, or reduced iron). Ferric orthophosphate is recommended for rice because it is relatively water insoluble and white in color [27]. In powder enrichment, a pre-blended mixture of vitamins and minerals is added to the rice. For parboiled rice, the premix is added soon after milling as the heat and moisture on the grain surface facilitates the powder adhering to the grain. Powder enrichment is less expensive than other types of enrichment; however, higher nutrient losses occur if the rice is rinsed before cooking. Twenty to hundred percent of the enrichment will wash off rice depending on the amount of water used and cooking time [27].

The second and most common type of enrichment is known as "grain" type, generally referred to in the industry as "premix." The vitamins and minerals are applied to the rice grain followed by coatings of a water insoluble substance so they will not rinse off. Usually, these premix grains have high concentrations of nutrients. These grains are then blended with un-enriched milled rice, usually at a ratio of 1 enriched grain to 200 un-enriched to attain the desired enrichment levels in the final product. Another method of grain type enrichment currently being tested is the development of enriched simulated/synthetic rice grains.

Vitamins and minerals are added to artificial grains made from rice flour extruded to form a rice-shaped kernel. These fortified grains are then mixed with regular milled rice to provide the target fortification levels in the final product. Artificial rice grains have provided an opportunity to increase the number of nutrients that can be added. Studies are continuing, as mentioned below, to investigate the feasibility of using this technology for vitamin A fortification. However, concerns are there on blending of the simulated grains with natural products and their consistency after cooking [28].

Another novel method of fortification is by parboiling the paddy. The production of parboiled rice is currently about >100 million tons annually, which is accounts for about half of the world rice crop, especially in Asia and Africa [6]. Processing facilities are also common in countries where parboiled rice is consumed such as India and Bangladesh. Briefly, the process of rice parboiling involves soaking the dehusked paddy rice, steaming and drying before milling to produce parboiled white rice, a preferred form among parboiled rice consumers [10]. The soaking time varies from 6 to 24 h with varying water temperature depending on the technique in each mill, leaving the time for Fe fortification can be implemented by spiking suitable form and concentration of Fe in the soaking water under the optimal soaking conditions. As a result, this does not require major changes to the existing parboiling rice production process and infrastructure.

Adding Fe to the paddy rice in the parboiling process significantly increases both total Fe concentration in the white rice grain and the bioavailable Fe fraction [55–56]. Under laboratory simulated parboiling conditions, Fe fortification has increased Fe concentrations in milled rice for up to 50 folds, compared to the background level of 4–7 mg Fe kg^{-1} dry weight, depending on the varieties and milling time. The fortified Fe effectively penetrated into the interior of the endosperm, which can be demonstrated by Fe localization staining with Perls' Prussian blue. The advantage of Fe penetration into the inner layer of the endosperm after fortification process ensure the retention of adequate Fe after polishing for optimum cooking qualities of rice grain, in contrast to the problem of significant Fe loss from milling of raw rice grains. This advantage also helps high Fe retention after rinsing prior to cooking, a common practice of rice consumers.

10.4.1 CURRENT PROGRAMS OF RICE ENRICHMENT AND FORTIFICATION

Enrichment of rice is only voluntary in United States, and even then most rice sold there is enriched. A recent regulation requires that folic acid also be added to enriched rice. Public health concern over low folate levels in the diets of young women and related increased risk of neural tube defects in infants born to foliate deficient mothers lead to this regulation. In Canada, enrichment of precooked rice is also voluntary. If the product is labeled enriched, then thiamin, niacin, and iron must be added. The addition of B_6, folic acid, and pantothenic acid is optional. Multi-nutrient enriched rice has been on the market since 1981 in Japan. Pantothenic acid, vitamin E, and calcium are added in addition to thiamin, niacin, riboflavin, and iron. In Japanese women, the enriched rice, known as Shingen (meaning "brown rice in the new age") is considered an important step in combating high rates of iron deficiency anemia [35].

Enrichment of rice with iron and thiamin in the Philippines has a long history. Although a law exists mandating the enrichment of all rice in the Philippines, it is currently not enforced. Attempts were made to fortify rice with vitamin A, but percent nutrient loss up to 20% due to washing was considered unacceptably high [20–47]. Studies have been carried out in the Philippines to fortify rice with the iron using ferrous sulfate. Studies are also underway to test the feasibility of marketing and distributing vitamin A-fortified rice in Indonesia, where vitamin deficiency is very common [42]. Fortified rice grains are made from rice flour by extrusion technology to form a rice kernel matching the appearance of local rice. These fortified grains are then blended with normal rice. Many such schemes are also run in India. Ultra rice marketed by PATH is one such one wherein vitamins and iron are packed in extruded rice grain (www.path.org). In India GAIN (Global Alliance for Improved Nutrition), a globally recognized U.N. based foundation has taken up fortification programs in Madhya Pradesh, Rajasthan, Andhra Pradesh and Orissa by fortifying wheat flour and biscuits.

10.4.2 FACTORS AFFECTING THE NUTRIENTS IN FORTIFIED RICE

Stability of the added nutrients in rice depends on enrichment process, storage conditions, washing procedure and cooking practices. Nutrient loss due to washing will be high in powder enriched rice where as coated rice has higher nutrient stability. Fortification of vitamin B, E, and minerals by Hoffmann-La Roche method and Wright method resulted in only 1% cooking loss [15]. However, vitamin A loss was high depending on the coating method [1]. Vitamin A adds to extruded rice resulted in a 25 % loss on storing and cooking [26]. Enrichment of artificial rice with both vitamin A and iron result in vitamin oxidation and discoloration of the products [46].

Bioavailability is another aspect because absorption of nutrient depends on the fortificant used. Bio-available forms of iron give unacceptable traits where as less bio-available compounds like ferric orthophosphate and ferric pyrophosphate does not affect the color and taste of cooked rice [65]. Fortification and consequent sensory quality with respect to the visual appeal, texture and taste also affect consumer acceptability [44]. Biological impacts of fortification depend largely on the region, parasitic load, malabsorption, etc. [36].

10.5   ADVANTAGES OF FOOD FORTIFICATION AS A STRATEGY TO COMBAT MICRO NUTRIENT MALNUTRITION (MNM)

Being a food-based approach, food fortification offers some advantages over other interventions aimed at preventing and controlling MNM. These include:

- if consumed on regular and frequent basis, fortified foods will maintain body stores of nutrients more efficiently and more effectively than willing intermittent supplements.
- fortified foods are also better at lowering the risk of the multiple deficiencies that can result from seasonal deficits in the food supply or a poor quality diet.
- it is an important advantage to growing children who need a sustained supply of micronutrients for growth and development, and to women of fertile age who need to enter periods of pregnancy and lactation with adequate nutrient stores.
- fortification does not require any alteration in the food habits.
- when properly regulated, fortification carries a minimal risk of chronic toxicity.
- fortification has been often more cost-effective than other strategies, especially if the technology already exists and if an appropriate food distribution system is in place

10.6   LIMITATIONS OF FORTIFICATION

Although it is recognized that food fortification can have an enormous positive impact on public health, there are, however, some limitations to this strategy for MNM control:

- While fortified foods contain increased amounts of selected micronutrients, they are not a substitute for a good quality diet that supplies adequate nutrition.
- Infants and young children, who consume relatively small amounts of food, are less likely to be able to obtain their recommended intakes of all micronutrients.
- Technological issues relating to food fortification have yet to be fully resolved, especially with regard to appropriate levels of nutrients, stability of fortificants, nutrient interactions, physical properties, as well as acceptability by consumers including cooking properties and taste.
- The nature of the food vehicle, and the fortificant, may limit the amount of fortificant that can be successfully added. For example, some iron fortificants change the color and flavor of many foods to which they are added.

10.7 CONDITIONS FOR SUCCESSFUL FORTIFICATION PROGRAMS

a. Political Support
b. Industry Support
c. Adequate Legislation
d. Consumer Acceptance
e. No Cultural or Other Objection
f. Availability of Micronutrients
g. Economical Sustainability

10.8 SUMMARY

Fortification is an important component to elevate the nutritional status of the population/community/group, etc. High prevalence of micronutrient deficiency diseases in developing countries can be overcome by fortification of various food products with minerals and vitamins. Various factors, including the cost effectiveness of the fortification in raising absorbable mineral intake in the targeted population, the palatability of the fortified food, and the etiology of deficiency diseases must be considered before initiating a fortification program. However, such a coordinated program must be firmly embedded with primary health care systems. A high degree of commitment from government and industry is required for the successful implementation of the program.

KEYWORDS

- anemia
- beverages
- bioavailability
- calcium pantothenate
- chronic
- coffee
- cognitive problems
- deficiency
- enrichment
- etiology
- feasibility
- folate
- food formula
- fortification
- heat-labile
- malnutrition
- micronutrients
- milling
- native
- niacin
- paddy
- parboiling
- precooked

- **premix**
- **pyridoxine**
- **riboflavin**
- **stimulated**
- **supplements**
- **sustainability**
- **thiamin**
- **vitamins**

REFERENCES

1. Anita Peil, Fred Barrett, Chokyun Rha, & Robert Langer (1982). Retention of Micronutrients by Polymer Coatings Used to Fortify Rice. Journal of Food Science, 47, 260–262.
2. Anonymous (1986). Tracing the marketing allure of calcium fortification. Food Eng. International, 11(12), 17–18.
3. Ballot, D. E., McPhail, A. P., & Bothwell, T. H. (1989). Fortification of curry powder with NaFe(III)EDTA: report of a controlled iron fortification trial. Am. J. Clin. Nutr., 49, 162–169.
4. Barret, F., & Ranum, P. (1985). Wheat and blended foods. In: Iron Fortification of Foods by Clydesdale, F. M., Wiemer, K. L., Eds., Orlando, FL: Academic Press, pages 75–109.
5. Bauernfeind, J. C. (1991). Foods considered for nutrient addition: fats and oils. In: Nutrient Additions to Food by Bauernfeind, J. C., and Lachance, P. A., Eds., Food and Nutrition Press, Connecticut, pp. 123–145.
6. Bhattacharya, K. R. (2004). Parboiling of Rice, American Association of Cereal Chemists, Inc., St. Paul, pp. 184–194.
7. Bierlen, R., Wailes, E., & Cramer, E. (1997). The MERCOSUR Rice Economy. Arkansas Experiment Station Bulletin, No. 954, University of Arkansas, pp. 234–243.
8. Brooke, C. L., & Cort, W. M. (1972). Vitamin A fortification of tea. Food Technology, 26(6), 50–52.
9. Burgy, H., Supersaxo, Z., & Selz, B. (1990). Iodine deficiency disorder in Switzerland one hundred years after Theodar Kocher's survey-A historical review with some new goiter prevalence data. Acta Endocrinological, 123, 577–590.
10. Choudhury, N. H. (1991). Parboiling and Consumer Demand of Parboiled Rice in South Asia, International Rice Research Institute, Manila, pp. 123–136.

11. Cocodrilli, G., & Shah, N. (1985). Beverages. In: Iron Fortification of Foods by Clydesdale, F. M., Wiemer, K. L., Eds., Orlando, FL: Academic Press, pp. 145–154.

12. Combs, G. F., Dexter, P. B., Horton, S. E., & Buescher, R. (1994). Micronutrient fortification and enrichment of P. L. 480 Title II commodities: recommendations for improvement. OMNI, Arlington, VA, pp. 234–254.

13. Cook, J. D., & Reuser, M. E. (1983). Iron fortification: an update. Amer. J. Clin. Nutr., 38, 648–659.

14. Cort, W. M., Borenstein, B., Harley J. H., Oscade, M., & Scheiner. (1976). Nutrient stability of fortified cereal products. Food Technology, 50, 52–61.

15. Cort, W. M., Borenstein, B., Harley, J. H., Osadca, M., & Scheiner, J. (1976). Nutrient stability of fortified cereal products. Food Technology, 30(4), 52–62.

16. De-Ritter, E., & Bauernfeind, J. C. (1991). Foods considered for nutrient addition: juices and beverages. In: Nutrient Additions to Food by Bauernfeind, J. C., Lachance, P. A., Eds., Food and Nutrition Press, Connecticut, pp. 324–354.

17. Disler, P. B., Lynch, S. R., & Charlton, R. W. (1975). Studies on the fortification of cane sugar with iron and ascorbic acid. Br. J. Nutr., 34, 141–148.

18. Edmonson, L. F., Douglas, F. W., & Avants, J. K. (1971). Enrichment of pasteurized whole milk with iron. J. Dairy Sci., 54, 1422–1426.

19. FAO/WHO. (1995). Amendment of the Codex standard for food grade salt to include iodization of salt. Codex Alimentarius Commission, CX/NFSDU 95/5.

20. Florentino, R. F., & M. R. Pedro. (1990). Rice Fortification in the Philippines. In: Combating Iron Deficiency through Food Fortification Technology. INACG, Washington, DC, pp. 112–123.

21. Garby, L. (1985). Condiments. In: Iron Fortification of Foods by Clydesdale, F. M., and Wiemer, K. L., Academic Press, pp. 165–170.

22. Garby, L., & Areekul, S. (1974). Iron supplementation in Thai fish sauce. Ann. Trop. Med. Parasitol, 68, 467–476.

23. Georg Steiger, Nadina, Muller – Fischer, Hector, Cori and Beatrice, Conde – Petit., (2014). Fortification of rice: Technologies and nutrients. Annals of the New York Academy of Sciences, 40, 1–11.

24. Hallberg, L., Brune, M., & Erlandsson, M. (1991). Calcium: effect of different amounts on non-heme and heme-iron absorption in humans. Am. J. Clin. Nut, 53, 112–119.

25. Hegenauer, J., Saltman, P., & Ludwig, D. (1979). Degradation of ascorbic acid (vitamin C) in iron-supplemented cow's milk. J. Dairy Sci, 62, 1037–1040.

26. Hernando, Flores, Nonete, B., Guerra Ana Clájdia, A., Cavalcanti Florisbela, A. C. S., Campos Maria Christina, NA., & Azevedo Andmarília, B. M. (1994). Bioavailability of Vitamin A in a Synthetic Rice Premix, J Food Science, 59, 371–372.

27. Hoffpauer, D. W., & Wright, S. L. (1994). Enrichment of Rice. In: Rice Science and Technology by Marshall, W. E., Wadsworth, J. I., Eds., Marcel Dekker, Inc.: New York, pp. 234–256.

28. Hoffpauer, D. W. (1992). Rice Enrichment for Today. Cereal Foods World, 37, 757–759.

29. Hunnel, J. W., Yasumatsu, K., & Moritaka, S. (1985). Iron Enrichment of Rice. In: Iron Fortification of Foods by Clydesdale, F. M., Wiemer, K. L., Eds., FL: Academic Press, Orlando, pages 234–254.

30. Hunnell, J. W., Yasumatsu, K., & Moritaka, S. (1985). Iron enrichment of rice. In: Iron Fortification of foods by Clydesdale, F. M., Wiemer, K. L., Eds., Orlando, FL: Academic Press, pages 124–136.

31. Hurrell, R. F., Furniss, D. E., Burn, J., Whittaker, P., Lynch, S. R., & Cook, J. D. (1989). Iron fortification of infant cereals. Am. J. Clin. Nut, 49(6), 1274–1282.

32. Hurrell, R. F., Furniss, D. E., & Burri, J. (1989). Iron fortification of infant cereals: a proposal for the use of ferrous fumarate or ferrous succinate. Am. J. Clin. Nutr, 49, 1274–1282.

33. Hurrell, R. F., & Jacob, S. (1996). The role of the food industry in iron nutrition: iron intake from industrial food products. In: Iron Nutrition in Health and Disease by Hallberg, L., and Asp. N., Eds., Lund, Sweden: Swedish Nutrition Foundation, pp. 339–347.

34. Hurrell, R. F., Lynch, S. R., & Trinidad, T. P. (1989). Iron absorption in humans as influenced by bovine milk proteins. Am. J. Clin. Nutr, 49, 546–552.

35. Hurrell, R. F., Reddy, M. B., & Dassenko, S. A. (1991). Ferrous fumarate fortification of a chocolate drink powder. Br. J. Nutr, 65, 271–283.

36. ILSI, (1996). Report of the ILSI Human Nutrition Institute's Fortification Working Group. ILSI, Washington, DC, pp. 11–13.

37. Juliano, B. O. (1993). Rice in Human Nutrition. FAO/IRRI, FAO, Food and Nutrition Series No.26. FAO, Rome, pp. 19–23.

38. Klug, S. L., Patrizio, F. J., & Einstman, W. J. (1973). Iron fortified soluble coffee method for preparing the same. US Patent 4006, 263.

39. Kurtz, F. E., Tamsma, A., & Pallansch, M. J. (1973). Effect of fortification with iron on susceptibility of skim milk and non fat Dry milk to oxidation. J. Dairy Sci, 56, 1139–1143.

40. Lachance, P. A. (1989). Nutritional responsibilities of the food companies in the next century. Food Technol., 43, 144–150.

41. Looker, A. C., Orwoll, E. S., & Johnston, C. C. (1999). Prevalence of low femoral bone density in older U. S. adults from NHANES III. J. Bone and Mineral Research, 12(11), 1761–1768.

42. Lotfi, M. (1997). Micronutrient Initiative. Personal Communication, pages 23–24.

43. Mannar, M. G. V. (1988). Salt iodination – Part 2. IDD Newsletter, 4(4), 11–16.

44. Misaki, M., & Yasumatsu, K. (1985). Rice Enrichment and Fortification. In: Rice Chemistry and Technology. Juliano., B. O., Eds, American Association of Cereal Chemists. Minnesota, pp. 244–265.

45. Morck, T. A., Lynch, S. R., & Cook, J. D. (1983). Inhibition of food iron absorption by coffee. Am. J. Clin. Nutr, 37, 416–420.

46. Murphy, P. A. (1996). Technology of vitamin A fortification in developing countries. Food Technology, 50, 69–74.

47. Murphy, P. A., Smith, B., Hauck, C., K., & Connor, K. O. (1992). Stabilization of Vitamin A in Synthetic Rice Premix. J. Food Science, 57, 437–444.

48. Nadiger, H. A., Krishnamachari, K. A. V. R., & Nadaminu, N. A. (1980). The use of common salt (sodium chloride) fortified with iron to control anemia: results of preliminary study. Br. J. Nutr. 43, 45–51.

49. Narasinga Rao, B. S. (1985). Salt. In: Iron Fortification of Foods by Clydesdale, F. M., Wiemer, K. L., Eds., Orlando, FL: Academic Press, pp. 155–164.

50. National Family Health Survey Report, July 17, 2008, pp. 23–25.
51. Nestel, P. (1993). Food fortification in developing countries. U. S. Agency for International Development, pp. 23–25.
52. O'Brien, A., & Roberton, D. (1993). In: The Technology of Vitamins in Food by Ottaway, P. B., Eds., Chapman Hall, Glasgow, pp. 54–68.
53. Parce, C. J. (1995). Micronutrient fortification of processed foods. Bulletin of Nutrition Foundation of the Philippines, 35(11), 1–3.
54. Pillaiyar, P. (1990). Rice Parboiling Research in India. Cereal Foods World, 35, 225–257.
55. Prom-u-Thai, C., Fukai, S., Godwin, D. I., Rerkasem, B., & Huang, L. (2008). Iron-Fortified Parboiled Rice—A Novel Solution to High Iron Density in Rice-Based Diets. Food Chemistry, 110, 390–398.
56. Prom-u-Thai, C., Glahn, P. R., Cheng, Z., Fukai, S., Rerkasem, B., & Huang, L. (2009). The Bioavailability of Iron Fortified in Whole Grain Parboiled Rice, Food Chemistry, 112(4), 982–986.
57. Prom-u-Thai, C., Longbin, H., Shu, F., & Benjavan, R. (2011). Iron Fortification in Parboiled Rice- A Rapid and Effective Tool for Delivering Iron Nutrition to Rice Consumers. Food & Nutrition Sciences, 2, 323–328.
58. Soekirman, A., & Jalal, F. (1991). Priorities in dealing with micronutrient problems in Indonesia. Proceedings of 'Ending the Hidden Hunger' (a policy conference on micronutrient malnutrition) pp. 71–88.
59. Subar, A. F., & Bowering, J. (1988). The contribution of enrichment and fortification to the nutrient intake of women. J. Am. Diet. Assn., 88, 1237–1245.
60. Theodore, H., & Tulchinsky, K., (2015). The Key Role of Government in Addressing the Pandemic of Micronutrient Deficiency Conditions in Southeast Asia. Nutrients, 2014–2017.
61. Victor, L., Fulgoni, N., & Rita, B. (2015). The Contribution of Fortified Ready-to-Eat Cereal to Vitamin and Mineral Intake in the U.S. Population, NHANES, Nutrients, 2007–2010.
62. Viteri, F. E., Alvarez, E., & Batres, R. (1995). Fortification of sugar with iron sodium ethylenediaminotetraacetate (NaFeEDTA) improves iron status in semirural Guatemalan populations. Am. J. Clin. Nutr., 61, 1153–1163.
63. Walter, T., Dallman, P. R., & Pizarro, F. (1993). Effectiveness of iron-fortified cereal in prevention of iron deficiency anemia. Pediatrics, 91, 976–982.
64. Wang, C. F., & King, R. L. (1973). Chemical and sensory evaluation of iron fortified milk. J. Food Sci., 38, 938–940.
65. Whittaker, P., & Dunkel, V. (1995). Iron, Magnesium, and Zinc Fortification of Food. In: Handbook of Metal-Ligand Interactions in Biological Fluids. Guy Berthon, and Marcel Dekker., Eds., Volume 1, Inc. New York, pp. 109–123.
66. Yip, R., Walsh, K. M., Goldfarb, M. G., & Binkin, N. J. (1987). Declining prevalence of anemia in childhood in a middle-class setting: a pediatric success story. Pediatrics, 80, 330–334.
67. Zoller, J. M., Wolinsky, I., & Paden, C. A. (1980). Fortification of non-staple food items with iron. Food Tech., 23, 38–47.

CHAPTER 11

FUNCTIONAL FOODS FROM THE INDIAN SUBCONTINENT

ARPITA DAS, ARIJIT NATH, and RUNU CHAKRABORTY

CONTENTS

11.1 INTRODUCTION

Diet and health relationship were initially proposed in the 4th century B.C. by Hippocrates. In 1999, European Community concerted Action on Functional Food Science in Europe and tightened the definition of 'functional food.' It declared a food as 'functional' if 'it is satisfactorily demonstrated to affect beneficially one or more target functions in the body, beyond adequate nutritional effects, in a way that is pertinent to either an improved state of health and well-being and/or reduction in risk of disease' [29]. People can take greater control of their health through the food choices they make, knowing that some food can provide specific health benefits. Examples can include fruits and vegetables, whole grains, fortified or enhanced food and beverages, and some dietary supplements. Biologically active components in functional food may impart health benefits or desirable physiological effects. Functional attributes of many traditional foods are being discovered, while new food products are being developed with beneficial components. Consumer interest in the relationship between diet and health has increased the demand for information about functional food. Rapid advances in science and technology, increasing healthcare costs, changes in food laws affecting label and product claims, an aging population, and rising interest in attaining wellness through diet are among the factors fuelling interest in functional food.

Indian traditional food is also recognized as functional food because of the presence of functional components such as body-healing chemicals, antioxidants, dietary fibers, and probiotics. These functional molecules help in weight management and blood sugar level balance and sustain immunity of the body. The functional properties of food are further enhanced by processing techniques such as sprouting, malting, and fermentation [19]. Dating back to Indian civilizations and Indian old literature, every community that lived in India had a clear and separate food belief system. Most of these have been influenced by Aryan beliefs and practices. According to Aryan belief, food was considered as a source of strength and a gift from God [1]. Generally cereals like lentil and rice were the combinations of complementary nutritional elements consumed by Aryans [36]. Cereals offer several challenges from nutrition point of view, especially the swelling of their starch upon cooking, the limited quantity and amino acid profile of their protein fraction, and the limited bioavailability of their mineral content due to relatively low mineral levels [23].

This chapter presents overview of functional foods in India.

11.2 STAPLE FOOD SOURCE OF INDIAN ORIGIN AS FUNCTIONAL FOOD

Cereals are an important economic commodity worldwide. Food ingredients from cereals with nutraceutical properties can contribute to health benefits to many people. Consumption of plant-based food, including fruits, vegetables and whole grains, cereals and nuts as well as intake of marine food plays a pivotal role in disease prevention and health promotion. Cereals like rice, wheat, maize, oats, etc. are now employed in preparation of food that are similar in appearance to conventional food and used in normal diet but have an added advantage of aiding physiological functions along with providing nutrition [31]. Traditionally, Indian food is classified into three main categories. Cooked vegetables, milk, fresh fruits, and honey are meant for the truly wise and are considered as *Satvika* food. Food that bring out the lowest, crass qualities of human behavior such as meat, liquor, garlic, and spicy and sour foods are classified as *Tamasika* food. Food that give enough energy to carryout daily work are categorized as *Rajsika* food [13].

In recent years, cereals have also been investigated regarding their potential use in developing functional food. Cereals are grown over 73% of the total world harvested area and contribute over 60% of the world food production providing dietary fiber, proteins, energy, minerals, and vitamins required for human health. Cereals are also good substrates for probiotics. The good growth of LAB in cereals suggests that the incorporation of a human-derived probiotic strain in a cereal substrate under controlled conditions would produce a fermented food with defined and steady characteristics, and possibly health-promoting properties combining the probiotic and prebiotic concept [9].

11.2.1 RICE

Rice bran contains primarily both soluble and insoluble fiber. Insoluble fiber adds bulk to gastrointestinal track in human causing more frequent stools that pass through the system more quickly, requiring less pressure to expel, and absorbing more bile acids and prevents their re-absorption to the body [46]. Rice bran lowers the serum cholesterol levels in the blood, lowers the level of bad low-density lipoprotein (LDL) and increases the level of good high-density lipoprotein (HDL) level, aids in cardiovascular health. LDL/HDL ratio is a reliable marker for coronary heart diseases, higher the ratio more will be the risk of coronary heart diseases [6]. Rice bran contains phosphorus, potassium, magnesium, calcium, manganese and other trace elements. Magnesium improves glycemic control and helps prevent insulin resistance. Rice bran also contain alpha lipoic acid, which can assist in metabolizing carbohydrates and fats thus called Metabolic antioxidants, which lowers glycemic index and controls body weight. Other than alpha lipoic acid, rice bran also contains energy boosting phytonutrients like CoQ10 and B vitamin including pangamic acid. They are vital for energy metabolism and electron transport in the mitochondria and serves as an effective intercellular antioxidant. Rice bran is potentially a valuable source of natural antioxidants such as tocopherols, tocotrienol and oryzanol [17].

Increased concern over the safety of synthetic antioxidants like butylated hydroanisole (BHA) and butylated hydroxytoluene (BHT) has increased the interest in finding effective and economical natural antioxidants. Antioxidants extracted from rice bran potentially could satisfy this demand. Protein quality of rice surpasses that of wheat and corn while it is just inferior to oats. Also, rice protein is hypoallergenic and contains good quantity of lysine. Thus it may act as a suitable ingredient for infant food formulations while adding variety to the restricted diets of children with food allergies [8, 18].

A global interest in rice and its fermented product is increasing due to their calorie value, unique quality characteristics and high acceptability [41]. Cereal/legume-based food are a major source of economical dietary energy and nutrients worldwide. Among the fermented food of India, *idli*, a fermented steamed product with a soft and spongy texture is a highly popular and widely consumed snack food in India. The lactic acid bacteria *Leuconostoc mesenteroides*, *Streptococcus fecalis*, *Lactobacillus delbrueckii*, *Lactobacillus fermenti*, *Lactobacillus lactis* and *Pediococcus cerevisiae* have been found to be responsible for the fermentation process [26, 27]. *Dosa* batter is similar to *idli* batter but the batter is thinner [42]. Dhokla is also similar to *idli*, but black gram is replaced by Bengal gram in the preparation. A mixture of rice and chickpea flour is also used as the substrate for the fermentation [5]. *Selroti* is another popular fermented rice-based ring shaped, spongy, pretzel like, deep fried food item commonly consumed in Sikkim and Darjeeling hills in India, Nepal and Bhutan [47]. The traditional semi-fermented food used by the Bhotiyas in Uttaranchal of India is called *sez*. It is made from rice, and is mostly used as snacks. Earlier, it was a delicacy and was prepared only during certain festivals [30].

11.2.2 WHEAT

Wheat is one of the major grains in the diet of vast number of the world's population and, therefore, can play an important role in the nutrition quality of the diet and human health. The bran and germ fraction of wheat are high in vitamins and minerals. Whole grains are a rich source of magnesium, a mineral that acts as a co-factor for more than 300 enzymes, including enzymes involved in the body's use of glucose and insulin secretion. The FDA permits food that contain at least 51% whole grains by weight (and are also low in fat, saturated fat, and cholesterol) to display a health claim stating consumption is linked to lower risk of heart disease and certain cancers [44]. Wheat antioxidants are mainly concentrated in bran layers and the amount of antioxidants depends largely on the grain variety, with red wheat generally containing higher levels than white wheat [20].

Jalebi is a sweetened fermented product made from refined wheat flour, curd and water. The fermented batter is deep fat fried in oil in spiral shapes and immersed in sugar syrup for few minutes. *Lactobacillus fermentum*, *L. Buchneri*, *Streptococcus lactis*, and *Saccharomyces cerevisiae* were found in the fermented batter [42]. Various types of traditional wheat based fermented snack food like Bhatura (white wheat flour product), kulcha (white wheat flour product), Nan (wheat flour product), are prepared indigenously in India. For the fermentation of those products, mainly *Saccharomyces cerevisiae* and lactic acid bacteria are used [34].

11.2.3 DAIRY PRODUCTS

Milk and dairy products have been associated with health benefits for many years containing bioactive peptides, probiotic bacteria, antioxidants, vitamins, specific proteins, oligosaccharides, organic acids, highly absorbable calcium, conjugated linoleic acid and other biologically active components with an array of bioactivities: modulating digestive and gastrointestinal functions, haemodynamics, controlling probiotic microbial growth and immunoregulation. Consumer's increasing interest for maintaining or improving their health by eating these specific food products has led to the development of many new functional dairy products. These dairy products contain many functional ingredients that decrease the absorption of cholesterol, can significantly reduce blood pressure, play role in the regulation of satiety, food intake and obesity-related metabolic disorders and may exert antimicrobial effects [7].

Probiotics have been defined as living bacteria and supportive substances that have beneficial effects on the host by improving the bacterial balance in the intestine [15]. This definition was later expanded to include living bacteria or mixed bacteria that have beneficial effects on the gastrointestinal and respiratory system of the host by improvement of the balance of intestinal flora [33]. *Lactobacillus* and *Bifidobacteria* are examples of genera of which some of the species are promising probiotics [32]. These microorganisms are gram-positive lactic acid producing bacteria that constitute a major part of the normal intestinal microflora in animals and humans [11, 39]. Probiotic bacteria are successfully incorporated in some dairy product like curd, yogurt, etc. Table 11.1 presents some functional foods which are available in India.

Potential probiotic lactic acid bacteria therefore attribute to these characteristics. They should be safe, viable in delivery vehicles, resistant to acid, tolerant to bile, have ability to produce antimicrobial substances, adhere to epithelial tissue, colonize the gastro-intestinal tract, stimulate a host immune response and influence metabolic activities such as vitamin production, cholesterol assimilation and lactose activity [40]. Food products with a short shelf-life (2–3 weeks) such as yogurt and fermented milks are the most common probiotic food available, although products with a longer shelf-life, such as probiotic Cheddar cheeses, have been developed more recently [28]. Dried preparations of live probiotic cultures are most convenient for long-term preservation and use in functional food applications. Freeze drying is the most frequently used method for the production of probiotic- containing powders, although the exposure of bacterial cells to the attenuating effects of freezing and dehydration can lead to cell injury and decreased viability in many cases [43].

Dahi or Indian yogurt is a lactic acid fermented product of cow or buffalo milk. It is consumed directly either as sweetened or as salted and spiced form [42]. *Dahi* is rich in lactic acid bacteria and demonstrates probiotic effect, which helps in intestinal health. Bacterial cultures help in controlling diarrhea in children [2]. The bioactive compounds produced by lactic acid bacteria such as diacetyl, hydrogen peroxide, and reuterin

TABLE 11.1 Some of the Functional Food Products Available in India

Name of the Product	Name of the Company	Base Product	Main Components
Aashirvaad atta with Methi	ITC-Aashirvaad	Whole wheat flour	Unique combination of fenugreek leaves and traditional spices, blended with whole wheat flour.
Elaichi tea	Twinings	Tea	Grade-A cardamom pods.
Farmlite	ITC-Sunfeast	Biscuit	Oats and wheat fiber with protein and iron.
Fruit yogurt	Mother Dairy	Yogurt	Lactobacillus, curd and fruits pulp
Marie light	ITC-Sunfeast	Biscuit	Enriched with natural wheat fiber with 0% Transfat and 0% Cholesterol.
NutriChoice 5 grain	Britannia	Biscuit	Oats, corn, ragi, rice (low in fat) and wheat.
Rice bran oil	Fortune	Oil	Naturally enriched with Oryzanol and Vitamin A, D, E.
Taj Mahal lemon tea	Brooke Bond	Black tea	Lemon.
Taj Tea-Masala	Brooke Bond	Black tea	Unique and balanced mixture of ginger, cardamom, nutmeg and cinnamon flavorings.
Top herb	Biskfarm	Biscuit	Enriched with herbs.

suppress the normal growth of undesirable flora, especially *E. coli, Bacillus subtilis,* and *Staphylococcus aureus* [35]. *Rabdi* is a milk-cereal based fermented product made from cooked maize flour and buttermilk. In addition with these, there are other fermented milk products like *paneer, shreekhand, misti dahi* are consumed in India. *Sandesh* is a protein-rich Indian milk product prepared by heat and acid coagulation. It has a characteristic aroma and is a rich source of vitamins A and D. Studies are being conducted to improve the nutritional value of traditional *sandesh* by incorporating herbs that adds to the antioxidant value of the product [4].

11.2.4 HERBS AND OTHERS

Herbs are as old as human civilization and they have provided a complete storehouse of remedies to cure acute and chronic diseases. The knowledge of herbals has accumulated over thousands of years provides effective means of ensuring health care [25]. Many edible plants, which are rich in specific constituents, are referred to as phytochemicals may have health promoting effects. These phytochemicals have the potential to be incorporated into food or food supplements as nutraceuticals. According to Dillard and German [12] the health promoting effects of phytochemicals and nutraceuticals and/or functional foods are due to a intricate mix of biochemical and cellular interactions which together promote overall health of the individual.

Garlic has a unique flavor and health-promoting functions which generally attributed to its rich content of sulfur-containing compounds, i.e., alliin, g-glutamylcysteine, and their derivatives. Processing a fresh and intact garlic bulb by crushing, grinding, or cutting induces the release of the vacuolar enzyme alliinase, which very quickly catalyzes alliin to allicin [3]. Garlic and its associated sulfur components are reported to suppress tumor incidence in breast, colon, skin, uterine, esophagus and lung cancers. A recent meta-analysis also showed that a high intake of garlic may be associated with decreased risks for stomach and colorectal cancer [24].

Ginger (*Zingiber officinale* Roscoe, Zingiberacae) is widely used around the world in food as a spice. Recently, it has been shown that (6)-gingerol is endowed with strong antioxidant action both under *in vivo* and *in vitro* conditions, in addition to strong anti-inflammatory and anti-apoptotic actions [6, 21].

Piper nigrum Linn. belongs to the family *Piperaceae* and its dried unripe fruit is used commonly as "black pepper." Phytochemical studies on black pepper have determined the presence of various minerals, vitamins (), polysaccharides (arabinose, rhamnose, galacturic acid), sterols (), fatty acids (linoleic acid), volatile oils (camphenes, pinenes), alkaloids (piperine, piperidine, piperolein, capsaicin, 2-dihydrocaspaicin), resins (chavicin), organic acids (hexadecanoic acid, octadecanoic acid), amides (pipnoohine, pipyahyine, guineensine, pipericide) and various phenolic compounds (benzamides, gallic acid, kaempferol, coumarins, quercetin) [14, 37, 38].

Turmeric, as a spice, is used by the Indian population for its color and flavor. In the Indian system of indigenous medicine it has also been prescribed for a few human diseases [10, 22] and is well known for its cosmetic properties [45]. These herbs are regularly used in every day's food as condiment.

Green tea, a time-honored Chinese herb, might be regarded as a functional food because of its inherent anti-oxidant, anti-inflammatory, antimicrobial and antimutagenic properties. They are attributed to its reservoir of polyphenols, particularly the catechin, epigallocatechin-3-gallate. Owing to these beneficial actions, this traditional beverage was used in the management of chronic systemic diseases including cancer. Recently, it has been emphasized that the host immuno-inflammatory reactions destroy the oral tissues to a greater extent than the microbial activity alone. Green tea with its wide spectrum of activities could be a healthy alternative for controlling these damaging reactions seen in oral diseases, specifically, chronic periodontitis, dental caries and oral cancer, which are a common occurrence in the elderly population [16].

11.3 CONCLUSIONS

The realization that food has a role beyond provision of energy and body forming substances has shifted scientific investigations with growing interest in the research and development of functional food. Whole-grain cereals protect the body against age-related diseases such as diabetes, cardiovascular diseases and some cancer due to the presence of fiber and micronutrients in the outer layer and germ fractions of the grain acting together to combat oxidative stress, inflammation, hyperglycaemia and carcinogenesis. Oxidative stress is associated with these metabolic diseases. Whole-grain cereals are a good source of vitamin E, folates, phenolic acids, zinc, iron, selenium, copper, manganese, carotenoids, phytic acid, lignins, lignans, and alkylresorcinols. Fermentation of cereal-based products with probiotic organisms has open up a new horizon of functional food along with the conventional probiotic enriched dairy products.

Presently there are few commercialized functional products in the Indian market. With India's strong tradition of consuming natural healthy food, the market of functional food is likely to double in the next five years. The highest growth is likely to be in sub categories such as energy drinks, enhanced shelf stable juices, probiotics, and omega fortified food and beverages. Products already in the market are food-containing probiotics like yogurt, Yakult and buttermilk. Probiotics produce a favorable environment for nutrient absorption, and promote the health of gastrointestinal tract. Fortified Juices, gluten free products, lactose free products, energy bars and fortified milk shakes, cereals with added fiber (Atta Mixes) Sprouted cereals /pulses, herb fortified bakery products are also equally consumed by the aware people as compared to conventional food. The development of new functional ingredients has the advantage that food manufacturers can add extra value to products the consumer is already familiar with. Indian consumers getting higher incomes are spending more on healthy food and India is expected to become the fifth largest consumer market in the world by 2025 from being the 12th largest currently.

11.4   SUMMARY

The term "functional food" (FF) refers to food supplements, containing nutrients or other substances (in a concentrated form) that promotes nutritional or physiological benefits. Within the last decade, consumers have made increasing reference to "functional food," recognizing the relationship between nutrition and health to the point of eschewing an overreliance on pharmaceuticals and regarding prescription drugs as often being unnecessary, too expensive, and un-safe and of dubious benefit once all the risks are considered. These functional food products are resulted of technological innovation at the processing level, such as multigrain biscuits, herb fortified tea and dairy products fermented with specific lactic acid bacteria.

This review chapter introduces the concepts of functional food in India and describes several traditional functional food across various regions of India. Food and nutrition science have moved from identifying and correcting nutritional deficiencies to design food that promote optimal health and reduce the risk of disease. In the era of globalization of the population and international food trading, health conscious citizens around the globe will benefit from the wealth of knowledge on traditional Indian and functional food of Indian origin.

KEYWORDS

- **antioxidants**
- **cereals**
- **fermented food**
- **functional food**
- **herbs**
- **India**
- **prebiotics**
- **probiotics**
- **rice**
- **tea**
- **wheat**

REFERENCES

1. Achaya, K. T. (1994). Indian food: a historical companion. Delhi: Oxford University Press, India.
2. Agarwal, K. N., & Bhasin, S. K. (2002). Feasibility studies to control acute diarrhea in children by feeding fermented milk preparations Actimel and Indian Dahi. Eur J Clin Nutr., 56(4), S56–S59.
3. Amagase, H., Petesch, B. L., Matsuura, H., Kasuga, S., & Itakura, Y. (2011). Intake of garlic and its bioactive components. J. Nutr., 131, 955S–962S.
4. Bandyopadhyay, M., Chakraborty, R., & Raychaudhuri, U. (2007). Incorporation of herbs into Sandesh, an Indian sweet dairy product, as a source of natural antioxidants. Int J Dairy Technol., 60, 228–233.
5. Battacharya, S., & Bhat, K. K. (1997). Steady shear rheology of rice blackgram suspensions and suitability of rheological models. J Food Eng., 32, 241–250.
6. Berger, A., Rein, D., Schafer, A., Monnard, I., Gremaud, G., Lambelet, P., & Bertoli C. (2005). Similar cholesterol lowering properties of rice bran oil, with varied γ-oryzanol, in mildly hypercholesterolemic men. Eur J. Nutr., 44(3), 163–173.
7. Bhat, Z. F., and Bhat, H. (2011). Milk and dairy products as functional foods: A Review. Int J. Dairy Sci., 6(1), 1–12.
8. Burks, A. W., & Helm, R. M., (1994). Hypoallerginicity of rice protein. In: Presented at the Annual Meeting of the American Association of Cereal Chemists. Nashville, TN.

9. Charalampopoulos, D., Wang, R., Pandiella, S. S., & Webb, C. (2002). Application of cereals and cereal components in functional foods: a review. Int J. Food Microbiol., 79, 131–141.

10. Chopra, R. N., Nayar, S. L., & Chopra, R. C. (1956). Glossary of Indian Plants. Council of Scientific and Industrial Research, New Delhi.

11. De Simone, C., Vesely, R., & Bianchi, S. B. (1993). The role of probiotics in modulation of the immune system in man and in animals. Int. J. Immunotherapy., 9, 23–28.

12. Dillard, C. J., & German, J. B. (2000). Review: Phytochemicals: nutraceuticals and human health. J. Sci. Food and Agr., 80, 1744–1756.

13. Dubey, K. G. (2010). The Indian cuisine. Delhi: PHI Learning Pvt. Ltd.

14. Duke, J. A., Bogenschutz-Godwin, M. J., Du Celliar, J., & Duke, P. A. K. (2002). Black pepper. In: Hand Book of Medicinal Herbs (2nd ed., pp. 98–99). Boca Raton: CRC Press Inc.

15. Fuler, R. (1991). Probiotics in human medicine. Gut., 32, 439–442.

16. Gaur, S., Agnihotri, R. (2014). Green tea: a novel functional food for the oral health of older adults. Geriatr Gerontol Int., 14(2), 238–250.

17. Godber, J. S., & Well, J. H. (1994). Rice bran: as a viable source of high value chemicals. La Agric., 37(2), 13–17.

18. Gurpreet, K. C., & Sogi, D. S. (2007). Functional properties of rice bran protein concentrates. J Food Eng., 79, 592–597.

19. Hotz, C., & Gibson, R. S. (2007). Traditional food-processing and preparation practices to enhance the bioavailability of micronutrients in plant-based diets. J Nutr., 137, 1097–1100.

20. Kim, K. H., Tsao. R., Yang, R., & Cui, S. W. (2006). Phenolic Acid Profiles and Antioxidant Activities of Sheat Bran Extracts and the Effect of Hydrolysis Conditions. Food Chem., 95, 466–473.

21. Kim, J. K., Kim, Y., Na, K. M., Surh, Y. J., & Kim, T. Y. (2007). [6]-Gingerol prevents UVB-induced ROS production and COX-2 expression in vitro and in vivo. Free Radical Research., 41, 603–614.

22. Kirtikar, K. P., & Basu, D. B. (1975). Indian Medicinal Plants, Vol. 1. Edited by B. Singh and M. P. Singh. Dehra Dun, India.

23. Nout, M. J. R. (2009). Rich Nutrition from the poorest—Cereal fermentation in Africa and Asia. Food Microbiol., 1–8.

24. Omar, S. H., & Al-Wabel, N. A. (2010). Organosulfur compounds and possible mechanism of garlic in cancer. Saudi Pharm J., 18, 51–58.

25. Patil, C. S. (2011). Current trends and future prospective of nutraceuticals in health promotion. Bioinfo., 1(1), 1–7.

26. Purushothaman, D., Dhanapal, N., & Rangaswami, G. (1993). Indian idli, dosa, dhokla, khaman, and related fermentations. In: Handbook of Indigenous Fermented Foods. Steinkraus, K. H. (ed) Marcel Dekker, New York, pp. 149–165.

27. Ramakrishnan, C. V. (1993). Indian idli, dosa, dhokla, khaman, and related fermentations. In: Handbook of Indigenous Fermented Foods. Steinkraus, K. H. (ed.) Marcel Dekker, New York, pp. 149–165.

28. Ross, R. P., Fitzgerald, G., Collins, K., & Stanton, C. (2002). Cheese delivering biocultures-probiotic cheese. Aust J Dairy Tech., 57, 71–78.

29. Ross, S. (2000). Functional foods: the Food and Drug Administration perspective. Am J Clin Nutr., 71, 1735S–1738S.

30. Roy, B., Kala, C. P., Farooquee, N. A., & Majila, B. S. (2004). Indigenous fermented food and beverages: A potential for economic development of the high altitude societies in Uttaranchal. J Hum Ecol., 15(1), 45–49.

31. Saikia, D., & Deka, S. C. (2011). Cereals: from staple food to nutraceuticals. Int. Food Res J., 18, 21–30.

32. Saito, T. (2004). Selection of useful probiotic lactic acid bacteria from Lactobacillus acidophilus group and their application to functional foods. Anim. Sci. J., 75, 1–13.

33. Salminen, S., Deighton, M. A., Benno, Y., & Gorbach, S. L. (1998). Lactic acid bacteria in health and disease. In: Salminen, S., von Wrights, A. (Eds.). Lactic acid bacteria: Microbiology and Functional Aspects, pp. 211–253. New York, Mercel Dekker Inc.

34. Sanjeev, K. S., & Sandhu, K. D. (1990). Indian fermented foods; microbiological and biochemical aspects. Ind J Microbial., 30, 135–157.

35. Sarkar, S., & Misra, A. (2001). Bio-preservation of milk and milk products. Indian Food Ind., 20, 74–77.

36. Sen, C. T. (2004). Food culture in India. Santa Barbara: Greenwood Publishing Group.

37. Siddiqui, B. S., Gulzar, T., Mahmood, A., Begum, S., Khan, B., & Afshan, F. (2004). New insecticidal amides from petroleum ether extract of dried Piper nigrum L. whole fruits. Chem Pharm Bull., 52, 1349–1352.

38. Siddiqui, B. S., Gulzar, T., Mahmood, A., Begum, S., Khan, B., Rasheed, M., Afshan, F., & Tariq, R. M. (2005). Phytochemical studies on the seed extract of Piper nigrum Linn. Nat Prod Res., 19, 703–712.

39. Smirnov, W., Reznik, S. R., & V'iunitskaia, V. A. (1993). The current concepts of the mechanisms of the therapeutic-prophylactic action of probiotics from bacteria in the genus Bacillus. Mikrobiolohichnyi Zhurnal., 55, 92–112.

40. Soomro, A. H., Massud, T., & Kiran, A. (2002). Role of lactic acid bacteria in food preservation and human health – A review. Pak. J. Nutr. 1(1), 20–24.

41. Steinkraus, K. H. (1994). Nutritional significance of fermented foods. Food Res Int., 27, 259–267.

42. Steinkraus, K. H. (1996). Handbook of Indigenous Fermented Foods. Marcel Dekker, New York.

43. Stanton, C., Desmond. C., Coakley, M., Collins, J. K., Fitzgerald, G., & Ross, R. P. (2003). Challenges facing development of probiotic containing functional foods. In: Handbook of Fermented Functional Foods. Edited by ER Farnworth. Boca Raton: CRC Press. pp. 27–58.

44. Van Dam, R. M., Hu. F. B., Rosenberg, L., Krishnan, S., & Palmer, J. R. (2006). Dietary calcium and magnesium, major food sources, and risk of type 2 diabetes in U. S. Black women. Diabetes Care. 29(10), 2238–2243.

45. Wealth of India. (1950). Raw Materials. Vol. II. Council of Scientific and Industrial Research, New Delhi, India.

46. Wells, J. H. (1993). Utilization of rice bran and oil in human diets. La Agric., 36(3), 5–8.

47. Yonzan, H., & Tamang, J. P. (2009). Traditional processing of Selroti – A cereal based ethnic fermented food of the Nepalis. Indian J Trad Know., 8(1), 110–114.

CHAPTER 12

PHYTOCHEMISTRY AND ANTI-OXIDATIVE EFFECTS OF *ASPALATHUS LINEARIS* HERBAL TEA – A REVIEW

ADEMOLA AYELESO, MUTIU IDOWU KAZEEM,
TAIWO AYELESO, and EMMANUEL MUKWEVHO

CONTENTS

12.1 INTRODUCTION

Rooibos (*Aspalathus linearis*) is a leguminous shrub native to the mountainous areas of the North Western Cape Province in South Africa. Rooibos belong to family Fabacae and it was first recorded by botanists in 1772 when they were introduced to the tea by the Khoi people [41]. The rooibos shrub can grow up to 2 meters in height with a long taproot that reaches as deep as 2 m (about 6 feet). As a legume, rooibos contains nodules of nitrogen-fixing bacteria on its roots; this characteristic helps the plant survive in the poor Cedarberg soils and minimizes the need for fertilizing commercial crops with nitrogen [31]. In 1994, Benjamin Ginsberg became interested in rooibos and realized its marketing potentials and profitability [42]. Although, rooibos is indigenous to south Africa, yet it is now been exported to and consumed in countries worldwide, including Germany, Japan, the Netherlands, England, Malaysia, South Korea, Poland, China, and the United States [41]. The major international market for rooibos is Germany (53%), followed by the Netherlands (11%), United Kingdom (7%), Japan (6%) and the USA (5%) [21].

The Khoi and San people of South Africa, who first discovered rooibos, crudely made the leaves and stems of the plant into a sweet, tasty herbal tea [21]. Generally, rooibos tea is produced by cutting its leaves and stem into about 5 mm length, fermenting by leaf enzymes and solar drying in a process similar to that of black tea or oolong tea [15]. Rooibos tea is normally either processed as green or fermented rooibos (Figure 12.1). The unfermented green rooibos tea was first introduced by a group of scientists in South Africa during the 1990s to achieve higher antioxidant level [40] and has since been in production as an alternative to its traditionally processed counterparts. When rooibos tea leaves are fermented, they turn into an orange red color and generally do have a sweeter taste [36].

Rooibos is not only consumed for enjoyment, as an alternative to oriental tea, but also for its possible medicinal properties and, till date, no adverse effects have been associated with its consumption. It has been reported to be safe for infants, children and pregnant women [14]. Rooibos tea is commonly used in the traditional medical system for treating abdominal colic, diarrhea, anemia, cardiopathy, eczema [13], asthma [11], inflammation, malignancies [32], nervous tension and allergies [9]. Rooibos is naturally caffeine free and contains very low levels of tannins [16] unlike other teas such as green tea. Owing to its safety and beneficial effects, rooibos tea is rapidly becoming popular as a health promoting tea. Infusions from rooibos are reported to exhibit antioxidative activity, which can be attributed to the presence of polyphenols. It has been reported that there is an inverse relationship between antioxidative status and incidence

FIGURE 12.1 Fermented rooibos herbal tea.

of human diseases such as cancer, diabetes mellitus, aging, neurodegenerative disease, and atherosclerosis [29]. The presence of the phenolic compounds with their antioxidative ability therefore account largely for the medicinal reports on rooibos tea.

This chapter reviews scope and technology of phytochemistry and antioxidative effects of *Aspalathus linearis* herbal tea in biological systems.

12.2 PHYTOCHEMISTRY

The presence of many chemical compounds has been reported in the rooibos leaves and stems. These compounds are discussed below:

12.2.1 FLAVONOIDS

Flavonoids (Figure 12.2) are molecules with a phenolic benzopyran structure and occur only in plants predominantly as glycosides. More than 4000 flavonoids have been identified in plants, and the list is constantly growing [18]. Flavonoids are extensively used in medicines, foods, textiles and cosmetics [8]. Several flavonoids have been detected and isolated from rooibos. The most abundant flavonoids in rooibos aspalathin (2, '3,4,4, '6'-pentahydroxy-3'-C-β-D-glucopyranosyldihydrochalcone)

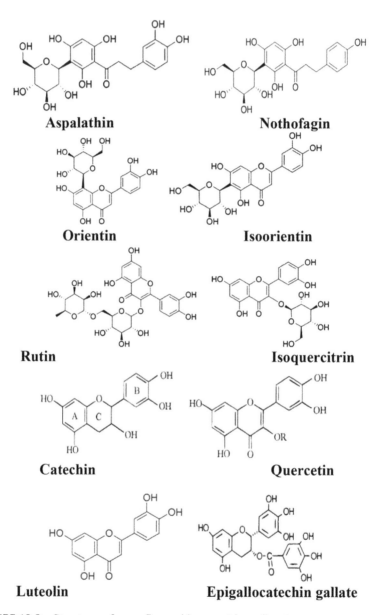

FIGURE 12.2   Structures of some flavonoids present in rooibos tea.

and nothofagin, which constituted about 10% and 1% respectively, of unfermented rooibos leaves [36]. Aspalathin is found only in rooibos while nothofagin is present only in rooibos and in red beech plant (*Nothofagus fusca*) [38]. Other flavonoids present in rooibos include isoorientin, orientin, vitexin, rutin, isovitexin, luteolin, quercetin, chrysoeriol, as well as isoquercitrin and hyperoside [9]. The presence of catechin and epigallocatechin gallate was also reported in rooibos tea [37]. Though fermentation alters the quantitative flavonoid composition of rooibos, all of them are still present in the fermented sample [10].

12.2.2   PHENOLIC ACIDS

These are compounds containing phenolic ring and an organic carboxylic acid (Figure 12.3). They occur in two forms, hydroxybenzoic acid and hydroxycinnamic acid [25]. The phenolic acid found in this plant includes protocatechuic acid, vanillic acid, *p*-coumaric acid, caffeic acid, ferullic acid, benzoic acid, *p*-hydroxybenzoic acid, and syringic acid [15, 34].

12.2.3   VOLATILE COMPOUNDS

Rooibos leaves and stem is also rich in volatile compounds which are of diverse classes and structures. They include guaiacol, 6-methyl-3,5-hepta-

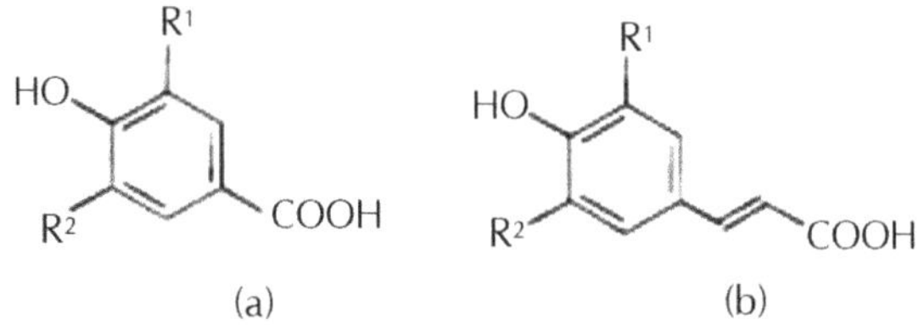

p-hydroxybenzoic acid: $R^1=R^2=H$ p-coumaric acid: $R^1=R^2=H$
Protocatechuic acid: $R^1=OH, R^2=H$ Caffeic acid: $R^1=OH, R^2=H$
Vanillic acid: $R^1=OCH_3, R^2=H$ Ferulic acid: $R^1=OCH_3, R^2=H$
Syringic acid: $R^1=R^2=OCH_3$

FIGURE 12.3   Hydroxy derivatives of (a) benzoic acid and (b) cinnamic acid present in rooibos tea.

diene-2-one, demascenone, geranylacetone and 6-methyl-5-hepten-2-one [23]. The presence of compounds like 4-butanolide, 2-phenylethanol, dihydroactinidiolide, β-demascenone, 3-methylbutanoic acid and 6,4,10-trimethylpentadecanone, was also reported in rooibos tea [23].

12.2.4   MINERALS

The minerals found in rooibos tea include iron, potassium, calcium, copper, zinc, magnesium, fluoride, manganese and sodium [22].

12.3   ANTIOXIDANT SYSTEM AND OXIDATIVE STRESS: OVERVIEW

An antioxidant is a molecule that inhibits the oxidation of other molecules by safely interact with reactive oxygen species (ROS) and nitrogen oxygen species (RNS), i.e., free radicals and quench them by terminating the chain reaction that is initiated by stealing of electrons thereby preventing oxidative stress (Figure 12.4). The antioxidant systems include both enzymatic antioxidants such as superoxide dismutase, catalase, glutathione peroxidase, and glutathione reductase) and non-enzymatic antioxidants which include vitamins A, C, and E, bioflavonoids, glutathione and minerals like copper, zinc, manganese, and selenium [30]. ROS are also produced by both endogenous metabolic functions and environmental stimuli such as ultraviolet light or other kinds of radiation [12]. Different examples of ROS include free radicals, i.e., superoxide (O_2-), hydroxyl (HO), peroxyl (RO_2-), hydroperoxyl (HRO_2-), and non-radical species such as hydrogen peroxide (H_2O_2), hydrochlorous acid (Hocl) and RNS include free radicals such as nitric oxide ($\cdot NO-$) and nitrogen dioxide ($\cdot NO_2-$) and non-radicals species like peroxynitrite (ONOO), nitrous oxide (HNO_2), and alkyl peroxynitrates (RONOO) [33].

Oxidative stress is an imbalance between the generation of ROS/RNS and the antioxidant defense capacity in the body [5, 33]. It is caused by increased production of free radicals and reduction in their removal by the body antioxidant system and this has critical health implications. The excessive production of ROS and RNS causes the structural deterioration

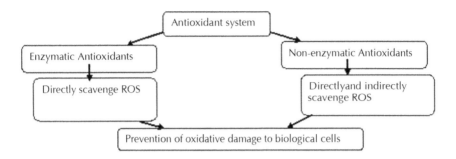

FIGURE 12.4   The classes of antioxidant system and the ways they operate.

of macromolecules (carbohydrates, proteins, lipids, and deoxyribonucleic acid, DNA), which leads to their instability and consequently loss of function [33]. Oxidative stress has been implicated in the pathogenesis of several diseases which include neurodegenerative disorders such as Alzheimer's and Parkinson's diseases, cardiovascular diseases, diabetes mellitus and cancer [1, 12].

12.4   ANTIOXIDATIVE ROLE OF ROOIBOS HERBAL TEA

The antioxidant enzymes are very good biochemical markers of stress and their elevated activity may confirm a potential for remediation [5]. Glutathione peroxidase (GPx) and glutathione reductase (GR) are important enzymes of the glutathione defense system, GPx catalyzes the reduction of hydrogen peroxide (H_2O_2) and lipid hydroperoxides using reduced glutathione (GSH) as a co-substrate, GR regenerates GSH from oxidized glutathione (GSSG) at the expense of NADPH [2]. Catalase is an antioxidant enzyme that catalyzes the decomposition of H_2O_2 to water and oxygen. Superoxide dismutase (SOD) is an antioxidant enzyme involved in the protection of cells by spontaneously dismutating $O_2\cdot-$ to H_2O_2. Following rooibos tea consumption in an animal model, a significant increase in liver GR, SOD and GPx activities in *tert*-butyl hydroperoxide-induced-oxidative stress was shown [1]. In another study, rooibos tea significantly increased SOD, GPx and GR activities in lipopolysaccharide (LPS)-induced liver injury [32]. Rooibos tea has been shown to significantly increased superoxide dismutase (SOD) in the serum of DSS-induced colitis [7]. In the liver of diabetic rats, supplementation with rooibos tea increased the activities of GPx and SOD [5]. The activities of catalase and SOD were increased in the sperm of rats due to supplementation with rooibos tea [4]. In another study, rooibos tea was also described as activator of catalase [35]. Sperm glutathione levels of rats consuming rooibos tea was also significantly increased and showed the tendency to lower the levels of ROS and lipid peroxidation in the epididymal sperm [4].

Treatment with rooibos tea increased antioxidant status in carbon tetrachloride (CCl_4)-damaged liver [24]. It was shown that the tea increased the level of alpha-tocopherol and reduced coenzyme $C_0Q_9H_2$ while reducing the level of malondialdehyde (MDA), an indicator of lipid peroxidation. The improved regeneration of coenzyme Q redox state may give explanation to the valuable effect of antioxidant therapy as effective hepatoprotector [24]. The tea has also marginally increased the oxygen radical absorbance capacity (ORAC) status in fumonisin B1-induced cancer model in rats [27] and plasma of diabetic rats [5]. Experimental animals subjected to chronic restraint or immobilization showed induction of oxidative stress in their brains and this was significantly attenuated by treatment with rooibos tea supplementation [19]. From their findings, it was concluded rooibos tea demonstrated the ability to reverse the increase in stress-related metabolites [5-hydroxyindoleacetic acid, 5-HIAA and free fatty acids (FFA)], prevent lipid peroxidation, restore stress-induced protein degradation, regulate glutathione metabolism (GSH and GSH/GSSG ratio), and modulate changes in the activities of antioxidant enzymes (SOD and CAT). The administration of rooibos tea to rats has also been reported to prevent age-related accumulation of lipid peroxides in several regions of the brain [20].

In a clinical study involving adults with cardiovascular disease, rooibos tea significantly decreased markers of lipid peroxidation such as conjugated dienes and thiobarbituric acid reactive substances (TBARS). In addition, GSH and GSH:GSSG ratio were both significantly increased following consumption of rooibos [28]. Rooibos tea has also partially prevented oxidative stress in streptozocin-induced diabetic rats especially by protecting the ocular (eye) membrane systems against peroxidation [39]. They found out that rooibos tea decreased advanced glycation end-products and MDA in plasma and in different tissues of diabetic rats, particularly malondialdehyde concentration in lens of diabetic rats. The decrease in reactive oxygen species and lipid peroxidation in the testes of experimental rats due to rooibos tea supplementation has also been reported [3]. They reported that rooibos tea could be rendering its protective effect against testicular tissue oxidative damage by possibly increasing the anti-oxidant defense mechanisms in rats, while reducing lipid peroxidation. Chronic administration of rooibos tea prevented age-related accumulation of lipid peroxides in several regions of rat brain and hence, protected the central nervous system (CNS) against damage. Supplementation with rooibos tea significantly decreased biomarkers of oxidative damage such as conjugated dienes and MDA levels in *t*-BHP induced oxidative stress in the liver of rats [1].

In another study, rooibos tea attenuated increased plasma and hepatic MDA, decreased liver and whole blood GSH:GSSG ratio [2]. Rooibos tea also exhibited protective activity against hepatic microsomal lipid peroxidation [26]. Using a *Caenorhabditis elegans* strain, rooibos tea and aspalathin, the most abundant flavonoid in rooibos were shown to inhibit acute oxidative damage caused by the superoxide anion radical generator, juglone, with aspalathin demonstrating a major role in improving the survival rate of *C. elegans* by stimulating the expression of stress and aging-related genes such as daf-16 and sod-3 [12].

12.5 CONCLUSIONS

Rooibos has good pharmacological activities which could be attributed to the presence of complex profile of antioxidants. The antioxidative potentials of the plant have been shown on several diseases which include cancer, diabetes mellitus, cardiovascular diseases and infertility. More studies are still needed especially on humans to further assess its medicinal benefits in the prevention and management of many other diseases. In addition, study on the mechanisms by which rooibos exerts its actions are also needed.

12.6 SUMMARY

Rooibos (pronounced ROY boss) herbal tea originates from the leaves and stems of the indigenous South African plant *Aspalathus linearis* which is grown on high mountain ranges. It is also known as Red-Bush. The presence of many chemical compounds has been reported in the rooibos leaves and stems. These compounds include flavonoids, phenolic acids and volatile compounds. Some of the polyphenolic compounds include aspalathin, catechin, isoquercitrin, luteolin, quercetin, rutin, caffeic acid, ferulic acid and vanillic acid. It has also been reported that rooibos is caffeine free and contains low tannins with a variety of minerals. The presence of these compounds is responsible for the various beneficial biological activities of rooibos tea. Due to the high prolife of antioxidants in rooibos tea, several experimental studies have confirmed its roles in the management of oxidative stress-mediated diseases. It is concluded that rooibos tea has good health promoting properties.

KEYWORDS

- antioxidants
- chronic diseases
- flavonoids
- free radicals
- oxidative stress
- phenolic acids
- phytochemicals
- rooibos tea
- South Africa

REFERENCES

1. Ajuwon, O. R., Katengua-Thamahane, E., Van Rooyen, J., Oguntibeju, O. O., & Marnewick, J. L. (2013). Protective effects of rooibos (*Aspalathus linearis*) and/or red palm oil (*Elaeis guineensis*) supplementation on tert-butyl hydroperoxide-

induced oxidative hepatotoxicity in wistar rats. Evidence-Based Complementary and Alternative Medicine, Article ID 984273, 19 pp.

2. Ajuwon, O. R., Oguntibeju, O. O., & Marnewick, J. L. (2014). Amelioration of lipopolysaccharide-induced liver injury by aqueous rooibos (*Aspalathus linearis*) extract via inhibition of pro-inflammatory cytokines and oxidative stress. BMC Complementary and Alternative Medicine, 14, 392.

3. *Awoniyi, D. O., Aboua, Y. G., Marnewick, J. L., du Plesis, S. S., & Brooks,* N. L. (2011). Protective effects of rooibos (*Aspalathus linearis*), green tea (*Camellia sinensis*) and commercial supplements on testicular tissue of oxidative stress-induced rats. African Journal of Biotechnology, 10, 17317–17322.

4. Awoniyi, D. O., Aboua, Y. G., Marnewick, J., & Brooks, N. The effects of rooibos (*Aspalathus linearis*), green tea (*Camellia sinensis*) and commercial rooibos and green tea supplements on epididymal sperm in oxidative stress-induced rats. Phytotherapy Research, 26, 1231–1239.

5. Ayeleso, A. O., Brooks, N. L., & Oguntibeju, O. O. (2013). Impact of dietary red palm oil (*Elaeis guineensis*) on liver architecture and antioxidant status in the blood and liver of male Wistar rats. Medical Technology SA, 27, 18–23.

6. Ayeleso, A., Brooks, N., & Oguntibeju, O. (2014). Modulation of antioxidant status in streptozotocin-induced diabetic male Wistar rats following intake of red palm oil and/or rooibos. Asian Pacific Journal of Tropical Medicine, 7, 536–544.

7. Baba, H., Ohtsuka, Y., Haruna. H., Lee, T., Nagata, S., Maeda, M., Yamashiro, Y., Shimizu, T. (2009). Studies of anti-inflammatory effects of Rooibos tea in rats. Pediatrics International, 51, 700–704.

8. Baranska, M., Schulz, H., Joubert, E., & Manley, M. (2006). In situ flavonoid analysis by FT-Raman spectroscopy: Identification, distribution, and quantification of aspalathin in green rooibos (*Aspalathus linearis*). Analytical Chemistry, 78, 7716–7721

9. Bramati, L., Minoggio, M., Gardana, C., Simonetti, P., Mauri, P., & Pietta, P. (2002). Quantitative characterization of flavonoid compounds in rooibos tea (*Aspalathus linearis*) by LC-UV/DAD. Journal of Agricultural and Food Chemistry, 50, 5513–5519.

10. Bramati, L., Aquilano, F., & Pietta, P. (2003). Unfermented rooibos tea: Quantitative characterization of flavonoids by HPLC-UV and determination of the total antioxidant activity. Journal of Agricultural and Food Chemistry, 51, 7472–7474.

11. Brown, D. (1995). Encyclopaedia of herbs and their uses. Dorling Kindersley, London, ISBN 0–7513–020–31, pp. 244.

12. Chen, W., Sudji, I. R., Wang, E., Joubert, E., van Wyk, B. E., Wink, M. (2013). Ameliorative effect of aspalathin from rooibos (Aspalathus linearis) on acute oxidative stress in Caenorhabditis elegans. Phytomedicine, 20, 380–386.

13. Duke, J. A., Bogenschutz-Godwin, M. J., Ducelliar, J., & Duke, P. A. K. (2002). Hand book of medicinal herbs, 2nd edn. CRC Press, Boca Raton, ISBN 0-8493-1284-1, pp. 612–613.

14. Erickson, L. (2003). Rooibos tea: Research into antioxidant and antimutagenic properties. The Journal of the American Botanical Council, 59, 34–45.

15. Ferreira, D., Marais, C., Steenkamp, J. A., & Joubert, E. (1995). Rooibos tea as a likely health food supplement. In Proceedings of Recent Development of Technologies on Fundamental Foods for Health; Korean Society of Food Science and Technology: Seoul, Korea, pp. 73–88.

16. Galasko, G. T. F., Furman, K. I., & Alberts, E., (1989). The caffeine contents of non-alcoholic beverages. Food and Chemical Toxicology, 27, 49–51.

17. Habu, T., Flath, R. A., Mon, T. R., & Morton, J. F. (1985). Volatile components of rooibos tea (*Aspalathus linearis*). Journal of Agricultural and Food Chemistry, 33, 249–254.

18. Harborne, J. B., & Williams, C. A. (2000). Advances in flavonoid research since 1992. Phytochemistry, 55, 481–504.

19. Hong, I. S., Lee, H. Y., & Kim, H. P. (2014). Anti-oxidative effects of rooibos tea (*Aspalathus linearis*) on immobilization-induced oxidative stress in rat brain. PLoS ONE, 9, e87061. doi:10.1371/journal.pone.0087061.

20. Inanami, O., Asanuma, T., Inukai, N., Jin, T., Shimokawa, S., Kasai, N., Nakano, M., Sato, F., & Kuwabara, M. (1995). The suppression of age-related accumulation of lipid peroxides in rat brain by administration of Rooibos tea (*Aspalathus linearis*). Neuroscience Letters, 196, 85–88.

21. Joubert, E., Gelderblom, W. C. A., Louw, A., & de Beer, D. (2008). South African herbal teas: Aspalathus linearis Cyclopia spp. and Athrixia phylicoides – A review. Journal of Ethnopharmacology 119, 376–412.

22. Kamen, B. (2000). Sippa cuppa Rooibos tea. Alternative Medicine, 75, 70–72

23. Kawakami, M., Kobayashi, A., & Kator, K. (1993). Volatile constituents of rooibos tea (*Aspalathus linearis*) as affected by extraction process. Journal of Agricultural and Food Chemistry, 633–636.

24. Kucharská, J., Uličná, O., Gvozdjáková, A., Sumbalová, Z., Vančová, O., Božek, P., Nakano, M., & Greksák, M. (2004). Regeneration of coenzyme Q9 redox state and inhibition of oxidative stress by rooibos tea (*Aspalathus linearis*) administration in carbon tretrachloride liver damage. Physiology Research, 53, 515–521.

25. Manach, C., Scalbert, A., Morand, C., Re'me'sy, C., & Jime'nez, L. (2004). Poly-phenols: food sources and bioavailability. American Journal of Clinical Nutrition, 79, 727–747.

26. Marnewick, J., Joubert, E., Joseph, S., Swanevelder, S., Swart, P., & Gelderblom, W. (2005). Inhibition of tumor promotion in mouse skin by extracts of rooibos (*Aspalathus linearis*) and honeybush (*Cyclopia intermedia*), unique South African herbal teas. Cancer Letters, 224, 193–202.

27. Marnewick, J. L., Van der Westhuizen, F. H., Joubert, E., Swanevelder, S., Swart P., Gelderblom, W. C. A. (2009). Chemoprotective properties of rooibos (*Aspalathus linearis*), honeybush (*Cyclopia intermedia*), green and black (*Camellia sinensis*) teas against cancer promotion induced by fumonisin B1 in rat liver. Food and Chemical Toxicology, 47, 220–229.

28. Marnewick, J. L., Rautenbach, F., Venter, I., Neethling, H., Blackhurst, D. M., Wol-marans, P., & Macharia, M. (2011). Effects of rooibos (*Aspalathus linearis*) on oxidative stress and biochemical parameters in adults at risk for cardiovascular disease. Journal of Ethnopharmacology, 133, 46–52.

29. Morales, G., Paredes, A., Sierra, P., & Loyola, L. A. (2008). Antioxidant activity of 50% aqueous-ethanol extract from Acantholippia deserticola. Biological Research, 41, 151–155.

30. Mukwevho, E., Ferreira, Z., & Ayeleso, A. (2014). Potential role of sulfur containing antioxidant systems in highly *oxidative environments. Molecules,* 19, 19376–19389.

31. Muofhe, M. L., & Dakora, F. D. (1999). Nitrogen nutrition in nodulated field plants of the shrub tea legume Aspalathus linearis assessed using 15N natural abundance. Plant and Soil, 209, 181–186.

32. Na, H. K., Mossanda, K. S., Lee, J. Y., et al. (2004). Inhibition of phorbol ester-induced COX-2 expression by some edible African plants. *Biofactors*, 21, 149–153.

33. Oyenihi, A. B., Ayeleso, A. O., Mukwevho, E., & Masola, B. (2014). Antioxidant strategies in the management of diabetic neuropathy. BioMed Research International, 2015, Article ID 515042, 15 pp., http://dx.doi.org/10.1155/2015/515042.

34. Rabe, C., Steenkamp, J. A., Joubert, E., Burger, J. F. W., Ferreira, D. (1994). Phenolic metabolites from rooibos tea (*Aspalathus linearis*). Phytochemistry, 35, 1559–1565.

35. Salkic, A., & Zeljkovic, S. C. (2015). Preliminary investigation of bioactivity of green tea (Camellia sinensis), rooibos (*Asphalatus linearis*), and yerba mate (*Ilex paraguariensis*). Journal of Herbs, Spices & Medicinal Plants, 21, 259–266.

36. Schulz, H., Joubert, E., & Sch¨utze, W. (2003). Quantification of quality parameters for reliable evaluation of green rooibos (*Aspalathus linearis*), European Food Research and Technology, 216, 539–543.

37. Snijman, P. W., Swanevelder, S., Joubert, E., Green, I. R., & Gelderblom, W. C. A. (2007). The antimutagenic activity of the major flavonoids of rooibos (*Aspalathus linearis*): Some dose–response effects on mutagen activation–flavonoid interactions. Mutation Research, 631, 111–123.

38. Snijman, P. W., Joubert, E., Ferreira, D., Li, X-C., Ding, Y., Green, I. R., & Wentzel C. A. Gelderblom. (2009). Antioxidant activity of the dihydrochalcones aspalathin and nothofagin and their corresponding flavones in relation to other rooibos (*Aspalathus linearis*) flavonoids, epigallocatechin gallate, and trolox. Journal of Agricultural and Food Chemistry, 57, 6678–6684.

39. Ulicna, O., Vancova, O., Bozek, P., Carsky, J., Sebekova, K., Boor, P., Nakano, M., & Greksák, M. (2006). Rooibos tea (*Aspalathus linearis*) partially prevents oxidative stress in streptozotocin-induced diabetic rats. Physiological Research, 55, 157–164.

40. Von Gadow, A. (1996). Antioxidant activity of rooibos tea (*Aspalathus linearis*). MSc in Food Science. University of Stellenbosch, Stellenbosch, South Africa.

41. WESGRO (2000). Western Cape Investment and Trade Promotion Agency, Cape Town, South Africa, Wesgro Background Report: The Rooibos Industry in the Western Cape. April.

42. Wilson, N. L. W. (2005). Cape Natural Tea Products and the U. S. Market: Rooibos rebels ready to raid. Review of Agricultural Economics, 27, 139–148.

APPENDIX A:
FOOD PROCESSING EQUIPMENTS
AND THEIR APPLICATIONS

Equipment name	Involved in operation	Application
Baker and blinder	Blind baking	Baking pastry, a pre filling process
Basting process	Basting	To moisten periodically with a liquid, such as melted butter or a sauce, especially while cooking
Belt conveyor	Conveying	Transportation of food product, food processing ingredients, etc.
Blancher	Blanching	Food substance (vegetable or fruit) is dipped into boiling water and removed after a brief interval, and then plunged into iced water or placed under cold running water to stop the cooking process
Blender	Blending	To combine or mix (different substances) so that the constituent parts are indistinguishable from one another: blended the flour, milk, etc.
Boiler, Evaporator, Kettle	Boiling	Rapid vaporization of excess liquid
Bottle/pouch filling Machines	Filling	Milk, juice, oil, etc. filling to pouch, bottles, cans, etc.
Braising equipments	Braising	Using combination of both moist and dry heat; generally food is first seared at a high temperature and then finished in a covered pot with a variable amount of liquid, that develop a specific flavor
Coddler	Coddling	Food is heated in water kept just below the boiling point
Dehusker	Husking	Removing husk cover of the food grains such as paddy, wheat, millet
Dicing equipments	Dicing	To cut food into small cubes

Equipment name	Involved in operation	Application
Dryer (Tray, Tunnel, Roller, Drum, Fluidized Bed, Spray, Pneumatic, Rotary, Trough, Bin, Belt, Vacuum, Freeze)	Drying	Process of removing excess moisture content of food material
Evaporator (Open Pans, Horizontal-tube, Vertical-tube, Plate, Long-tube, Forced-circulation)	Evaporation	Removing excess water or moisture
Fermenter	Fermentation	Enrichment of the diet through development of a diversity of flavors, aromas, and textures in food substrates, Preservation of substantial amounts of food through lactic acid, alcohol, acetic acid, and alkaline fermentations, Biological enrichment of food substrates with protein, essential amino acids, and vitamins, Elimination of antinutrients, A decrease in cooking time and fuel requirement
Fruit and vegetable slicer	Cutting	Cut into small pieces to make salad or make suitable for cooking
Fryer	Frying	Cooking food in oil or another fat
	Deep frying	Food is submerged in hot oil or fat
	Hot salt frying	Hot slat is used as heating medium
	Hot sand frying	Hot sand is used as heating medium
	Pan frying	Cooking food in a pan using a small amount of cooking oil or fat as a heat transfer agent and to keep the food from sticking
Grater	Grating	To rub cheese, vegetables, etc., against a rough or sharp surface in order to break them into small pieces
Griller	Grilling	A form of cooking that involves dry heat applied to the surface of food, commonly from above or below
Grinder	Milling	The act or process of grinding, especially grinding grain, spices seeds, cereals into flour or meal

Equipment name	Involved in operation	Application
Infuser	Infusion	Process of extracting chemical compounds or flavors from plant material in a solvent such as water, oil or alcohol, by steeping
Juice or milk pasteurizer	Pasteurization	Milk or fruit juice pasteurization
Juicer	Juice making	Preparation of juice from fruits
Julienning processing	Julienning	To cut vegetables into fine strips.
Kneader	Kneading	To mix and work into a uniform mass, as by folding, pressing, and stretching with the hands: kneading dough
Microwave oven	Microwaving	Heats foods quickly and efficiently, a microwave oven does not brown bread or bake food
Mixer	Mixing	To combine or blend into one mass or mixture. Generally mix the dry ingredients first.
Oven, Hot ashes, hot stones	Baking	Prolonged cooking of food by dry heat acting by *convection*
Parallel flow or counter flow or cross flow heat exchanger	Heat-transfer	Milk and juice pasteurization
Peeler	Peeling	Remove the outer peel of the fruits and vegetables
Pickling method	Pickling	Process of preserving or expanding the lifespan of food by either anaerobic fermentation in brine or immersion in vinegar. The resulting food is called a pickle. The pickling affect the food's texture and flavor
Poacher	Poaching	Process of gently simmering food in liquid, generally milk, stock
Presser Cooker	Cooking	Cooking in a sealed vessel that does not permit air or liquids to escape below a preset pressure, which allows the liquid in the pot to rise to a higher temperature before boiling
Pressure cooker	Cooking	Cooking at low pressure, helps in fast cooking
Refrigerator	Cooling and freezing	For storing perishable food items such milk, prepared food items, etc.

Equipment name	Involved in operation	Application
Roaster	Roasting	Cooking method that uses dry heat, whether an open flame, oven, or other heat source. It causes caramelization or Maillard browning of the surface of the food
Refrigerator	Refrigeration	Freezing and chilling
Salting process	Salting	Most bacteria, fungi and other pathogenic organisms cannot survive in a highly salty environment, due to the hypertonic nature of salt. Any living cell in such an environment will become dehydrated through osmosis and die or become temporarily inactivated. It is one of the oldest methods of preserving food.
Seasoning process	Seasoning	Salt, herbs, or spices added to food to enhance the flavor
Seed metering system	Metering	To regulated the feed rate
Separator	Cleaning	Removing dust, broken, stones, etc.
Simmers	Simmering	Cooking of food in hot liquids kept at or just below the boiling point of water but higher than poaching temperature
Slicer machine	Cutting/slicing	Fruits and salads slices making
Smoker	Hot smoking	Process of flavoring, cooking, preserving food by exposing it to the smoke from burning or smoldering plant materials (wood). It helps in cooking and flavoring the food.
Sniper	Snipping	Cut medicinal herbs, plants, fruits and vegetables with scissors
Souring process	Souring	Having a taste characteristic of that produced by acids; sharp, tart, tangy, food items is made acid or rancid by fermentation then it may be having characteristics of fermentation or rancidity
Sprouting process	Sprouting	Having good health benefits as functional food
Steamer	Steaming	Cooking the food by heat which is carried by steam
Steeper	Steeping	Saturation of a food in a liquid solvent to extract a soluble ingredient into the solvent
Stewing Devices	Stewing	Food is cooked in liquid and served in the resultant gravy
Sugaring process	Sugaring	Food preservation one of the method

REFERENCES

1. Das, H. (2005). Food Processing Operation Analysis. Agricultural and Food Engineering Department, IIT Kharagpur (India). Press: Asian Books Private Limited, Ansari Road, Darya Ganj, New Delhi.
2. Geankoplis, C J. (1999). Transport Processes and Unit Operations. Prentice-Hall of India, Private Limited, New Delhi, pp. 200–243.
3. Sahay, K. M., & Singh, K. K. (2004). Unit Operations of Agricultural Processing. Vikash Publishing House Pvt. limited, New Delhi, p. 13.
4. *Unit Operations in Food Processing*, (1983). R. L. Earle (Ed.), NZIFST (The New Zealand Institute of Food Science & Technology), Inc.

INDEX

9 781774 630419